understanding statistics

an informal introduction
for the behavioral sciences

understanding
statistics
an informal introduction
for the behavioral sciences

R. L. D. Wright

University of Victoria

Under the general editorship of
Jerome Kagan
Harvard University

Harcourt Brace Jovanovich, Inc.
New York Chicago San Francisco Atlanta

To the memory of my teacher and friend
Edwin Stephens Waycott Belyea
and one other

Interior cartoons: Tony Hall
Cover: designed from drawings by Yvonne Steiner
Technical illustrations: Bert Schneider

ISBN: 0-15-592877-5

Library of Congress Catalog Card Number: 75-37066

Printed in the United States of America

preface

This book is an introduction to statistics for students of psychology, the other behavioral sciences, and education. Statistics has the reputation of being boring and difficult, an idea that probably stems from preoccupation with the finer details of methodology to the exclusion of the simple but fascinating core concepts. *Understanding Statistics* omits some of the finer points and minor qualifications of advanced statistics and emphasizes conceptual clarity, but every effort has been made to prepare the ground for the study of more advanced statistics.

The book's chapters should be read in order, because new points tend to build on a grasp of previous ones. However, the chapter on probability may be skipped by students who are already familiar with this material. Coverage includes all standard topics in statistics through two-way analysis of variance. Set theory, Bayesian statistics, and those topics in probability theory not needed here have been omitted.

Every book must have some special emphasis to set it apart from others in ways the author considers important. This book emphasizes the following:

● Use of words and visual imagery rather than mathematics to establish the important basic concepts of statistics. Mathematics has its role, but for most students the initial phases of understanding involve nonarithmetic insights. Algebra is used only in optional derivations of formulas.

● Use of a light approach whenever possible. This is not to belittle the importance of the material, but to ease the student's burden as much as possible. Moreover, the author believes that if something is worth doing, it is worth doing enjoyably.

● Concentration on the basic concepts of statistics, with secondary emphasis on their mechanical application. Thus, the arithmetic used is minimal, and understanding of concepts is emphasized more than memorization.

● Frequent examples of the application of statistics to the behavioral sciences and education. Statistics is an invention of people, an activity of people, directed toward human ends. The author's experience has shown that students learn best those things with which they can identify.

● Use of Self-Tests and end-of-chapter problems as learning aids, but with minimal mechanical calculations. Each page of Self-Test questions is followed by a set of carefully discussed answers, which allow students to evaluate their performance before proceeding to the next page of questions. The Self-Tests and problems have been chosen to improve the student's ability to use statistical concepts rather than just to provide practice in computation. Additional problems are included in *Using Statistics,* the study guide that accompanies this book.

Understanding Statistics is the creation of many people whose contributions I must acknowledge. At the beginning were my own teachers, A. H. Shephard and the late Edwin Belyea, who planted the idea that statistics could be elegant and enjoyable. A vast number of students in my statistics course at the University of Victoria applied the final polish to many of my rough ideas; outstanding among them were Peter Johnson, Ardis Bicknell-Brown, and James B. Brown. Many of my colleagues are the unwitting donors of ideas that I incorporated to good effect.

The manuscript, in one or more versions, was read thoroughly by Thomas B. Hamilton of Seattle University, David H. Krantz of the University of Michigan, Lawrence Lapin of San José State University, and Leroy Wolins of Iowa State University. Their comments and suggestions have been comprehensive and helpful, and I thank them all. The staff of Harcourt Brace Jovanovich, Inc., have also been helpful, supportive, patient, and determined that you shall have the best possible statistics book. I thank the publishing team of Judith Greissman, Judy Burke, Gail Lemkowitz, Abbi Winograd, Nancy Kirsh, Marian Griffith, and Kenzi Sugihara for all their efforts. I am also grateful to Tony Hall, whose original cartoons have enlivened the presentation.

Every author is indebted to a few people who defy categorization but who deserve thanks for all sorts of miscellaneous support. For me, these people are my wife, Iris, who took time from her own professional career to lend support; Kevin, Heather, and Edward Wright; Randi Masters; my parents, Robert Hamilton Wright and Kathleen Joan Wright; and Doris Balch.

I am grateful to the Literary Executor of the late Sir Ronald A. Fisher, F.R.S., to Dr. Frank Yates, F.R.S., and to Longman Group Ltd., London, for permission to reprint Tables III and IV from their book *Statistical Tables for Biological, Agricultural and Medical Research* (6th edition, 1974.)

Lastly, I thank my employer, the University of Victoria, for being the kind of place, with the kind of people, that makes possible the conception of a book like this one and the time to write it.

R. L. D. Wright

contents

4 averages: measuring togetherness   90

5 variability: measuring apartness   111

6 percentiles and standard scores   143

7 probability: measuring likelihood   170

8 samples and populations   211

9 correlation: measuring relatedness 239

10 regression: measuring predictability 278

|| experiments: mathematics meets reality 303

|2 the *t* test 341

|3 analysis of variance and the *F* test 371

14 how to do two experiments at once 402

15 nonparametric statistics 425

chapter 1
some argument, some arithmetic, and a little psychotherapy

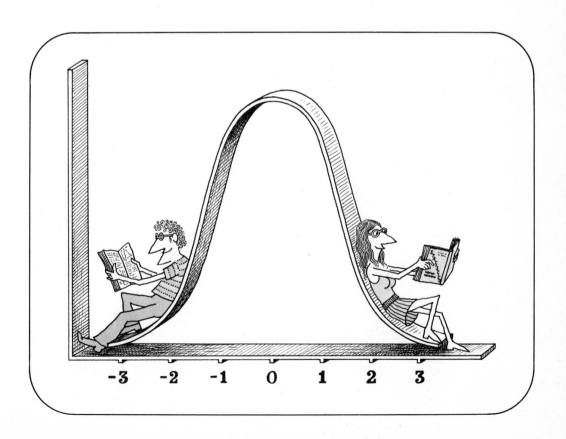

This book was written in response to a widespread need on the part of students. It was teaching statistics to students who were not mathematics or statistics majors that drew my attention to this need. Although I am a psychologist who knew very little about statistics, one day I suddenly found myself having to teach it to a roomful of students who knew, if possible, even less about it. To make matters worse, statistics is, as everyone knows, a branch of applied mathematics, but I soon found that nobody in the room, myself included, knew much mathematics. Things looked pretty hopeless.

However, the students seemed to like the course, and an amazing number of them went on to more advanced courses and did quite well, so there seemed to be no question that they had actually been learning statistics rather than just goofing off for a semester. The suspicion dawned on me that statistics might really be easier than previously suspected, and the present book is offered as evidence that it is.

How can this be possible? Because, as it turns out, statistical *concepts* are plausible, simple, and appealing, even if statistical *formalities* (or formulas) and *techniques* are difficult and boring. Once you have a good conceptual grasp of the topic, you have it pretty well made. What's more, the formalities and techniques then become tame and cooperative, fitting easily into your grand scheme of things. You wonder: Is this really true? Let us see if you qualify as the typical student in a course like this one.

- Math background is shaky; motivation is shakier.
- Basically, you doubt that you were *born* with adequate mathematical ability.
- Certainly you doubt *now* that you have the time, energy, or inclination to do statistics.
- Lacking spiritual communion with the statistics nut, you feel resentful, guilty, even a bit stupid when confronted with the need to learn statistics. It all seems hopelessly beyond you and therefore (such is psychology) beneath your dignity.
- But you *are* a student, even a serious one if only somebody would give you the chance! Maybe you're inclined, even if only tentatively, to stay with it.

This book is intended to ease much of the pain of doing just that.

Unfortunately, textbooks tend to justify their own existence by treating their subject matter as very serious and horribly difficult. (This tendency is called pedantry.) The Pedants Union typically ignores

2

#001   21-11-2016 6:16PM
Item(s) checked out to p1178927x.

TLE: UNDERSTANDING STATISTICS: AN INFO
THOR: WRIGHT, R L D
RCODE: 77123101
E DATE: 12-12-16

Please keep this receipt

books that treat serious topics in happy ways. But this is not a book for pedants—it is a book for the typical student just described.

To many people, mathematical thinking is a hardship, but in fact it doesn't have to be. One probable source of this problem is that mathematics is often badly taught. It's likely that you have not learned to read, to vocalize, the *language* of mathematics. Only after you can decipher something can you understand it. For example, can you speak the following phrase? Can you understand its meaning?

$$s_{\bar{x}} = \sqrt{\frac{\sum X^2 - (\sum X^2)}{n^2(n - 1)}}$$

Very likely not, and foreign languages can be terrifying, like nightmares, if people assume that you can understand and speak them when you really can't. It's embarrassing to confess such ignorance—like having to ask what your boss's name is after you've been on a job six months. Such embarrassment does not mean you are stupid or incompetent, and certainly not that you are *incapable* of learning.

Humanity's quest for understanding nature has been encouraged by the finding that great basic truths, great insights, are essentially very simple. The *basic* concepts of statistics are exceedingly simple; they could hardly have been possible otherwise. Such bafflement as does arise in the student occurs as a result of two types of strategic error: (1) trying to replace basic understanding of fundamentals with detailed and overqualified descriptions of details—like trying to decorate the house before you've built it; (2) trying to learn the finished formalities of a subject, as laid out in full mathematical grandeur, rather than sharing the groping, the ephemeral intuitions, the dreams, and sometimes the hallucinations, of those whose insights have changed the world. Insight always precedes theory; insight always involves the grand picture first, and the qualifications and little details afterward. The great chemist Kekulé, the father of organic chemistry, "discovered" the secret of organic molecules while dozing atop a London bus. "Let us learn to dream," he said later, "and then we may be able to discover the truth. But let us beware of publishing our dreams until we have put them to the test of the waking understanding." His priorities were correct. Archimedes "discovered" specific gravity while almost afloat in his bath, but he certainly did not run skinny through the streets of Athens shouting "Eureka! $P_{G+W} = W_2/(W_1/P_G + W_2 - W_1)$!" The things we understand best, and use most comfortably, are the

Figure 1.1
Insight ALWAYS precedes formal theory (*by courtesy of American Scientist*).

things we are allowed to understand and only later subject to the fussiness of formula making.

It is that "Eureka!" feeling, that "gut comprehension," that we want to encourage in you. Mathematical intricacies are secondary; in some places, we will avoid them altogether. You don't need to be a statistician or a mathematician *already* to read this book.

So what *do* you need for this book? Not very much:

1 You should be able to read and understand English. But if you can't, there are better things to learn than statistics.
2 Then, you must actually read the book. Just *having* it won't do. Knowledge exists in people, not books.

3 You should know how to study. This may be your main problem anyway. Some wisdom on this point later.

4 You should be normally intelligent. Having been admitted to college and gotten far enough to need this course, you probably are, so don't worry about it.

5 You should know how to **add**, **subtract**, **multiply**, and **divide**. Yes, that's all; as in the fifth grade. But now, you can even use desk (or shirt-pocket) calculators (which aren't even essential), which you couldn't use in the fifth grade.

For all of the above, you are, to a great extent, on your own. However, we can help you with two of the items: the matter of study habits, and the arithmetic.

how to study and how not to

It's a familiar scene: the night before the important quiz. You are at your desk, cleared for action, the TV silenced and the roommate evicted (downstairs, probably, carousing). Belt loosened, shoes off, feet carefully flat on floor. A deep breath; the book (THE Book) reverently opened, possibly for the first time, at Chapter One, Page 1. It's all in there, all 453 pages of it. Let's see . . . at 40 pages an hour, less than one page per minute. Not bad at all, you'll be finished by 3:00 AM, easy. It's all in there, just got to move it from the page to the mind, but don't miss anything. . . .

You have been reading Page 1, Paragraph 1, Chapter 1. Now you're on Page 2. What have you been reading? A little stab of panic . . . you can't remember! Frowning slightly, eyes boring holes in page, you turn back to the beginning. Concentration, spurred on by slowly growing concern. . . .

Three hours later: Chapter 2, Page 32. Eyes hurt, slight headache. Glance at watch . . . physical panic–pain: 421 pages left, but one-third of time gone. Plunge onward, skimming now. Forget about learning everything: get the essentials. . . . Later: time for short break. Rise, stagger glassy-eyed downstairs. . . . "Man! Have I been *working*??!!"

Sound familiar? It is to most students. Unfortunately, though heroic, it is *not* studying. Next morning, you teeter into the examination room; exhausted; open the exam—ZONK. Mind goes blank. End of scene.

Studying is rather like a proper combination of eating and exercise. You need each to balance the other, and the food ideally alternates with the exercise regularly and constantly. In study, the analogy to feeding is the ingestion of information; the analogy to exercise is the integration of the new information with the old and the use of the new information in altering slightly your grasp of the subject matter as a whole. In study, then, *time* is of the essence. You need lots of it, but in small, balanced, daily doses. The new material ingested in these doses will continue to digest, to rearrange itself, to settle in comfortably, during the periods between study. *Go for basic comprehension, not just facts.* Comprehend, and the facts will be much easier to remember.

It often helps to make an outline of points as you go along. Ensure that you do so in a way that stresses the **organization** of the points. Don't memorize the pages; instead try to understand the message. Keep on integrating new points with old knowledge. Keep asking yourself, "How does this tie in with that?" Move slowly, comfortably—there's lots of time. Knowledge is fascinating for its own sake; why bother if you don't like it? Study so as to really appreciate, not just to pass the examination. Anyway, the exam should care for itself: the only question will be whether your grade will be a high A or a low A.

Stop periodically. The mind needs to burp sometimes. You needn't rest too long—a minute or two every ten minutes. If you can find another person studying the same material, interact. Ten minutes of discussion (maybe spelled *argument*) for every ten minutes of reading may not be enough. Test yourself constantly as you go; invent little exercises and then do them. Remember that comprehension is not the same as highlighting your text. If it helps, remember how much you are paying to be a student; get a good feeling about being one. Think happy.

Examinations should almost be casual. If you really understand it all, what is there to terrify you? Questions you can't answer from memory? Then stop and think a while. If nothing else, you can answer with wisdom rather than knowledge. Wisdom might be defined as How to Handle Ignorance Creatively.

In any course, especially this one, try to analyze your basic attitudes. Is this a game, between yourself and the professor (The Enemy), in which your main effort is The Great Con Job, rather than (as it were) giving in and learning what there is to be learned? *Really* learning it? This is often quite a problem in unpopular courses like statistics, where the unpopularity tends to generalize to the teacher. Confrontation aside, all professors were once students, and many of them understand your dilemmas better than you might think. Try to give him or her, a fair chance.

Finally, keep a sense of proportion. If going to school isn't fun, then either make it fun or leave school. There are other ways to Fame and Fortune. But if you stay, make it fun, keep a reasonable perspective on the day-by-day (or course-by-course) difficulties, and keep a sense of humor. Growth toward the profession of Capable Thinker has not always been easy, but never awful, either. For the point of view of a career, consider your college experience a lengthy training in how to use your mind effectively, in which the opportunity to think logically and quantitatively (from this course) adds to, rather than detracts from, your development as an Educated Person.

arithmetic: a therapeutic review

Numbers are fun if you have the right attitude toward them. Given the chance, they are often quite friendly. For example, there is a very simple way of telling quickly whether a number, even a very large one, is evenly divisible by 9, without taking the trouble to divide it by 9. Simply add all the digits of the number, and if the answer has more than one digit, add all of *them,* and so on until only one digit remains. If the answer is 9, then the original number is an even multiple of 9; otherwise it is not.

$$81: 8 + 1 = 9$$
$$54: 5 + 4 = 9$$
$$5301: 5 + 3 + 0 + 1 = 9$$
$$7812: 7 + 8 + 1 + 2 = 18; \quad 1 + 8 = 9$$
$$673227: 6 + 7 + 3 + 2 + 2 + 7 = 27; \quad 2 + 7 = 9$$

Isn't that friendly? If you want to know whether a number is evenly divisible by 3, use the same method, but now the total may be 3, 6, or 9. You don't *have* to enjoy playing with numbers, but it's not difficult and it helps. If you are already an expert with numbers, you may feel like skipping this section.

addition and subtraction

These are the basic operations in all arithmetic, and you already know how to do them, in principle if not in accuracy. The basic notion in addition is the *procedure* of combining specified quantities to give a new, larger quantity. Subtraction is the opposite. The operation is commonly set up as an **identity**, or **equation**, in which the way the **quantities** are to be combined is denoted by means of **operators**. Examples of operators are such signs as + and −. Thus

$$1 + \quad 2 + \quad 3 = 6$$
$$8 + \quad 1 - \quad 9 = 0$$
$$2 + (-1) + (+4) = 5$$

↑ ↑ ↑   ↑   ↑   ↑ ↑

quantity operator quantity operator quantity identity quantity

Here you will note that the sign, + or −, can be either an operator or a part of the quantity. The rules of procedure are the same, but it helps to maintain the distinction. In this light, the operations like

$8 - -3 = 11$	*Note:* Examples like these will "scan" properly in your
just as $8 + \quad 3 = 11$	reading if you pronounce them as though they were
and $\quad 7 + -2 = 5$	written out. Thus, "eight minus minus three equals
just as $7 - \quad 2 = 5$	eleven, just as eight plus three equals eleven."

cease to be mere mathematical busywork, and take on some sense.

Why force this issue of **operators** as opposed to **quantities**? Because a lot of confusion results from failure to make this distinction. It is hardly a difficult distinction: any child who can cook is able to distinguish between "amounts" (two cups, one pound) and "directions" (add, stir, simmer) in the recipe book.

Here are some rules for addition and subtraction.

1 A number is assumed to be positive (+) when there is no sign in front of it. Hence, 2 means +2.
2 You can arrange a set of numbers being added in any order you wish; the answer won't be affected. This is not always true in arithmetic, but it always holds for addition (whether of positive or of negative numbers).
3 The operation of subtracting an already negative number is equivalent to adding a number that is positive. If you subtract $3.00 from what you owe, that's like adding $3.00 to your assets.
4 Adding a negative number is equivalent to subtracting a positive number. Adding a $3.00 debt to your books is like subtracting $3.00 from your pocket.

5 Sometimes, you must subtract a larger number from a smaller one, both numbers being positive. 5 minus 8 = what? If the temperature is 5 degrees and drops 8 degrees overnight, the new temperature is −3 degrees. Arithmetically in such cases, merely substract the smaller from the larger number and make the answer negative. Thus 5 − 8 = −3, while 8 − 5 = +3.

A frequent dilemma is the role of 0 (zero), in mathematical operations. The question seems to be whether 0 is really a *number,* in the sense of a quantity; is a zero quantity still "something"? It helps greatly here if numbers are visualized as lying on a scale centered on 0, with positive numbers extending to the right and negative ones to the left:

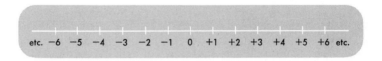

Then, while the sign (+ or −) denotes the **location** on the scale (left or right of 0), the *operation* of addition refers to the **direction of movement** on the scale. To add a positive number starting anywhere, you move to the right; to subtract a positive number (the opposite of addition), you move to the left. If the number being added is negative, this is like subtracting; if the number being subtracted is negative, this is like adding. In all of these instances it is possible to cross the 0 point, and a glance at the scale above should show that 0 is a numeral to be counted just like any other when you're stepping off any added or subtracted value. Supposing that the scale above refers to inches, find the number of inches between −4 and +3, counting along the scale. Then 0 is counted. Actually, 0 does have a unique function in addition. It is the point on the sequence from which another number is measured or counted to give the notion of **quantity**, pertaining to that number. That is, quantity is conceived as *distance from 0* on the scale, not directly as merely a *point* on the scale. In statistics a prime concern is to convert measurements so that 0 is the point from which comparisons are made, when the data don't have this feature initially.

absolute numbers

Sometimes, fortunately rarely, a number is important only for showing quantity, with no regard to sign. Such a number is termed an **absolute**

number and is written with a vertical bar on each side, e.g., $|6|$ (best read as "the absolute value of six").

In one sense, absolute numbers are treated as positive, inasmuch as they do denote quantity. Quantity alone doth not arithmetic make. However, when absolute values are submitted to arithmetical analysis, problems do arise. An excellent example is the mean deviation, considered in Chapter 4. When the mean deviation is calculated, the absolute differences between scores and their mean are treated uniformly as being positive. The result is a value that can literally be called the average distance of scores from the mean. But the mean itself is defined as being at the point of balance where that average distance is 0. A paradox is presented. It is resolved by the fact that an important arithmetical rule was suspended during the calculation of the mean deviation, namely, that scores have not only a *distance* away from their mean, but also a *direction* in which that distance is located.

Direction as well as distance is implicit in how we learn our basic arithmetic in grammar school. No six-year-old would mistake a person in debt for $1,000,000 as a millionaire. It is interesting to speculate that such lack of confusion, from the very earliest stages of infant development, is due to the earlier development of directional sense, and a relatively later development of distance concepts. A baby will accurately reach toward an object that is ten feet away. The concept of anything being a certain distance away, but in no particular direction, does not square with our experiences, so such concepts as absolute values are alien to our ways of thinking.

Let us return to the subject of **operators**. We have another problem to overcome. When we are adding we use the + sign between the numbers being added, so the latter must actually be present. What happens when we want to denote addition in general, without the specific numbers actually being present? The answer is easy: you say (or are told): "George, add up those numbers." This is quite equivalent, but much more convenient than saying: "George, find the sum of 2 and 3 and 9 and 7 and 5 and 8 and 11 and. . . ." It would be nice to have a convenient shorthand symbol to replace the + sign for such general instructions. We have such a symbol: it is the Greek letter Σ, sigma (the Greek for S), which symbolizes the instruction to **sum**. It can also mean **the sum of**. . . . Pronounce the Σ as "sum" and *not* as "that funny-looking E." It isn't an E. Only students of Greek literature need call it **sigma**. A second nice thing would be to have some symbol for the values to be added. You could be told to add "those scores," but the expression

$$\Sigma \text{ those scores}$$

isn't very dignified. Nor is it very efficient. So we have the tradition of letting some letter in the alphabet stand for the scores that are to be added. Usually the capital letter X is employed. Occasionally we have two sets of scores that must be treated separately, so the first set is X and the second is Y. So

$$\Sigma X \quad \text{is the same as} \quad X_1 + X_2 + X_3 + \ldots + X_n$$

and

$$\Sigma Y \quad \text{is the same as} \quad Y_1 + Y_2 + Y_3 + \ldots + Y_n.$$

Note that the subscripts refer to the first, second, third, ... and nth scores in the series, where n is the total *number* of scores. Since n can be any value, including very large ones, the ΣX notation rescues us from the potentially horrible fate of listing all the scores.

We'll find the same principle extended to the addition of other values as well. For instance, ΣX^2 means to summate the squared values of X, just as ΣY^2 means to. ... You've got it!!

An important therapeutic point for those in need: We are only doing simple arithmetic, adding numbers. The symbols X, Y, x, y (the latter two are different from the former two, as we'll see) do not stand for *unknowns* as they do in pure algebra. Formulas in this book do not get *solved* as they do in algebra. Statistics involves formulas merely as abbreviated arithmetical instructions. And, arithmetic may be a bore, but it's no terror.

equality

The equals sign ($=$) is already very familiar to you, and nothing could be easier than the concept of equality. There are, however, at least three different shades of meaning embodied in the sign $=$.

1 $=$ can mean that two sets of values *are* equal; that is the notion of **identity** mentioned earlier.

2 $=$ can denote an "if ... then" situation. *If* you add these values of X, *then* you will have a given value for the total.

3 $=$ can imply a situation of **probability** rather than certainty. Tossing a coin gives a 50:50 chance, or a 0.50 probability, of getting heads: If we let p stand for the probability, then we can write

$$P_{\text{heads}} = 0.50$$

But what you *get* doesn't equal 0.50; you *get* either a head or a tail. Such situations arise quite frequently in statistics. They are easily recognized, and you will become quite accustomed to them. Sometimes you may see other signs besides =. These are as follows.

≠ A slashed equals sign. It means "does not equal." 3 ≠ 5, as you have always known.

≅ or ≈ An equals sign with a wave over it, or a "watery equals" sign, constructed of two wavy lines. Both mean "approximately equals."

→ An arrow. It means "becomes almost equal to" or "approaches." For example, if you add all possible positive numbers, 1 + 2 + 3 + 4 + 5 + . . ., the total will approach infinity (written as ∞).

One final remark about equality. As treated here, the term refers to *quantitative,* or numerical, equivalence. We exclude the other connotations, such as merit, value judgment, and quality. 10 cats = 10 dogs, in number, but not in talent for barking.

multiplication and division

Once again, these are things you once learned in school, although possibly you are not much with it, not any more. Forget the mechanical details: for those you can use a calculating machine, or buy a slide rule. We must assume that you can do very simple multiplication and division in your head, or on paper.

Consider the expression $4 \times 3 = 12$. We see two numbers, 4 and 3, which, combined in a certain way, give a product, 12. Does this mesh with what we already know about **quantities** being combined by means of **operators**? If so, what are the quantities and what are the operators in the expression $4 \times 3 = 12$?

Let's translate it into a problem of addition.

$$4 + 4 + 4 = 12$$

Four, taken three times, equals twelve. You will note that while four and twelve are quantities, the three is not. Three is a type of operator: an instruction to "take the quantity 4 three times." Generally, multiplication may be understood as a shorthand for repeated addition of a number to itself, a specified number of times. Note that the above simple problem could also be written

$$3 + 3 + 3 + 3 = 12$$

Here, the positions of the quantity and the operator have been reversed, without any effect on the product. $4 \times 3 = 3 \times 4 = 12$. Such a situation, in which the order of arithmetic calculation is immaterial to the result, is termed **commutative**. Sometimes, however, calculations must proceed in a definite order. Some such situations are noted below.

The symbols for multiplication scarcely need concern us. The procedure is denoted by $\times$, by a dot $\cdot$, or by no symbol at all. Depending upon context, 4×4 can be written as $(4) \cdot (4)$ or $(4)(4)$. The latter convention is quite frequent in statistical formulas.

Division as a topic is also familiar to you; it might be described as the reverse of multiplication. It is more difficult, however, to make it intuitive. Division is the exact reverse of multiplication. Dividing 12 by 3 means subtracting 3 from 12 repeatedly until nothing remains. So

$$12 - 3 - 3 - 3 - 3 = 0$$

The result, or quotient, is the number of times 3 must be subtracted: in this example, 4. The operation is subtraction, and the quotient is the number of times the operation was performed. This example works equally well when 12 is divided by 4 and the quotient is 3.

A SLASHED EQUALS SIGN IN THE MAKING

multiplying and dividing negative numbers

If you followed the earlier discussion about addition and subtraction of positive and negative numbers, multiplication and division should be an easy extension of the same logic. For example,

$$4 \times -3 = -3 + -3 + -3 + -3 = -12$$

and

$$-3 \times -3 = \quad 0 - -3 - -3 - -3 = +9$$

The actual working rule is even simpler: If the number of minus signs in the expression is even, the product is positive; otherwise, it is negative. Exactly the same rule pertains to division: If the dividend and divisor are both negative, the quotient is positive. If only one of the two is negative, then the quotient is negative also.

raising numbers to powers: squaring and taking square roots

By raising to a power, we mean only that a number is multiplied by itself a specified number of times. The total number of times is called the **power** of that number. For example:

$$2 \times 2 \times 2 \times 2 \times 2 = 32$$

This is 2 taken five times, or raised to the power 5, and for convenience we may write instead

$$2^5 = 32$$

The **power**, as written here in superscript, is also called the **exponent**. By far the most common power to which we raise a number is 2. The **second power** of a number is also called the **square** of that number: thus, 2^2 (two squared) = 4; 3^2 (three squared) = 9, etc. It should be apparent that the square of any number must be positive. For example,

$$5^2 = 25$$

and, as well, $(-5)^2 = -5 \times -5 = +25$, because the number of minus signs in the expression is even: two. The expression

$$-5 \times +5 = -25$$

is not an example of squaring, because -5 and $+5$ are not the same number. The main usefulness of squaring in statistics is this very feature: negative numbers can be made positive by squaring them, provided that ways can be found to unsquare them again when necessary.

Remember that an exponent applies only to the numeral or value immediately before it. So $3s^2$ is $3 \times s \times s$, not $3s \times 3s$. The two results are different! If you want to get the latter, use parentheses: $(3s)^2$.

Square roots are the reverse of squares, but more difficult intuitively and much harder to calculate. Intuitively, try to think of a square root of a given number X as "that number which, if squared, would give the value X." It should be apparent that every number must hypothetically have two square roots: one positive and one negative. Usually we concern ourselves only with the positive one. Negative numbers cannot have square roots; there is no way that a square can be negative. However, a negative number can be multiplied by -1, which doesn't affect its value, only its sign, and you could find the square root of the result. There must be some good reason, however, if you are to multiply by -1. It can't be done just for convenience.

How are square roots calculated? Here are eight ways:

1 Work it by hand, arithmetically. This is bothersome but, fortunately, unnecessary.

2 Refer to a Table of Squares and Square Roots. There is one at the back of this book.

3 Use a Table of Common Logarithms. Your instructor would be delighted to show you how.

4 Use the square root button, probably marked $\sqrt{}$, on the nearest electronic calculator.

5 A slide rule, for less than \$5.00, will give you an excellent tool for taking square roots, as well as a good mathematical image.

6 The value of a square root may be *approximated* by squaring numbers that you guess are close to it, and seeing how close your squared values fall to the problem value. You want the square root of 30? 5^2 is 25 and 6^2 is 36, so the square root of 30 must be somewhere between 5 and 6. Now, why not move even closer by squaring 5.5? And so on.

7 Keep your intelligence at work. Do you actually need to find the square root? Example: on an examination you might be asked to find the product of $\sqrt{2} \times \sqrt{8}$. Looks bad. But if rewritten as $\sqrt{2 \times 8}$, the square root can now be done mentally. This clever stratagem is possible because order of multiplication and division is commutative, and taking square roots is a special variety of division. But remember that order of multiplication/division and addition/subtraction is not commutative, so $\sqrt{2} + \sqrt{8}$ or $\sqrt{2} - \sqrt{8}$ must be worked out the

hard way. Tests and examinations often segregate sheep and goats by such a gimmick.

8 If all else fails, ask the student next to you. Square roots have been the basis for some beautiful friendships.

factorials

Sometimes you may have seen exclamation marks ! mixed in with messy-looking formulas. Probably you felt surprised that anyone, even a math freak, could get so enthusiastic about all that.

We must admit, sadly, that ! denotes not enthusiasm but merely another operator. Suppose that you wanted to multiply

$$9 \times 8 \times 7 \times 6 \times 5 \times 4 \times 3 \times 2 \times 1$$

How would you express this problem except by writing it out in full? As for the problem of summation, a briefer expression is needed. It is called **factorial**, and its notation is !. So the above problem can be read as "nine factorial" or "factorial nine" and would be written 9!.

$$9! = 9 \times 8 \times 7 \times 6 \times 5 \times 4 \times 3 \times 2 \times 1$$

Factorials are of some importance in statistics, particularly in connection with the notion of probability. More on this in a later chapter.

combining addition, subtraction, multiplication, and division

We take it for granted that actions must occur in a proper order. This is true in cooking, dressing, and arithmetic. And, as in the other two, arithmetic sequences are just a matter of logic, a little memory, and experience. Unlike cooking and dressing, though, arithmetic provides definite rules in advance. They are quite simple, if arbitrary.

1 In the absence of guides to the contrary, multiplication and division are done before addition and subtraction. Thus,

$$4 + 3 \times 5 - 4 \div 2$$
equals
$$4 + 15 - 2 = 17$$

No other order of solution will give the correct answer.

2 An exponent attached to a quantity takes precedence over any other operation on that quantity. Thus

$$5 \times 3^2$$

would equal

$$5 \times 9 = 45$$

and not

$$15^2 = 225$$

3 Exceptions to the preceding rules are made clear by the use of brackets, or by the layout of the formula. Hence,

$$(5 \times 3)^2 \neq 5 \times 3^2$$

because

$$15^2 \neq 5 \times 9$$

or

$$225 \neq 45$$

As another example,

$$\Sigma X^2 \neq [\Sigma X]^2$$

in that the left-hand term means to square every value of X, *then* total the squared values, and the right-hand term means that the sum of X values is to be found first and then squared. The two answers will *not* be the same. Because this is a rather important point, let us work a short problem to illustrate.

> The values of X will be 2, 3, and 5. Working first to get ΣX^2, we first square each score to get 4, 9, and 25. Adding these squares gives a total of 38. We now calculate the value of $[\Sigma X]^2$ by adding the scores to give 10 and then squaring this total to give 100. Clearly, 38 is not the same answer as 100.

The above few lines may seem bothersome, but they are important. Make a point of understanding them now; the trouble saved later will reward you. In case you are still puzzled let's use an analogy: making pancakes. If the directions say to add milk to the flour and then beat, you'll find that first beating the flour and then adding milk will give you some of the worst pancakes on record. The good thing about expressing such sequential notions by means of "formula shorthand" is the complete removal of ambiguity—more than most cookbooks can claim. To repeat the rule on brackets, always work out the innermost bracketed quantities first, and work outwards.

About formulas: To most students who have a bad feeling about mathematics, formulas are The Enemy. Arrogant, inscrutable, impregnable. Certainly couched in an undecipherable foreign language that is too difficult to care about.

On the other hand, you have been dealing with formulas for most of your life, most recently over the past few pages. Possibly one's attitude depends more on terminology than on issues. A recipe is a formula; if it's for refueling a baby, it's even *called* a formula. But we all use recipes, knitting pattern books, and telephone numbers quite happily. All are examples of formulas; all consist of coded instructions for the fulfillment of some particular purpose. Already you can follow the formula that says, "Square these numbers, and then find the total of the squares." Since high school you have known how to calculate the "**average**" or **arithmetic mean** of a set of numbers, and now you can even understand the meaning of $\Sigma X/n$, which is the same thing written as a formula (n is the total number of quantities being averaged). If you can understand the meaning of the averaging formula above, you can understand any formula in this book, as well as most others in elementary statistics.

If formulas always look very hostile to you personally, it is probably a holdover from your days of analytic algebra (if you ever had it) or the stories you heard about it from other people (if you didn't). Perhaps you have often felt that mathematics is a gratuitous, useless, irrelevant, difficult, unnecessary bore with absolutely no good points. Especially algebra, which doesn't even have the sense to involve numbers (which you *might* understand), but fools around with letters instead. Even if you did cope with mathematics, you may have unpleasant memories of solving for X, for Y, for quadratic components, for linear components, for slope—and often getting them wrong. You definitely probably deserve to have bad memories.

Fundamentally, such difficulties involve *algebra* rather than *arithmetic*. Simple arithmetic may be dull and annoying, but at least it's familiar and manageable. Now note: using and understanding basic statistics requires *only* basic arithmetic ability. This is not to say that algebra is irrelevant, only that it is unecessary for *our* purposes. Statistical formulas *can* be played with, using algebra, and students who are so inclined will enjoy doing it. Algebra *is* necessary for developing special-purpose **working version formulas** from the simpler **theoretical** versions. But assuming that you are prepared to accept our word for formula equivalents, you can forget about algebra. If you really want to convince yourself that one formula is the same as another, you can do so by substituting simple invented arithmetic data into them and seeing that they work.

counting versus measuring

When you count objects, you arrive at a total that is absolutely precise. It may be wrong—inaccurate—but it will be exact (this distinction between *accuracy* and *precision* should be noted); for example, there are exactly 23 letters in the heading to this paragraph. Not 23.0 or 23.0000000—such a use of decimal places is not wrong, but it is unnecessary and grotesque. Such numbers, which are achieved by the process of **counting**, are termed **integers**. In statistical work as treated so far, the best example of an integer is n, the number of scores in a sum or the number of subjects in an experiment (or even both of these if one subject gives one score).

By comparison, if you measure something, you are not likely to find an exact total. Your notebook is not *exactly* 11 inches long; the measurement will vary depending upon how much care, or how accurate a ruler, you use in making it. In this example it would not always be adequate to say "the measurement is 11 inches," because for some purposes this approximation would not be close enough. Perform very exact measurement and you may find 10.9758365 inches, but the decimal does not end there. You have merely stopped there, with a close approximation of the length, but an approximation nonetheless.

Suppose that the measurement *were* exactly 11.0000000 inches. Would it suffice to write merely 11 inches? Perhaps. You may not care about seven-decimal precision. But if you do, you should include all the decimal places justified by the precision of your measure, even if the decimals are 0. The figure 11 does not signify the same as the figure 11.0000000, and neither signifies the same thing as 11.0. The differences lie in the degree of precision attained: the first value, 11, merely means somewhere between 10.5 and 11.5; the second is far more precise—ten million times more precise—and the third is also more precise, but only ten times more. Measurement would cease to have much point if some degree of precision were not specified. How much precision depends strictly upon your purposes.

Such a situation is rather unsettling to some people. Suddenly it appears that measurement cannot arrive at the exact "truth" about things being measured. This is true. However, for practical purposes approximation can be made to suffice. In statistics generally we find that "truth" is necessarily replaced by "suitably precise approximation." Integers, which *are* absolutely precise, are the only exception.

Even integers retain precision only if you do nothing more to them. Integers added to or subtracted from other integers will still give integers, but should you *divide* an integer by another integer, the quotient may not be an integer. For example, the integer 5 when divided by 3 gives the quotient 1.66666. . . . If we round this number off (and scientists always do), whether as 1.67 or 1.666667 or as 1.66666-666667, we lose some precision. Wisdom dictates that we keep enough decimal places for our own purposes, whatever they are. In fact, the same rules apply here as pertain to measured quantities. Ratios have the same properties as quantities.

Why are these matters important to our purposes? They are not directly important. But perhaps you should always use some arbitrary number of decimal places (you choose them) when working with quantities or ratios, but not when you are using integers.

significant figures

Discussion of decimals, which can be carried out an infinite distance from the decimal point, raises the question, "How far is far enough but not too far?" There are several simple rules.

1 In the original data measurements, be as precise as possible. This means that your precision of measurement should be at least ten times (one decimal place) finer than the variations in the data themselves. Otherwise, the data variations are going to be lost. You can't pull out small fleas with a coarse comb.
2 In recording your data, use the number of decimal places justified by your measurement. If you measured length to the nearest millimeter, then don't talk about tenths of millimeters.
3 In analyzing your data, i.e., performing arithmetical operations on them, do not round off any numbers if you can avoid doing so. This could mean that you wind up with a dozen decimal places in a product of three quantities, if each had four places to start with. Naturally, you should use all the decimal places available from your data. Why go to the trouble of gaining precision and then not using it?
4 Once only, at the very end of your calculations, may you round off excess decimal places. Perhaps you will want to retain one or two more than were embodied in the data.

You may protest that arithmetic is hard enough without having to hang on to dozens of decimal places. Nonsense. With any luck or artistry you'll be using a calculating machine anyway.

The term **significant figures** may now be more understandable. By this term is meant figures which, if discarded, would remove data from a calculation—data that are valid and possibly important. You should note too that significant figures do not include only decimal places. To the left of the decimal place, all figures are significant as far as the left-most value that is other than 0. For example, there are ten significant figures in the number

100014.0010

Perhaps you recall from your previous encounters with mathematics that the last 0 above should not be construed as significant. From a purely mathematical standpoint it is not. However, if the last decimal place refers to actual *measurement,* i.e., to four decimal places of precision, then it is a significant figure. Pure mathematics differs somewhat from the mathematics of measurement.

Rounding off may be accomplished in two ways.

1 Rounding up or down to the nearest value. Thus, 2.18 becomes 2.2, and 2.13 becomes 2.1, if we are rounding to one decimal place.

2 Merely dropping the excess numbers to the right of the cut-off point. Thus 2.18 and 2.13, rounded to one decimal place, would both become merely 2.1.

The choice between these two methods should perhaps be discussed with your instructor. A case could be made for either. The usual method is the first one. Its only problem is what to do when, for example, 2.15 needs rounding to one decimal place. The rounded-off digit, 5, is equidistant from between 2.10 and 2.20. The conventional answer here is to round toward the *even* number. 2.2. is even and 2.1 is odd, so 2.15 becomes 2.2. 2.25 would also become 2.2, not 2.3.

The second method is what your calculating machine does. Actually, if you have enough decimal places to start with, the choice of method does not make much difference.

summary

The intent of this book is to give an intuitive, not mathematical, approach to statistics. The main formulas given will have to be interpretable by you (and you already know a bit about this), but algebra games are out. There is a penalty: there are many ways of expressing the same formula, and you may not recognize (not right away) strange versions in unfamiliar statistics books. However, if you can accept our word for it, we'll give you the appropriate **working formula** for each **theoretical formula** whenever required. You should be careful to distinguish between the use of *algebra* in tinkering with formulas, and the use of *arithmetic* in using them. If you find, as you should, that the arithmetic is uncomplicated, then yours is not to reason why. You may enjoy playing with the algebra, too, but it isn't necessary in this course.

Having come this far, you are likely to be doubting the earlier implication of "statistics *without* mathematics." So far it's been *all* mathematics. So here is the truth: now that you have gotten through the foregoing, that's all the mathematics there is. Nothing that follows will require more. Now you can relax and enjoy it.

Possibly a review would be in order first. The basic concepts that you should know are

1 The distinction between **quantities** and **integers**.
2 The notion of **operators** as distinct from the above.
3 The significance of **arithmetic sign**, + or −.
4 The rules for addition and subtraction.
5 **Absolute numbers** and why you should avoid them.
6 The symbols for **squaring**, **square roots**, **summation**, and **factorial**.
7 How arithmetic sign is determined by multiplication, division, and and squaring.
8 How to find square roots.
9 **Commutativity** and its implications, including examples.
10 **Significant figures** and how to keep orderly in your calculations.

As both a study method and a morale-improving self-test, you might want to make the above into a hypothetical examination and try to answer verbally (to another student) or even in writing. You might even maneuver a real mathematician into checking on you. But first you should acquaint him or her with your purposes and with this chapter. Otherwise he or she may expect you to know more mathematics than is needed for this course. At any rate, do not proceed until you have a good feeling about the foregoing mathematics.

statistics and the behavioral sciences: a meeting of minds?

Just a glance through the table of contents of the usual statistics text shows you such words as percentile, variance, standard deviation, correlation, t test, regression, and much else. Your initial reaction to such terms, and to the approach underlying them, may be negative, and this is understandable; having elected to enter the social sciences, you may not be pleased to find yourself embroiled in mathematics.

The social sciences (such as anthropology, psychology, and sociology) seem at first glance to be more a matter of description and insight than of statistical measurement. Biosocial areas such as ecology, nursing, and medicine seem a bit more plausibly connected to statistical methods, but still not very importantly so. Economics and education seem even closer to the use of numbers, but still not as intimately as the "hard" sciences like chemistry, physics, astrophysics, and engineering. Students who enter the latter expect to study mathematics; those who enter the social and behavioral sciences do not.

The beginning student of psychology may see it as largely common sense, dignified by a diploma and a white coat. Sociology students often have an initial idea that they will shortly become social workers. Many people entering the professional areas of nursing, medicine, and education expect to need only memorized facts and the insightful application of them. Such impressions are incorrect, although understandable. As will be seen throughout this book, statistical procedures are applicable and necessary in these disciplines. They offer a way of making decisions based on a lot of data that may not point clearly, at first glance, to any decision at all. Such decisions are often needed in these fields.

In some very fundamental ways, the physical sciences and social sciences are similar. A major similarity is their reliance on sound methodology, critical judgment, and disciplined thinking. Unfortunately, popular media have dramatized the *results* of scientific methodology without equal time for its requisite discipline. How many ten-year-olds dream of physics in terms of calculus? Very few; but in terms of a space mission to Mars, lots of children have stars in their eyes. The same contrast holds for the social sciences. The glamor of doing field research, of helping the victims of poverty or illness, and of teaching are well known, but the methodology and discipline needed to do these things successfully are not. The essential methodological foundations of all sciences tend to be invisible.

Yet the mastery of *methodology* is what distinguishes the scientist from the interested and informed lay person. In the case of the social sciences, where there pre-exists so much ill-founded popular belief (about race, social classes, the way human minds work, and so on), strong methodologies are necessary if we are to avoid biased, hasty, unwarranted conclusions. Certain habits and procedures must be employed, more or less formally, if the quicksands of illusion are to be removed from the underlying bedrock of reality.

The habits and procedures are of two main sorts: *research design* and *statistics*. The former involves the various principles by which we seek to avoid attributing effects to the wrong causes; more on this in a future chapter. The latter, statistics, guides the finer details of design and analyzes the results of the research. There is no way that bad research design can be remedied by statistics; both must be adequate if we are to avoid goofy, persuasive, but incorrect conclusions. Since the average person lacks these techniques, it is hardly surprising that some very strange beliefs are to be found in human society. Our quest for knowledge about nature (including ourselves) often resembles a contest between humanity and nature: we try to understand nature, and nature tries (usually successfully) to mislead us. Nature has the advantage of having designed us as very credulous beings, whose credulity may be complicated further by a basic need for mystery going beyond workaday explanations. People should not be blamed for being the way nature made them, which is very intelligent and capable of abstract thinking. But, frequently, abstract thinking requires collaboration with self-disciplined tests of reality. And all too often, the latter have been unavailable. Our history is resplendent with beliefs in magic, witchcraft, flat earth, gods made in the image of humanity, racial supremacy and inferiority, ad infinitum and ad nauseum. We haven't many advantages in this unequal contest, but statistics is one of them.

Statistics is like any potent technique; it must be used with knowledge and wisdom. Otherwise it will mislead instead of enlighten. You may ask, if we get lies as well as truth, why regard statistics as having any advantage over other methods of discovering truth? Because statistics is *our* weapon, not nature's, and *if done properly will not mislead us*. Nature gets the last blow, however, having designed us so that we find statistical thinking unpleasant—until we get used to employing it, anyway. Like it or not, the social and behavioral sciences must use it. Here are some examples of how statistical thinking is relevant—no, *essential*—to them.

example 1

Imagine yourself a respected and progressive educator. You believe that free study should produce better learning than the lock-step formal discipline still found in some schools, but mere belief is not evidence. You need evidence. Somebody points out that your belief is *obviously* correct. *You still need evidence.* (Five hundred years ago, certain women were *obviously* witches, too.) Somebody else points out that another research team, Gugelpranz and Dewlap, have already found that your belief is correct. You look up their article. Sure enough, they found that Hawaiian ninth graders (who, in a Maui high school, were studying the history of the hula by rather free-study methods) achieved significantly higher GPAs[1] than did an eighth-grade class in Hanover, Minnesota (where very formal methods were being employed in the study of the influence of Bismarck on the Austrian Question of 1848).[2] However, you *suspect* that other variables, like age, geographic location, culture, and subject matter, might have produced or influenced the difference in average grade point. When several possible causes are all mixed up, as above, so that an effect cannot clearly be pinned to one of them, they are said to be **confounded**. A study in which confounding has occurred had better be ignored. So you *still* need evidence.

Therefore, you find a very large high school in Victoria, British Columbia, and concern yourself with the ninth grade only. You ascertain that assignment to the various classes within the grade is random, so that the classes are roughly equivalent. You find an instance in which one teacher is responsible for two mathematics classes. You arrange for one class to be taught by the free-study method and the other by the regimented method. You find that the grades are initially, on the average, the same for the two classes. You decide that your measure of performance will be the individual grades on a given mathematics test at the conclusion of the semester. Except for the study method, the two classes are as identical as possible.

Notice that your experiment is free of **confounding**. If you find a

---

[1] For those of you who know it by a different name, that's Grade-Point Average.

[2] As you may suspect, the Gugelpranz and Dewlap study was published in the *Journal of Confounded Behavioral Science*. Fortunately, still imaginary.

difference between the two classes, these differences cannot be attributed to initial GPA differences: you know they are similar. The students are within the same school, of the same age, from the same geographic region, and have the same teacher (who is equally skilled in both methods of teaching). Therefore, a difference between the two classes at the end of the semester cannot be due to these variables, which are held constant. A difference, if present, must be attributed either to sheer accident or to the teaching methods. Statistics gives ways (in later chapters) of judging the likelihood of any such differences being merely accidental; if the likelihood is very low, we would be left with only one viable conclusion: that teaching method did have an effect upon learning, at least for students in ninth grade, in this school, with this teacher, and at this moment in history.

You'll note that we have been considering only research design so far in this example. But now the experiment is over; we have all the **data**, the students' **scores**. The word **data** is plural, from the Latin word meaning "given" (as a noun). Data are *given* by an experiment and arrived at by means of conceptualizing some observation along a numerical scale. Numerical data are termed **scores**. Scores, therefore, are the result of **measurement**.

We look at the scores from the two groups, free study and regimented study, expecting (or at least hoping) to find better learning in the former. All we have is a bunch of numbers, some high, some low, and (probably) most medium. There are high and low scores in each group; there is no clear-cut obvious picture. But unless we can get a clear-cut conclusion, the experiment will have been a waste of time. So **statistical analysis** of the data is necessary.

The human mind has the disadvantage of being unable to process a lot of information at once. You can't take in more than a few scores simultaneously, let alone estimate their relation to other scores that you haven't taken in. It is necessary to *summarize* the data, so that their essential meaning is expressed in just a few numbers. There are two general methods of summarization:

1 Pictorial summarization in graphs. You are already somewhat familiar with these.
2 Mathematical summarization, in which many scores are democratically represented by just one or two.

Discussion of graphs will be left to a later chapter. So will our main consideration of **descriptive statistics**, statistics that serve to summarize data. Here, we are only providing an overview.

The most familiar descriptive statistic is the **average**. While there are several kinds of average (each, confusingly, usually giving a different value), they all seek (by different rules) to elect some single score to represent a group of scores. Usually such an average score is somewhere in the middle of those to be represented; hence the calculation of such averages is sometimes termed the **measurement of central tendency**. You already know the most frequently used type of average, known more technically as the **arithmetic mean**. It is obtained by adding the numbers to be averaged and then dividing the total by the quantity of scores, or more briefly, as $\Sigma X/n$, where n refers to the number of scores.

A second problem in summarizing data is to ascertain *how closely* the average value represents its constituent scores. There are really two questions here.

1 Was the best type of average (for your purposes) used?
2 How closely are the constituent scores approximated by the average?

As discussed more fully later, an incorrect choice of average can give a totally wrong picture, and incorrect choices are far from rare. Sometimes they are deliberate. Figures don't lie, but liars figure. Turning to the second question, an example may be illustrative; suppose that in the free-study group the scores ranged from 14 to 99 with a mean of 80. This average represents scores over a range of 86 (99.5 − 13.5 =86). In the regimented group the scores ranged from 49 to 58, with a mean of 55. Here the average represents scores over a range of only 10 (58.5 − 48.5 = 10), and hence in the latter instance the average would seem to be a much closer representative of all its scores. A measure of *how closely* scores cluster around their average is therefore a second type of descriptive statistic, and there are several choices among various **measures of variability**, discussed in a future chapter.

Let us apply this logic to our hypothetical experiment. We have seen that the free-study group has a mean of 80, much higher than the regimented group mean of 55. This looks promising. However, you can see that there is a very great overlap between the scores of the two groups. If you don't see this, stop and think until you do. Is the amount of overlap so great as to rule out what we seek, namely, *consistent* (give or take a little) superiority in the free-study group? A glance at Figure 1.2 illustrates this question.

Descriptive statistics now become the raw material for **statistical analysis**, which involves choosing among several **tests**, such as the *t* **test**, *F* **test**, and **chi-square test**. They will be discussed more fully

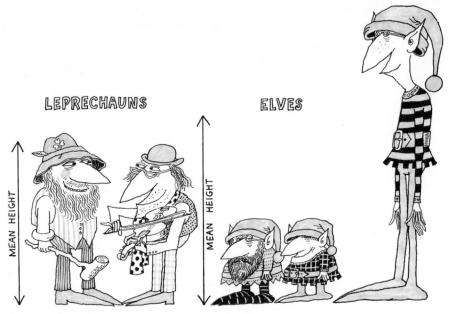

Figure 1.2
Judging from their *mean* heights only, we might conclude that elves were generally (or even necessarily) taller than leprechauns.

when you are ready for them. In common, but by various methods, statistical tests allow us to assess how likely it is that an experimental difference (as between the two means, 80 and 55) is merely a *chance* difference. By *chance* is meant the various unpredictable, unsystematic accidents that occur in sampling (selecting subjects) and conducting the experiment. The likelihood of such accidents causing a given difference between means can be ascertained more or less accurately. If it is less than 1% likely that the difference is due to chance (and this is determined by comparing the means with their variabilities), then we would prefer to reject chance as the cause of the difference. The odds would favor the experimental hypothesis instead, not as a "proven fact" but as an "overwhelming probability"; at least for this experiment, with these subjects, at this time in history. If you want to **replicate** the experiment (do it over again), then all power to you. Your original hypothesis (as a Respected Educator) is supported, and you are now regarded as a Respected Statistician as well.

example 2

Imagine that the time is the future and you are an educator. The following problem is handed to you. Certain children in your school are being carefully followed by social agencies because of broken-home backgrounds or other psychosocial handicaps. Every semester you are asked to report the progress of these "high-risk" children, relative to their classmates in the many classes involved. Different teachers, different subject matter, different standards of assessment, and even different numerical scoring systems are involved. In short, to report a child's mathematical prowess by giving a score of 88 is to exercise oneself in futility. Eighty-eight could be out of any value; it could represent any level of performance from appalling to outstanding. Letter grades are too imprecise. Qualitative verbal reports aren't objective enough. In short, you want to have some standard system for stating a score in its relation to the group of which it is a member. Fortunately, there are several easy ways of doing this—fortunately, because this is a very frequent problem in the social sciences and education. The more important ways are discussed in Chapter 6, but the most easily comprehended method may be touched on to illustrate the present problem. It really is quite easy. You ask each teacher confidentially to rank order the children in his or her class, with 1 being the lowest-ranking student, 2 being the next, and so on, until the best student has the highest rank. Suppose that high-risk student Jimmy has a rank of 9 in a class of 25 children. He is nine twenty-fifths, or thirty-six one hundredths, or 36% of the way from the worst to the best student in his class. The value 36 is called his percentile, and you can report it to the agency. No confidences regarding other students are betrayed, yet Jimmy has been given a numerical score relative to the other students. Percentiles have the same meaning irrespective of the characteristics of the raw scores from which they come. And other types of standard score, even better than percentiles, will be considered in some detail later.

example 3

People are immensely interested in **relationships**—mostly human, but sometimes mathematical. Maybe sometimes both. The author has often wondered whether there is any relationship between how close to the front of the lecture room a student sits and the grade the student achieves in the course. Casually, I notice sometimes that the best students seem to sit near the front (closest, as it were, to the fountain of knowledge). Several years ago, I wondered whether the idea could be tested. Suppose I could show that final grade (in the 10-point system used at the University of Victoria) decreased from A+ to F in step with the ten rows of seats in my classroom. If I could show this, goodbye to the need for setting and scoring tests, and arguing about them afterward. All I would need to know would be each student's seat position, and I would know his or her grade. If such a relationship did exist and could be expressed in terms of numerical scores, it would be an example of the statistical notion of **correlation**. Using the correlation to predict the grade of a new student who chose to sit in the fifth row is termed **regression**. Unfortunately, the relationship between position and grade was not very good, though there was a tendency for the best students to sit in the first three rows and the worst to sit in the back row. There was a slight correlation, not a perfect one. The "fountain of knowledge" theory took a nose dive. But for prediction purposes, *some* relationship is better than none, to put me somewhere between dead certainly and sheer guesswork. I now try to ensure that the rear third of the room keeps a happy look on its collective face, because regression suggests that these will be the problem students.

example 4 (or is it 1 again?)

Recall, if you haven't repressed it already, the first example in this series. You found (hypothetically) that your experimental subjects (ninth graders in Victoria, B. C.) learned better with free learning than under regimentation. At least, you are 99% sure that they did, and 1% fearful that the whole thing is an accident. You probably used 80 students as subjects, altogether. But who ever heard of Victoria, B. C.? Your real concern is to apply the conclusion to *all* students, including those who are yet unborn. You have actually studied 80 people out of many billion; to generalize seems risky. But to go to all that work, and then restrict the results to only 80 people would seem ridiculous. There must be something between risk and ridicule. How do you apply experimental findings to people in general? This is the business of **statistical inference**, which is closely similar to **statistical analysis,** mentioned above. Statistical inference might be called How To Overgeneralize and Not Get Caught.

Now you might glance again at the Table of Contents of this and/or any other textbook on basic statistics. The topics should now seem a little more comprehensible and applicable to your interests, whether in psychology, sociology, education, public health, or another field.

Authors are very fond of putting lengthy arithmetic exercises at the end of chapters, in the naive (and mistaken) belief that students will find them challenging, helpful, and soothing. The present author, a realist, will therefore renounce this tradition—for the most part. Every chapter will have a necessary minimum of exercises. None are totally repetitions of others; all are designed to establish or illustrate a point. None are mere drill. Some give practice in the art of taking examinations. A few are in the guise of games, which may actually be played or merely imagined, both to equal advantage. Please attempt them. They will be ridiculously easy arithmetically, put there for the sole purpose of letting you do them easily and thus realize that you still know what's going on.

So work awhile on the following, going back and reviewing where necessary.

1 With the overview behind you, do you still want to be here?
2 Do you have what you need to stay and succeed?
3 Can you add? How about subtract? Divide? The population curve already proves that you can multiply.
4 Do you know the difference between quantities, integers, ratios, and operators?
5 Is the summation sign Σ Greek to you?
6 What is the meaning of *commutative*?
7 What is The Word about using decimal places?
8 Do your promise to behave yourself while *rounding off* your calculations?
9 Think about the following, as fondly as possible (you will have to know them and love them sooner or later):

RESEARCH DESIGN

CONFOUNDING

DATA and SCORES

DESCRIPTIVE STATISTICS

CENTRAL TENDENCY

ARITHMETIC MEAN

VARIABILITY

STATISTICAL ANALYSIS

STANDARD SCORES

PERCENTILES
CORRELATION
REGRESSION
STATISTICAL INFERENCE

Are you wondering whether to treat the word *statistics* as singular or plural? Here is (are?) some help. When the word *statistics* refers to the field of study, the mathematical specialty, etc., it is singular. You would say, "Statistics *is* tommyrot," or "Statistics *is* great!" When talking about an arithmetic mean (which is a *statistic*) or a correlation (which is another *statistic*), you would use the singular when referring only to a single *statistic* ("*That's* a surprising statistic!"). If you are referring to several *statistics* together, of course you would use the plural.

exerci∫e

At the present, you and your fellow students are probably unacquainted with each other. You know a fair bit about what you want from your curriculum and life, if not from this particular course involving statistics. The purpose of this exercise is for you to clarify your own ideas and to help other students do likewise.

With classmates (Instructor, please help here if necessary), organize an informal debating group with about six other students. Individually and in a group, consider the substance of this first chapter. Debate it pro and con, as freely as you can contrive. Just as an exercise, take the parts of both protagonist and antagonist—argue both ways. If you need a formal resolution to debate, let it be something like
Resolved: **That statistical thinking is central and crucial to the discovery and explanation of facts in the science of psychology.**
After all that, you will be in one of the following states of mind.

a You are strongly and unalterably opposed, intellectually and emotionally, to the resolution give above.

b You have become just as strongly fervent about the essential role of this new and exciting field of statistics just laid before you.

c You are somewhere in between, who knows exactly where?

Take time to ruminate and write your thoughts, possibly to be handed in if your instructor wants to understand your ideas and feelings. But at any rate, do it for yourself. Knowing your own ideas and feelings, at all times, is part of the course.

chapter 2
dimensions, variables, and measurement

Normally we take measurement for granted. Length, weight, time, and cost are naturally conceived of in such units as feet, pounds, hours, and dollars. Without these measures the concepts they stand for would be quite hazy; any concept that has a measure to go with it is fairly clear-cut. Clarity of concept and availability of measure seem to go hand in hand. The more sophisticated the measure, the clearer the concept generally is.

Unfortunately, to some people, the concepts found in the social sciences are not very clear-cut. However, to work effectively with such concepts, we need a definite and clear understanding of them. Such understanding must come from experience, and the experience must involve either active or mental manipulation of the concept. The most precise conceptual manipulations are possible when a measure exists for the conceptual quality. As the noted physicist Lord Rutherford remarked, "One cannot fully understand something unless he can measure it."

An example may illustrate this relationship among concept, measure, and experience. The Pygmy people of the Ituri Forest in central Africa spend their lives in dense jungle where vision is limited to a few feet. The concept of linear perspective (decreasing apparent size with increasing distance of an object) is therefore poorly developed. The Pygmies, who are not unintelligent, thus have a poor concept of distance as estimated visually. Taken from the forest to the grasslands, one Pygmy interpreted a herd of buffalo in the distance as insects, seen close up. This illusion resolved itself with further experience, in which a familiar Pygmy concept, movement toward the object for a period of time, served as a kind of measure of the several miles separating the subject from the animals.[1] The Pygmy's concept of distance became much clearer when a familiar measure, movement over a period of time, was applied, and gradually the new or undeveloped concept of perspective also became more familiar as a result.

The nonscientist sometimes has similar trouble with social science concepts, such as motivation, group cohesiveness, and conformity. One may have a general idea of the concept but be unable to express clearly what it means, and the idea of *measuring* a fuzzy concept seems ridiculous. But it may be fuzzy *because* it has no measure! Adoption of an acceptable measure depends upon practice and experience, as did the Pygmy's growing use of perspective as an index of distance.

[1] For a further account, see Colin M. Turnbull, *The Forest People,* Simon & Schuster, N.Y., 1961, Chapter 14.

The adoption of new concepts, or of new measures for old ones, is usually difficult. When Kenge, the Pygmy, was told that the insects were buffaloes, he became emotional and accused his friend of lying. Quite intelligent North Americans sometimes behave similarly; the idea of measuring motivation is taken quite lightly. And such psychological and sociological concepts involve ideas in which vagueness is a necessary characteristic. Concepts that are built around their measurements, such as length and temperature, don't suffer from this problem. Temperature, in degrees, seems to be a more solid concept than motivation.

Difficulties with concepts give rise to difficulties in the everyday world. Modern business executives are skilled at manipulating money (measured in dollars), and go to great expense in air conditioning their employees' offices to 68 degrees. Degrees! These seem like tangible, important things in keeping the troops happy. But problems of morale and motivation are typically missed or misinterpreted, and hence mishandled. Morale problems are usually countered by pay raises, improvements in working conditions, and other tangible, measurable things. People often act as if what's measurable exists and what's not measurable does not. It's interesting to wonder what the world would be like if we could measure love as easily and confidently as we measure office temperatures.

The apparent tangibility of the natural sciences, physics, and chemistry arises from the fact that their basic concepts are defined in terms of measurement. But tangibility (literally, "touchability") is not really an issue. *Gravity* to the physicist, *valence* to the chemist, are no more tangible than *motivation* to the psychologist. There is popular acceptance of measurement in the physical sciences, however. The social sciences are becoming measurement oriented, but tend to achieve public alienation as a consequence.

is humanity measurable?

Our discussion clearly is leading toward the point that human characteristics such as motivation may well be as measurable as any physical phenomena, but some legitimate problems do exist and we should consider them first.

Objections to measurement of human beings show an interesting paradox in that physical measurement is taken for granted (you feel easy about your shoe size), but psychological measurement sometimes seems to be an invasion of privacy. We hear claims that psychological measurement is in violation of human rights. In a society that is rapidly perfecting techniques of genocide, such sweeping accusations have a curious flavor. But in some specific instances, these attitudes have some merit, stemming from two sources.

1 A societal history in which psychological measurement has been, at times, inept, exploitive, and careless of the individual person. A good example is the use of widescale IQ screening of children, using rather unreliable measures, in ways that adversely affect individual freedom of opportunity. Certainly there is room for concern, but our ammunition should be reserved for *misuse* of measurement rather than measurement itself.

2 The belief that the human mind is too complex, too subtle, and too mysterious to be measurable, or an indivisible, qualitative whole that cannot be analyzed into measurable parts. This belief is founded in some basic facets of our culture: our needs for individuality, privacy, and personal uniqueness. Many of us believe (in various ways) that the human mind is the image of God's, and God is unmeasurable. Such beliefs are both good and understandable, but they miss the main point.

While it's true that humanity (in the abstract) is philosophically unmeasurable, individual people possess many *characteristics* that are measurable. Such things as reading speed, arithmetic ability, emotionality, creativity, persistence, and a host of other things are expressed in our behavior. And behavior can be observed; and what can be observed can also be measured. In principle, any observable human characteristic can be measured, provided that some suitable measuring scale is handy.

Such measurable characteristics are called **dimensions**. Just as our bodies have physical dimensions, our minds have psychological ones. And just as our bodies can never be perfectly described by the sum of their physical dimensions, our minds can never be encompassed by even the most painstaking listing of psychological ones. Our individual needs for uniqueness and existential integrity remain intact. But a *dimensional* approach to human characteristics can be very useful. A clinical worker with responsibility for rehabilitation of a brain-injured patient must have some idea of his or her specific goals. To desire "recovery" is not, in itself, enough. To seek specific improvement in attention span, spatial memory, verbal fluency, and emotional sta-

Figure 2.1
One way of scaling love . . . perhaps best reserved for those who love scaling. (*Reprinted through the courtesy of Mell Lazarus and Field Newspaper Syndicate. Copyright Field Newspaper Syndicate.*)

bility is preferable. These may be the patient's *dimensions* most impaired by the brain injury. And they are capable of being measured; hence, the clinical worker will be able to measure the effects of his or her treatments, rather than rely on clinical intuition. This is merely one out of countless examples; others may be found in education, sociology, psychology, psychiatry, and human-oriented research in general.

the scaling of psychological dimensions

In this section we shall consider how scales of measurement can be fitted to social, psychological, or behavioral dimensions. By *scaling* we mean the use of numbers to measure or at least describe some aspect of experience. When we *scale* temperature in degrees Celsius, we can then *measure* by using that scale. It doesn't matter that the scale is arbitrary. The three temperature scales commonly used, Fahrenheit, Celsius, and Kelvin, are equally arbitrary, though only the first one named is familiar to most Americans. Familiarity *does* make a difference in the general acceptability of any scale, as we've seen already.

When a dimension is scaled and then observed or studied in terms of the scale, it becomes a **variable**. The use of this term, *variable,* is obvious when we consider that people vary in all of the ways that let us conceive of dimensions in the first place. (A human *constant,* such as having two arms, is rarely interesting.) A dimension is a measurable concept; a variable is the measure of that concept by means of some arbitrary scale. A dimension can be scaled in many ways, and therefore gives rise to many possible variables. The dimension of *motivation,* for example, can give rise to these variables:

1 The number of *hours* that a person will persist in trying to solve a difficult problem;
2 The number of *volts* that a hungry rat will tolerate in order to reach food in a maze;
3 The number of *dollars* that a person will bid at an auction for some desired item.

The possible list is endless. In each example here, the scaling unit for the variable is italicized. "Motivation" remains a dimension that can have many scales and give rise to many variables. Other dimensions can similarly give rise to many variables, limited only by the creativity and ingenuity of the researcher.

various procedures for scaling

While the *purpose* of scaling (the rendering of ideas into variables) is straightforward, the *procedures* of scaling vary according to the needs, purposes, and ingenuities of the researcher. Basically, you can choose among three methods: measuring, comparing, and categorizing.

measuring

Measurement is attaching a number to an observation according to some consistent rule. In principle, it is familiar to everybody. We discussed it above in connection with various physical phenomena such as temperature and distance. Now let us look at the measuring scales themselves and their essential characteristics.

A measuring scale involves two principles: an orderly method for making an observation, and an arbitrary (though often rational) subdivision of that observation into standardized, equal-sized, numbered units. Look at measures of length as an example.

The orderly observation is done by laying a ruler or tape measure alongside the distance to be measured. The arbitrary scale units may be inches or centimeters, yards, furlongs, rods, chains, meters, or miles. Each such unit is adopted through agreement and custom. All the units have two features in common. (1) They start at 0, denoting 0 distance from the starting point. (2) Their total increases from 0 in perfect step with the magnitude of the thing being measured, while remaining individually constant in size. This is ensured by having the units engraved upon the tape measure, and the tape measure backed up by national or international standards. Thus, an inch is an inch wherever you find it.

This simple procedure is an example of **ratio scaling**. The numerical sequence of equal-size scale units (1 inch, 2 inches, 3 inches, . . .) corresponds to increasing amount of distance. We can say with confidence that 10 centimeters is *twice* 5 centimeters, because 10 is twice 5. The numbers on the scale may legitimately form a ratio that pertains to the actual amounts of distance measured.

Ratio scaling gives the most sophisticated form of measurement, but it is difficult to achieve in the social sciences. There are three reasons why. (1) We are uncertain as to the "goodness of fit" between the number sequence 1, 2, 3, 4, etc., and the characteristic being measured. There may be jumps and gaps in any but the simplest psychological attributes. (2) We cannot agree on the necessary external scale—tape measure—for any but the simplest human characteristics. (3) Even if the preceding problems could be solved, there is still a problem with getting the 0 point of the scale to coincide with a complete absence of the characteristic being measured. Take IQ, for instance; it is barely possible that a person might scale 0 on an IQ test, but would this mean a *total* absence of intelligence? Such a person would still have *some* ability, even if only the ability to keep on breathing. This would reflect *some* intelligence.

Measurement can be done with scales whose 0 points do *not* coincide with a true 0 quantity of the attribute. If such scales allow comparison between any two objects or persons measured with them, and if the scale units have fixed size, we have *interval scaling*. This is a slight step down from ratio scaling. A good example of interval scaling is the

Fahrenheit temperature scale, in which 0°F does *not* indicate a total absence of heat (it is merely the coldest temperature achievable by mixing salt and ice). The scale units, or degrees, are each defined as 1/212 of the tememperature range between 0° and the boiling point of water. Highly arbitrary? So is the Celsius (or centigrade) system, in which 0 still does not indicate the total absence of heat, but merely the freezing point of pure water. Both of these familiar temperature scales are interval scales. A ratio scale for temperature is found with the Absolute or Kelvin scale, in which 0 indicates total absence of heat energy, at −459°F.

Sometimes we give the name **cardinal scaling** to ratio and interval scaling taken together, inasmuch as they employ cardinal numbers to denote the amount of whatever is being measured.

Interval scaling gets its name from the fact that given sizes of interval are of equal size along the scale. Five inches have the same length wherever you find them, and a 10-degree rise in temperature means the same thing no matter what temperature you start from.

Interval scales are rare in the social sciences for the same reasons as the first two given above for the scarcity of ratio scales.

A third type of measurement employs what might be called **plastic interval scaling**. Here the measuring scale is marked into intervals of plastic, or variable, size. An example is IQ, or intelligence quotient, where "one point of IQ" covers a greater span of mental ability at the extremes than it does in the center of the scale. There are many other examples in the social sciences of such plastic interval scaling. Whenever you do anything that gets a score from a person and you are not certain that the scale units all denote equal amounts of the quantity being measured, you have this type of scale.

In summary, measuring involves a comparison of the phenomenon to be measured with a calibrated measuring scale. If the scale has known equal-sized units corresponding with equal-sized increases in the phenomenon itself, it is a cardinal scale. If the cardinal scale has a 0 point corresponding with *none* of the thing being measured, it is a ratio scale; otherwise it is an interval scale. If the scale units correspond with variable amounts of the thing being measured, you have a plastic interval scale.

comparing

A person is shopping for a new set of wheels and wants the cheapest available. The choice is among a buggy, a bicycle, and a behemoth.

Now, the price tags are, respectively, $1,000, $100, and $6,000, a case of plastic interval scaling of transportation value. But there is no need to look at the prices themselves, except to rank-order the three choices. This involves comparing each choice with the others and winding up with the bicycle as cheapest of the three.

This is an example of **ordinal scaling**, arranging and describing things in rank order of dollar value (in this example). It seems clear that anything that can be measured can also be rank-ordered, but it is often possible to rank-order easily when true measurement is difficult. This makes ordinal scaling very useful in the social sciences. For example, school children can be ranked according to spelling ability, families according to socioeconomic level, professors according to effectiveness. External scales of measurement are thus unnecessary. Should such a scale be available, it would permit plastic interval scaling instead.

A limitation of ordinal scaling is its reliance upon comparison within a given group. The bicycle is cheapest only if the group of alternatives does not contain anything for less than $100. A straight-A student from a weak college is not necessarily as good as a straight-A student from a strong one. The two students have their A grades with reference to different groups. This limitation is not found in cardinal scaling, since the standard of comparison, the scale, would be constant.

categorizing

Suppose that our researcher is working in a hospital laboratory, in charge of blood typing. Many people have Type O blood, some have Type A, fewer have Type B, fewer still have Type AB. These are categories into which the blood of all human beings can be sorted. Blood type is one of the many dimensions possessed by the body.

Although blood type is a variable, it really can't logically be measured, or even ranked. No one blood type is better than the others. They differ qualitatively rather than quantitatively. But numbers can still be used as shorthand labels for qualitatively different categories. In many hospitals these four types are labeled, for the computer, as 1, 2, 3, and 4. Football players are labeled similarly by numbers on their sweaters. Social security numbers and the labels on classroom doors employ numbers to *identify* rather than to *measure*. Your telephone number is another instance. This use of numbers to label, identify, or classify, is called **nominal scaling** (from the Latin *nominare,* to name).

Nominal scaling is widespread in the social sciences. Socioeconomic level, sex, occupation, medical diagnosis, psychiatric classification, and questionnaire-response categories can all be labeled numerically.

You might ask why. After all, there is current public reaction against being "treated as a number," so why not use names rather than numbers to identify different categories? There are at least three reasons: (1) The number serves as a convenient shorthand; it's briefer than a verbal label. (2) If the categories are numbered upward from 1, the highest number also gives the total number of categories, which is sometimes helpful. (3) The growing use of computers and automatic data processing requires brief, clear, unambiguous labels for things, and computers are accustomed to using numbers.

Classification is the first step in science, and categorizing is therefore necessary. It is important to remember that the end result of classification is not perfect description, since only one or two dimensions can be classified at a time without the number of categories becoming too huge. Imagine the quantity of classifications if we were to look, simultaneously, at age, sex, educational level, IQ, blood type, political party, income level, blood pressure. . . . Impossible. You need not fear that the classifiers will ever pigeon-hole you completely.

variables and empirical research

The reason for doing statistics is to summarize and understand research information, some coming from experiments, some from controlled observation in which the researcher does not actually manipulate matters, and some from critical and repeated field observations in which control is difficult or impossible. Provided that the observations are scaled somehow, statistical analysis is possible. We saw some examples of it in the previous chapter.

The highest level of empirical investigation is the **experiment**, characterized by manipulation of a possibly causal variable and observation and measurement of the result on some other variable. For example, a researcher might wonder whether level of anxiety has any effect upon performance. The latter may depend upon the former, and consequently the latter is termed the **dependent variable** of the experiment. The former, level of anxiety, may be arbitrarily varied by the experimenter (as by tranquilizing one group of subjects and scaring another group) and is thus the **independent variable**. But so far, only a general idea has been expressed. This is a **hypothesis** or, more exactly, a **conceptual hypothesis**.

The researcher must now decide how to measure both anxiety and performance. In the simplest possible experiment, which involves only two groups, he might form two groups as suggested above, calling one "Group I—Tranquilized" and the other "Group II—Scared." The measure of performance might be the length of time required for each subject to complete correctly some given task, an instance of cardinal scaling. With all other possible variables held constant, any change in the dependent variable may be attributed to the independent variable. Of course, the effect of chance must be ruled out in interpreting any difference between the two groups, but the statistical analysis will do this. Thus, the conceptual hypothesis has been translated into an experimental hypothesis: "Will tranquilizing one group of subjects and scaring another group be followed by a difference between the times required for the two groups to complete this given task?" Notice that the experimental hypothesis is much more explicit (measurement oriented) than the conceptual hypothesis. And finally, the experimenter is able to decide upon the correctness of the experimental hypothesis on the basis of the dependent variable measurements. By cautious and tentative generalization, he may then have a few clues about his conceptual hypothesis as well.

Other loose variables might have affected the dependent variable and been responsible for the results. If these other variables were to be held constant between the two groups, they could *not* affect the results differentially. But sometimes they are not held constant. They are of two kinds: **subject variables** and **extraneous procedural variables**. The first type pertains to built-in characteristics of the subjects, their various dimensions as discussed earlier. If performance is aided by intelligence and if one of the groups has more intelligent subjects than the other, then they will give different performance results that might wrongly be attributed to anxiety. When two variables (such as anxiety and IQ) are mixed up in this way, they are **confounded** with each other.

The second type, **extraneous procedural variables**, pertain to things that the experimenter does or allows to happen that affect the dependent variable. The tranquilizer used might have contained a powerful sedative, making the subjects sleepy and thus decreasing ability to perform. This effect may be wrongly attributed to low anxiety rather than to the sleepiness: another case of confounding.

In social science research it is often difficult, immoral, or impossible to manipulate an independent variable. For example, there are limits beyond which experimental increases in anxiety cannot go. Experiments are not possible when manipulations of independent variables are not possible. However, it may still be possible to observe relationships that occur naturally. A psychiatrist might note a relationship between levels of naturally occurring anxiety and performance. If these variables are scaled and measured, their **correlation** (as we already know) can be calculated. We would never be sure about which variable *caused* which, or even whether either of them was the causal agent. Correlated variables thus cannot be thought of as dependent or independent; perhaps *interdependent* would be a good word.

Sometimes the refinements of experimentation and correlation procedures are left alone in favor of *mere observation*. Whereas experiments can establish cause and effect, and correlation can show the presence of relationships, a lot of empirical research occurs for the purpose of *finding facts*. Facts, in themselves, are usually qualitative rather than quantitative, but scaling and measurement are still very convenient in organizing information. Imagine a large building full of filing cabinets, each cabinet full of "facts," which, properly organized, might contain some important message, like the basic mechanisms of schizophrenia or cancer. Until such "facts" are retrievable, they really aren't facts at all. Retrievability requires organization and summarization, functions for which computers are beautifully suited; but computers operate quantitatively, so all those facts must be scaled somehow, distilled, boiled down, and presented in some way that can be encompassed by the human mind. This process of summarization is the purpose of **descriptive statistics**, outlined in the following few chapters.

A creative person looking at statistically summarized information will probably get ideas from it. There is no mechanical way of getting hypotheses from collections of facts; it requires originality. So don't let anybody tell you that good science is creatively sterile.

By this time your ideas of the relationships among scaling, measurement, dimensions, and variables should be pretty clear. But there is a danger awaiting the enthusiastic convert to empirical methods: the danger of becoming *too* blindly convinced. This happened fifty years ago to many psychologists who were followers of behaviorism. Behavior, the empirically observable, was *everything*; the very existence of mental processes, cognition, and other inner "events" was challenged. This was going too far. Not surprisingly, raw behaviorism quickly ran out of fresh ideas.

We should always remember that dimensions are abstractions made by the observant creative mind and that variables are chosen arbitrarily or for expediency. They are chosen because they work, not because they are perfect. Not even the largest collection of variables, used simultaneously, can fully encompass the meaning of most phenomena. For example, the dimension *motivation* can give rise to numerous variables, such as speed of rat-running in a maze, or duration of human persistence on a hopeless task. Such variables, dependent variables in this case, have the advantage of being measurable, but the disadvantage of being narrow. Practically all psychological and social experiments occur in quite select circumstances; care is needed in extending any results to "life in general." Such necessary care is no criticism of an empirical approach. The only other choice would be pure mysticism, which has its drawbacks too. Obviously, we are stuck between a limited methodology and no methodology.

data

Data is a plural Latin word meaning facts or figures from which conclusions can be drawn. As we saw above, a building that is full of filing cabinets full of potential facts does not contain *data* until those facts are summarized or known to be summarizable. Thus, data must practically always be numerical in style. They are the end point of observation, scaling, measurement, and experimentation and the beginning point of statistical analysis leading to some kind of conclusion.

There is a frequent tendency to treat the word "data" as singular, as in "*this* data *is* . . .," instead of "*these* data *are*. . . ." Usage makes for acceptability, and you can take your choice. The singular form of data is **datum**, or **score**. In this book we'll be pedantic and plural with data.

Since data are the starting point for scientific conclusions, we must be very careful with them. No conclusion can be better than the data leading to it. Computer scientists have a word for it: *gigo,* meaning "garbage in, garbage out." Let us look now at some problems that may occur with empirical data.

validity

Valid data are data that reflect reality. The biologist observing microorganisms, classifying and counting them, is obtaining data. The schizophrenic patient who is hearing imaginary voices, classifying and counting them, is not. In an experiment, data must pertain to the effect of the independent variable the investigator had in mind, not to some other procedural variable he or she hadn't thought of. All too often, data are **invalid**. The scores do not reflect the reality the researcher was trying to measure. There can be three reasons for this.

(1) the problem of "what is reality?"

The mind does funny things sometimes. Those who have taken LSD certainly know this, and they would probably stress that inner reality is one thing and that external reality is something else. For better or worse, data must pertain only to the latter, external reality. Why?

It's accepted that science must deal only with consistent phenomena in nature. After all, its objective is to organize and understand, and therefore some degree of consistency is needed; the first step in science is to find it. There is no "science of unique events." Therefore, since most personal experience is unique, science is restricted to those aspects of experience that are shared by all people at any given time and place. The fact of such sharing can be established by what the observers say and do. Thus, for example, you and your classmates are able to agree on the length, width, and thickness of this book. Such measures as you might make can be agreed upon and would yield data. Whether the book is a *motivating* one is another matter, because you will have individual ideas as to what this word means. If, however, you decide to define and measure "motivating power" somehow, your agreement and your procedures of scaling will permit data to forthcome.

Defining something in terms of some agreed-upon method of observation and measurement is called **operational definition**. "Operational" because of the agreed-upon operations, or procedures, used in doing the observing. So if you (and your friends) agree that "motivating power" is to be defined as "what percentage of students learn to like statistics" from any book, that's one operational definition of "motivating power." Of course, there could also be others.

The validity of data depends in part, therefore, on the presence of some agreed-upon operational definition underlying them. Without it, you'll be lost in a sea of conflicting value judgements populated by creatures of various uncoordinated private worlds.

Operational definitions have become a fetish in some areas of psychology and other social sciences. It all started in physics, where it is now recognized as useful but not ideal. Reality goes beyond what can be operationally defined. This is now accepted by most scientists. We now know that science is only one among many approaches to making sense of our realities.

(2) inappropriate choice of measuring scale

Validity will suffer if the data have resulted from an inappropriate measuring scale, even if good operational definition has been accomplished in principle. Obviously, there will always be a range of choice in possible scales, some relatively good, others relatively poor. Intelligent creativity is needed to pick the best one. We can imagine an extreme example of picking a poor scale: in a study by the author some years ago, an experimental antidepressant drug was administered to several dozen depressed psychiatric patients. The index of "clinical improvement" was a rating scale used by the attending nurses. Disappointingly, very few of the patients showed "improvement" in their ratings. A deeper look into the situation revealed that their *depressions* did improve, but that many other clinical aspects of their illnesses got worse as a result of greater energy released through lifting of the depressions. Net result: the *patients* did not get better, though their depressions did. A somewhat improved rating scale was the answer to the problem, directing attention more closely to the target symptoms of depression. The medication, after all, was intended to relieve only this, not all of the patients' life problems as well. For the latter problems, psychotherapy and various methods of social activation were found effective.

(3) lapses in experimental procedure

Even if there is a good operational definition and a good choice of scale, invalid data can arise when the experiment or study involves faulty methodology. A good example is the so-called "placebo effect," when a patient may recover from an illness as a result of getting an impressive pill, medication, or treatment that is actually worthless physiologically. In the above mentioned drug study it was necessary to give half of the patients an inert concoction that exactly resembled the antidepressant drug. This provided a baseline for determining the effects of the *drug,* as against merely getting a *pill.* Quite a lot of amateur research suffers from poor methodology and invalid data, because some of the controls needed are indeed very subtle.

reliability

Reliability is the same as *consistency*. A steel tape measure will always give the same (or closely similar) readings of the length of a log. An elastic rubber tape measure applied to the same log would give widely variable readings, as a result of varying tension during each measurement. The steel measure is reliable; the rubber one is not. And between the two extremes, there can be intermediate degrees of reliability, as with dressmakers' cloth tape measures, which are less exact than steel ones but not as awful as rubber ones.

Data can lack reliability for two possible reasons.

(1) unreliable measures

This is what we saw in the preceding paragraph. The tape measure illustration is obvious to the point of being ridiculous. But there is a *real* problem in the social sciences, which tend to lack such things as physical yardsticks, gauges, and dials, which are enjoyed by engineers and aerospace technicians. The personnel manager who is trying to evaluate a job recruit may use a rating scale, a checklist of aptitudes, or something similar requiring an input of judgment on his or her part. But the manager's judgment will probably be affected by circumstances and moods, so it will be variable in the same sense that the rubber tape measure is. Such a problem is typical in psychological measurement, sociological surveys, and pupil evaluation by teachers in classrooms. It is not anybody's fault, but it *is* something to be aware of. Extensive training can often reduce such "variable user-input" unreliability to acceptable levels.

(2) variability in what is being measured

There might be great instability in the thing, or whatever, that is being measured. Taking the measurement only once or a few times probably won't be sufficient to allow any reliable conclusion about the *average,* or most typical, value. For example, if you wanted to determine the average height of leprechauns and managed to catch five of the little people, the odds are poor that they will average exactly the average height of leprechauns in general.

Reliability problems can be minimized in three ways: (1) adopting systematic *procedures* of measurement, known to be as reliable as possible; (2) *applying* such procedures in practiced and consistent ways (a lot of experience and practice is often necessary, as with intelligence testing); (3) applying the measuring procedure *to as many cases as possible* before any conclusion is drawn. If you could catch 500 or 5000 leprechauns, you'd get a more reliable picture of them than if you had settled for only five. This is a basic point. In pages to follow, we'll find that statistical procedures give firmer and firmer answers as the number of cases or size of samples increases.

usefulness of data

We should never forget that data and statistics are intended to serve a purpose or to answer a question, whether theoretical or practical. But we live in an era of tremendous data-gathering prowess, in an age of incredible computer sophistication. Far more data are collected than are ever analyzed, simply because such data are *possible* to collect. There is a real problem even to find room for storing such gratuitous information, which, once stored, is unlikely to see the light of day again, because fresh useless information is continuing to flow in at an even greater rate.

Straight thinking and creative question-asking are more important than mere data. So try to develop the former and not worship the latter. It's notable that the world's great discoveries were made by clever thinkers, not by mechanical hoarders of facts. Data are only valuable within the context of a question, though it often happens that questions arise when alert people look at surprising data. As Einstein said, "chance favors the prepared mind."

Sometimes, we become involved with questions for which it seems impossible to collect data. The problem may be merely to reformulate the question; turning a question into a hypothesis and then into an experiment does require mental discipline, training, and experience as well as creativity. But once in a while, a question will defy all efforts. When this occurs it is well to consider critically whether you are asking a *real* (as opposed to *metaphysical*) question. Real questions refer to observable, describable, measurable phenomena; metaphysical ones do not. Some metaphysical questions may be important (e.g., Does God exist? Is humanity "good"? Is the soul the same as the mind?), but they are not subject to scientific investigation. Therefore they are not subject to data-oriented, statistical approaches. Science is fine for scientific problems, but not all problems are scientific.

an inspirational last thought

The eminent mathematician Jean Baptiste Fourier once remarked that ". . . the chief attribute of mathematical analysis is clarity . . . mathematics has no symbols to express confused ideas."

active mental summary

1 Spend some time thinking about the concept of *dimension*. As an exercise, think of a dozen dimensions pertaining to your field of interest.

2 Try to find variables that would serve to measure each of your dimensions. Try to identify the level of scaling your variables involve.

3 Review the following terms. They stand for concepts central to this chapter.

MEASUREMENT-DEFINED CONCEPTS (what other kind is there??)
DIMENSIONS
SCALING
VARIABLE
   INDEPENDENT VARIABLE
   DEPENDENT VARIABLE
   SUBJECT VARIABLE
   EXTRANEOUS PROCEDURAL VARIABLE
MEASURING
   CARDINAL SCALING
     RATIO SCALE
     INTERVAL SCALE
     PLASTIC INTERVAL SCALE
COMPARING—ORDINAL SCALING
CATEGORIZING—NOMINAL SCALING
DATA
   VALIDITY
   RELIABILITY
REAL and METAPHYSICAL QUESTIONS

4 Debate this one: "We can't measure the mind, but we *can* measure what the mind does."

self-test on chapters 1 and 2

Directions

On this and following multiple-choice questions, you may encircle (lightly, in pencil) each correct choice. Any number may be correct within a question. Score yourself 3 points for correct circling; penalize yourself 2 points for incorrect circling; penalize yourself one point for not circling a correct choice at all. When you have finished a page, turn over the right-hand margin of the page to find the answers and get instant feedback. The remaining questions appear on the following right-hand pages.

Example: In working through this test, you should

   a encircle in ink

+3 (b) credit three points for correct

−2 (c) cheat by turning page over
           beforehand

−1 d think carefully about the principles
           involved in each question

1 A value achieved only by counting is termed a

   a factorial

   b sum

   c integer

   d quotient

   e result of a noncommutative operation

   f result of a commutative operation

2 An experimenter wishes to determine the effect of alcohol upon time judgment. In this experiment, the independent variable could be

   a the amount of alcohol consumed.

   b number of subjects tested.

   c the reading on the timer after subject stops it.

   d time elapsed between alcohol ingestion and testing.

   e whether alcohol or water was administered via the subject's stomach catheter.

SCORING AND ANALYSIS of Self Test questions on the other side of this leaf. Align the left-hand margin of the scoring side with the choices on the test side.

If you have made errors, read and digest the comments provided.

Answers to Question 1

a F A factorial is a serial multiplication with multiplier diminishing by 1, e.g., $3 \times 2 \times 1$.

b T Yes, you would obviously get the sum of the number of things counted; if the count were 10, then the total number of things counted must have been 10.

c T Yes, an integer (whole number without fractions attached) occurs if you count on an all-or-none basis.

d F No, a quotient is what you get by *dividing*.

e F You will get the same total count in no matter what order you count
f T your cookies; therefore, the process of counting must be commutative.

Answers to Question 2

a T The amount of alcohol is set by the experimenter, to see what the effect of varying it is.

b F This is irrelevant to hypothesis.

c F Timer reading could be the *dependent* variable!

d F Since alcohol takes time after ingestion to reach the bloodstream, this *could* be an independent variable; but a very awkward one, confounded with individual variability in rate of absorption. Altogether, a poor enough choice to be unacceptable.

e T This could also be an independent variable.

3 We may control unwanted, extraneous variables in an experiment by

   a ignoring them.

   b eliminating their sources.

   c holding their sources constant.

   d concentrating their sources into one experimental group.

   e randomizing their sources among all experimental groups.

   f confounding them with the independent variable.

4 Correctly rounding off the value 5.61237, we could get

   a 5.61000

   b 5.612

   c 5.61240

   d 5.6124

   e 5.6123

   f 5.0

   g 5

   h 6

5 Which of the following are most likely to be *confounded* when looking at a group of school children?

   a age and sex

   b grade level and age

   c sex and years of schooling

   d sports ability and sex

   e years of schooling and age

Answers to Question 3

a F   Ignorance is *not* bliss in this instance.

b T   Yes, either. If eliminated or held constant, they cannot cause the de-
c T   pendent variable to vary; a player can't make the ball move by remaining motionless.

d F   No! If this were to happen, you couldn't distinguish the effect of the extraneous variable from the independent variable. This is confounding.

e T   Yes; this way its effects are randomized and therefore equalized (one hopes) between the groups, and thus they won't differentially affect any one group.

f F   If you circled this one and were actually awake, go back and reread the whole chapter.

Answers to Question 4

a F   Wrong. The last three 0s indicate that *nothing* is there, when really 237 is there.

b T   Dropped term is less than 5, so correct.

c F   That last 0 is incorrect; the precision it suggests isn't really there.

d T   Either is correct, so long as your rule is consistent: rounding (prefer-
e T   able), or merely dropping unwanted digits (not the best idea).

f F   Wrong; if one decimal place is to be included, it should show 5.6, not 5.0.

g T   As with choices (d) and (e), either could be correct.
h T

Answers to Question 5

a F   Age and sex are quite unrelated to each other.

b T   Yes; older children are in the higher grades.

c F   **d** F   Not likely unless you're in a sexist school district.

e T   Generally speaking, the older you are, the more schooling you've had.

6 Suppose that the individual personalities, etc., of every human being could be exhaustively and exactly specified by reference to only 12 dimensions, each of which could take on only one of 10 possible values for any individual. Theoretically, how many human beings could exist before any two had to be *exactly* alike? (Note: The world in 1973 had a population just under 4,000,000,000—4 billion.)

a $10 \times 12 = 120$

b 12 raised to the tenth power = 61,917,362,000

c 10 raised to the twelfth power = 1,000,000,000,000, or one quadrillion

d about 250 times the present world population

e the answer is indeterminable

7 From among the alternatives offered the most *valid* operational definition of aggressiveness in children would be

a opinions of teachers.

b number of fights initiated per eight-hour school day.

c psychologist's evaluation report.

d distractibility in class.

e the children's feelings of aggressiveness.

f how frustrating the children's home situations are.

8 The difficulty that typically afflicts the development of operational definitions and cardinal scale units is

a that operational definitions don't get at the essence of any concept.

b that operational definitions are in fact meaningless, for the reason mentioned in (a) above.

c that an operational definition is highly arbitrary.

d that operational definitions do not permit assessment of their own reliability

e that most people are emotionally unwilling to see their fuzzy ideas made concrete.

f The applicability of any of (a) through (e) to the acceptability of cardinal measurement is parallel to their validity as arguments against operational definitions.

Answers to Question 6

a F **b** F $12^{10} \neq 10^{12} \neq 10 \times 12$

c T (c) is one correct answer. If each of twelve scales can give ten possible values, the total number of different outcomes must be $10 \times 10 \times 10 \times 10 \times 10 \times 10 \times 10 \times 10 \times 10 \times 10 \times 10 \times 10$. A simpler example: flip a penny, a nickel, and a dime. Each may come up heads or tails, i.e., two possible outcomes. The total number would then be $2 \times 2 \times 2 = 8$; the same principle applies in both cases.

d T One quadrillion is about 250 times the present world population!

e F

Answers to Question 7

a F Teachers' attitudes and childrens' aggressiveness may not coincide. And can "opinion" be measured?

b F This is certainly operational, and you are looking at an aspect of aggressiveness itself. If it's representative, it should be valid.

c T Same problem as with (a). Perhaps psychologists' reports have a bit more class, but when possible, look at the phenomenon itself instead of some other person's interpretation of it.

d F Is distractibility the same thing as aggressiveness? If so, fine. But if not, you have merely replaced one question with another. And usually these two concepts are held to be quite different.

e F You don't know how the child feels, only what the child does.

f F How would you operationally define and measure *this*? And anyway, it's not the same thing as what we're looking for; many sweet children come from frustrating backgrounds.

Answers to Question 8

a T Yes, this is a difficulty. But how much are we bothered by the limitations of a simple thermometer reading when we try to encompass all the connotations of "temperature"? Having limitations does not mean being useless.

b F Meaningless? Don't throw out the baby with the bathwater!

c F Arbitrary, yes. But so what? The inch was originally set arbitrarily as the width of King Henry VIII's thumb.

d F The nice thing about operational definitions is that you *can* subject them to the necessary scrutiny.

e T? You know, this could be true! What do you think?

f T This is quite true, because cardinal measurement is the ideal end result of getting operationally defined data.

9 Matching question. In the usual way, try to match the items on the right with the items on the left. Count one for each correct.

Integers	( )	**a** nominally scaled categories
Commutative	( )	**b** Σ
Sum—add up	( )	**c** achieved only by counting
Factorial	( )	**d** $9 \times 8 \times 7 \times 6 \times 5 \times 4 \times 3 \times 2 \times 1$
Data	( )	**e** How old you are
Independent variable	( )	**f** order of arithmetic operations doesn't matter
Dependent variable	( )	**g** intangible
Male/female	( )	**h** variable whose values are set by experimenter
Ratio scaled	( )	**i** scores

EVALUATION

Determine your score; the highest possible is 72. Follow this guide:

60 to 72	Excellent
24 to 59	OK
0 to 23	Review carefully where needed
Negative score	Ascertain what is wrong before going any further. See hints on page 58.

Answers to Question 9

c (b) and (i) are possible, but (c) is best.

f (f) is the only choice.

b (b) What else is so obvious?

d (d) is the only one. We made this one easy for you!

i (i) Data consist of scores.

h (h) is the best alternative, but (a) is possible inasmuch as many independent variables are nominally scaled, e.g., whether drugs or psychotherapy are more therapeutic for treating goitre.

i (i) again; the dependent variable comes out as scores.

a (a) Male/female is an example of nominal scaling.

e (e) is ratio scaled; a person of age 10 is twice as old as a person 5 years old.

Note: Are you upset because some alternatives were correct more than once, and because one was not used at all? Don't be: this was deliberate. Life in the rough is like that; indeed, a major problem is to find the confidence needed to ignore distracting irrelevancies and stand on our considered convictions, even if they don't add up instantly to a neat picture.

HINTS ON REMEDIAL STUDY

1 Could your difficulty be anxiety or hostility? If, so, remember that the book, the author, your professor, and your school are basically on your side, so you probably won't regret it if you relax and enjoy things a bit.

2 Are you trying to memorize the book? If, so, is it because you don't *understand* it? Come on, try to be friendlier than that. This book is not difficult, and you *can* understand it.

3 Once a point is understood, don't let it go flabby again by just letting it sit. Explore it, use it, twist it, and see what its ramifications are. Relate it to other points. State it in different ways from the book's wording.

4 When an examination question looks strange, stop and figure out what principle is being referred to. It should be well into your mind, but maybe in very different terminology. Spread your thinking;

you're a person, not a computer. You are being asked to *think,* not just *respond.*

5 It's customary for examinations and quizzes to adopt the two-valued logic of *right* and *wrong.* It's equally customary for real life *not* to do so. You'll have noticed that some of the items in the foregoing self-test (and in the future ones too) are messy, in the sense that a "correct" answer is really impossible, and that a good "score" is thereby impossible too. Don't fret. These self-tests are not here to squash you. They are here to exercise you, and real-life exercise is seldom neat.

*s*ome ordinary problem*s*

1 Think of operational definitions for the following.
   a hostility
   b dependence
   c persuasiveness

2 Work out the following bits of arithmetic, paying particular attention to significant figures
   a $6!/(3.000 \times 10.000) =$
   b $X = 2, 7, 4, 7, 5, 3.$
   $$\sum X =$$
   $$\sum X^2 =$$
   c $0.003 \times -2.45 \times (-3.1)^2 =$

3 State the most sophisticated form of scaling involved in the following measures.

   a the U.S. Postal Zip Code system.
   b the letter grades assigned in your school.
   c the "counting off" from the left when you are lined up with the rest of your squad (platoon, chain gang) on the field (parade ground, rock pile).
   d the different heights of submarine mountains whose peaks form the Hawaiian Island chain, as shown in most atlases.
   e the numbering on the pages of this book.
   f the temperatures of patients in a hospital.

4 The following formulas omit parentheses because the rules don't *really* require them; however, place parentheses and brackets in such a way as to reinforce proper application of the rules

a $\sum X^2/N - \bar{X}^2$
b $5^2 + 11 \div 72/2^{3!}$
c $105 \div 8 - 5^2/8 - 1$
d $M! - N! \div M^2 - N^2$, where $M = 6$ and $N = 3$

Work out the answers arithmetically for (b) through (d).

5 With the four items in the preceding question, place parentheses or brackets in ways different from the rules intrinsically implied within the expressions (and reinforced by you above), and work out the answers. Are they the same as the answers in Question 4?

6 Under what circumstances does $[\Sigma X^2/N]$ equal $[(\Sigma X^2)/N]$? This will become an important issue later, so spend some time thinking it out.

7 A researcher studying the dietary habits of Eskimos gets data from a total of six Eskimo families, all from the west coast of Greenland. Comment on the validity and reliability problems that may affect the data.

8 List the essential characteristics of the various levels of scaling considered in this chapter.

9 The surgical records of a doctor are reviewed and all of the patients are classified into categories "improved" and "not improved." What level of scaling is this? There are 7547 patients in the first category and 3327 in the second. What level of scaling do we have here?

10 Validity and reliability are different concepts, independent of each other. What are the consequences of having data which are

a very reliable but not valid?
b valid but not very reliable?

chapter 3
all
about
graphs

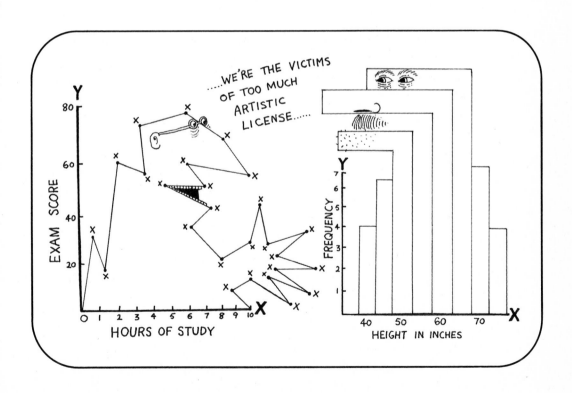

A graph is a picture of reality as seen in a particular way. Its purpose is to display data, to allow the viewer to discern at a glance those facts, those relationships, that are to be abstracted from their background of unsorted information. A well-constructed graph may contain more information than many words. Any object may, however, be portrayed in a huge variety of ways, each showing a different facet of reality. If one model were to be portrayed by Rubens, by Karsh, by Dali, by Goya, and by Picasso, the various end products would differ according to the artists' various purposes.

Similarly with graphs. There are different ways of graphing a given set of data, and they will probably encourage different interpretations of the data. In order to portray information without misleading the reader, proper selection of the style of presentation is needed. If you want to use graphs effectively and resist their misuse as instruments of confusion, you have to know what you're doing. It's almost certain that you are already familiar with graphs from the consumer angle, but actually producing a graph is another matter.

meeting graphs

For the purposes of this book, a graph is meant to display an independent variable, a dependent variable (whose value *depends on* the value of the independent variable), and the actual results of an entire experiment on the relationship between them. This is the objective to keep in mind. Only a very few principles are necessary; once you understand them, you will find graph-making very easy.

The term **graph** is really short for **graphic figure.** With subtle one-upmanship, psychologists rarely use the word **graph;** they use the term **figure** instead. A figure is any sort of drawing that gives a graphic message. Try to use the term **figure** exclusively.

The basic characteristic of a graphic figure is that variables are scaled (marked off) along two straight lines called **coordinate axes**, one horizontal and one vertical. In line with mathematical custom, the horizontal coordinate axis is termed the X **axis**, or the **abscissa** (pronounced ab-SIS-sa). The vertical coordinate axis is the X **axis**, or **ordinate**. The two coordinate axes are placed at right angles to each other, usually crossing toward the lower left-hand part of the figure; see Figure 3.1. The X axis and the Y axis, together, are called the (plural) **axes** (pronounced "acks-eeze").

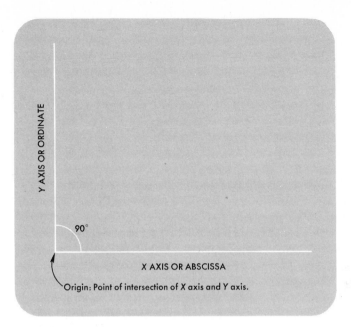

Figure 3.1
The skeleton of a graphic figure.

The different values of the independent variable are marked off along the X axis. If the independent variable is nominally scaled, it can be laid out along the X axis in any order. If it is ordinally or cardinally scaled, the lowest values are placed to the left (where the axes intersect at the **origin**), and the values increase toward the right. Except in unusual circumstances, the numbering along the axis is evenly spaced. If you have laid out the X axis properly, every possible value of the independent variable may be located unambiguously somewhere along it. For example, in Figure 3.2, the independent variable of a hypothetical experiment (in which *hours of study* were experimentally manipulated, each student studying for 0, 1, 2, 4, or 8 hours before a given examination) is marked off on a horizontal axis. As in Figure 3.2, when the zero point on the abscissa represents an actual experimental level, it may be moved slightly to the right of the origin, for clarity. Also note the ratio scaling of the abscissa numbering: the distance from 0 to 4 is half the distance from 0 to 8.

On the Y axis we mark off values of the dependent variable, which are scores of some kind. Scores are usually positive, and the value 0

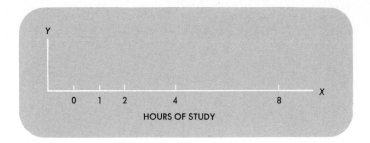

Figure 3.2
A sample X axis, showing the independent variable scaled.

is usually placed at the origin, with higher values marked at appropriate heights along the Y axis. When there are negative as well as positive scores, they are arranged from the largest negative number, up through 0, to the largest positive number. The abscissa can run out from 0, from above 0 (part way up the ordinate), or from a negative location on the ordinate, below all the Y scores. (The latter would be quite unusual, but clarity sometimes demands it.) Every possible score within the range of the data must be represented by some position along the Y axis. The ordinate should be scaled so as to permit this. For example, in the experiment on hours of study mentioned above, the lowest score might have been 6, the highest 91. Therefore it would be reasonable to have the ordinate scaled from 0 (at the origin) to 100 (at the top). On the other hand the height of the ordinate should not be too much greater than the possible range of the data. Here, for example, it should not go from 0 to 1000, because no score values were greater than 100.

Now let us suppose that the students who did not study at all had an average (mean) score of 16 on the exam; that those who studied 1 hour had a mean of 35; that those who studied 2 hours had a mean score of 72; that those who studied 4 hours had a mean score of 80; and that those who studied 8 hours had a mean score of 85. These data can be represented by dots placed as in Figure 3.3. Here the dashed lines (which are omitted from real graphs except in very unusual circumstances) are drawn perpendicular to the coordinate axes, starting at the actual values of the independent and dependent variables; then a dot is placed at the intersection of two dashed lines. The dot above 0 on the X axis and to the right of 16 on the Y axis shows the value of the dependent variable (mean test score) obtained from a particular value of the independent variable (study time). This principle of plotting a graph is known as the **rule of right angles**.

From Figure 3.3 you can see at a glance the relationship between study time and examination score, at least for those values of the independent variable that were actually tried. The scores that correspond to levels of the independent variable that were actually tried out are called **empirical** scores (from the Greek *empeirikos*, "experienced"). But what might have been the mean score for a group of students who had studied for 6 hours? The study time of 6 hours was not tried out.

What would you assume, if you had to assume *something*? Probably that the result for 6 hours would be somewhere between the results for 4 and 8 hours. The best guess (best because it makes the fewest assumptions about what happens between any two empirical points) would be on a straight line connecting the mean scores of the 4-hour and 8-hour groups. The exact point would be directly above the value 6 hours, had it been included on the Y axis. This procedure is termed **linear interpolation**. Empirical points are *always* connected by straight lines; to use curved lines is to add unknown amounts of

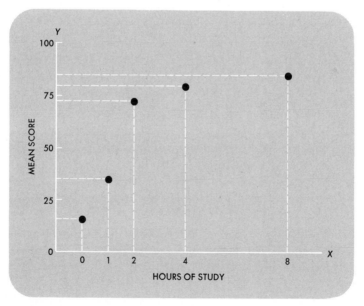

Figure 3.3
Hypothetical relationship between hours of study and examination score, which are, respectively, the independent and dependent variable. Each plotted point is the average (mean) of several individual scores achieved after study for the number of hours shown.

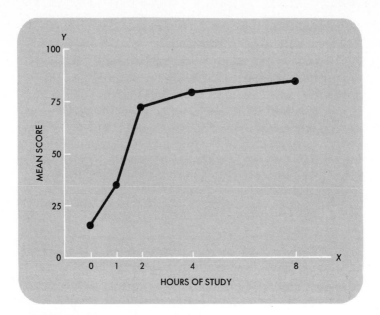

Figure 3.4
Hypothetical data from Figure 3.3, shown as an empirical curve. Note
the interpolation lines, which are drawn straight.

imagination to the data. The completed graphic figure is shown in
Figure 3.4. From this figure, a mixture of empirical data and linear
interpolation will give a pretty good estimate of the mean score to be
expected from any number of study hours (up to 8) under the experimental conditions of this hypothetical study.

Be careful about using curves for empirical data extrapolation.
Curves generally denote *theoretically expected* results, *not* obtained
results plus extrapolation. Of course, empirical data and theoretical
curves *can* be shown together, if that's your deliberate intention. It's
often interesting to compare the theoretically expected results with
those obtained empirically.

As shown in Figure 3.4, labels for the coordinate axes should be clear,
printed in block letters lying parallel to the appropriate axes, and the
plotted (empirical) **points** should stand out clearly from the extrapolation lines. Such graphs should be drawn on (or traced over) squared
graph paper to ensure that the rule of right angles is being followed
exactly. Some esthetic judgment is called for in graph making: figures
should "look good," with reasonable economy of room but without
undue crowding. When a report is being typed, a figure should appear

on its own separate page and have a reasonable margin with enough space within it to be easily read.

There are four major types of graphic figures: **frequency distributions**, **correlation scattergrams**, **data curves**, and **theoretical functions**. Each comes in several varieties and has some unique features.

data curves

The example considered above is an instance of a graph portraying data, a **data curve**. At a glance we can observe the effect of the independent variable, hours of study, on the dependent variable, the mean score obtained. Note that the independent variable is always placed on the abscissa (X axis) and the dependent variable on the ordinate (Y axis). But note too that a **data curve** does not contain any curves: only points and straight lines! The overall effect may approximate a curve, though.

A graph can do more than convey information about the relationship between two variables, as the one in Figure 3.4 did. It can also show relationships among relationships. Suppose that the initial hypothetical experiment involved study of a *verbal subject*, such as French vocabulary. The results, graphed in Figure 3.4, show that performance continues to improve as study time increases to 8 hours. But now, let us suppose that the study-time experiment was also done with a *nonverbal skill*, perhaps playing the flute; another set of empirical data has been collected, and it can be plotted on the same graph and connected by interpolation lines, as shown in Figure 3.5. For the sake of simplicity, the dependent variable is measured on the scale of the ordinate (ranging from 1 to 100). To avoid confusion, the two curves are drawn in different styles, one with open circles and the other with solid circles for the plotted values, and one with dashed lines and the other with solid lines for the interpolations. Graphs like this must be clearly labeled. At a glance, this figure shows that while study for up to 8 hours increasingly improves performance of the verbal task, the nonverbal performance is improved by study only up to 4 hours, after which performance quality decreases. The graph shows the data and their relationships instantly and precisely, making it easy to grasp the entire depicted result.

Graphic figures are nice in that individual needs and artistic preferences may dictate style to a large extent. However, wisdom suggests

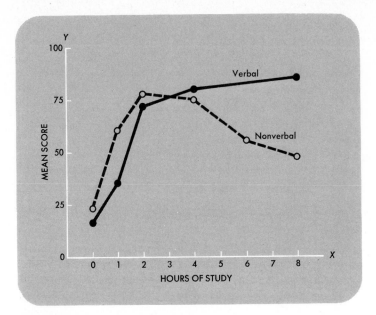

Figure 3.5
Hypothetical data from Figure 3.4, but with data obtained from a different type of task also added. Note that the two curves are easily distinguished.

that convention be followed except for unusual instances. Otherwise people might think you're ignorant rather than artistic.

Some don'ts: Don't use colors to distinguish the parts of your graphs. They won't reproduce very well in black-and-white illustrations when (and if) your work is published (dream on!).

Don't clutter your graphs; if you do, they will not be very legible or instructive. Four curves per graph should be a rough maximum. If you need to illustrate more than four relationships, duplicate the graph side by side with its partners.

Label coordinate axes as instructively as possible. Frequently we give experimental groups rather arbitrary names, such as A, B, and C, which convey no information about the independent variables they represent. At least, let the labels for the various groups suggest what they stand for. In Figure 3.5 the groups are labeled according to amount of study for the examination; no additional verbal explanation of the meaning of the points on the axes is needed.

Formal figures are drawn with India ink, by means of drafting instruments, and lettered with the aid of a stencil. Informally, ruled lines and freehand lettering in ball-point pen are usually sufficient.

correlation scattergrams

The topic of **correlation**, the analysis of relationships, will be covered more fully in a later chapter, so again we give only an illustration here. Let us consider an adaptation of the previous one, on study time and grades. This time, instead of manipulating study time, suppose we merely *ask* 20 students how long they studied and then determine their test scores. Now, instead of an independent and a dependent variable, we simply have two variables that may not be related at all (although we may suspect that they are). Let us scale one of these variables on the abscissa and the other on the ordinate of a graph. Then, using the rule of right angles, we can place a dot in the body of the figure for each student; the dot will be placed above the student's score on the abscissa and to the right of the same student's score on the ordinate. When all scores are entered, there is a scattering of plotted points, each point representing one pair of observations from one student. Figure 3.6 shows such a **scattergram**.

The scattergram has two special features to be noted. The variables may be shown on either coordinate axis, since neither is the independent variable. Moreover, interpolation lines are not used to connect the plotted points. A scattergram simply shows all the data collected on the two variables. Lines are sometimes to be seen on scattergrams, but let's leave them for a later chapter.

theoretical functions

Sometimes graphic figures are used to depict relationships that theoretically ought to exist, quite apart from any empirical data. A theoretical *learning curve* makes a good example. According to the theories of learning fathered by C. L. Hull in the 1940's, learning increases by a constant fraction of the amount remaining to be learned, up to some upper limit; each learning trial results in an improvement, but as trials go on, the amount of improvement becomes smaller and smaller. In psychology there are numerous instances of such mathematical models of behavior, which utilize graphic curves obtained from mathematical formulas. Since psychological theory is best kept as simple as possible, and since theoretical curves generally picture some postulated process

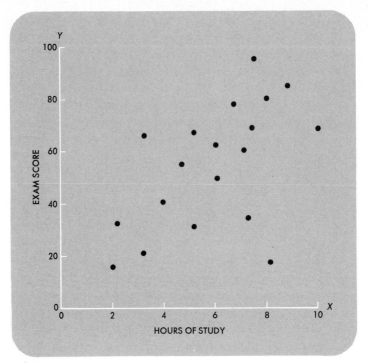

Figure 3.6
Hypothetical data shown as a scattergram. Each plotted point is one
subject, so there is no averaging of scores in this type of graph. Here can
be seen a tendency for more hours of study to result in a higher exam
score, which is consistent with the data shown in Figures 3.4, 3.5, and
3.6. In a scattergram it is not possible to combine (without confusion)
the scores from two or more groups, as was done in Figure 3.5.

or phenomenon, the formulas and curves are usually kept as simple as
possible too. Figure 3.7 shows a learning curve of the Hullian variety,
along with its mathematical formula.

Theoretical curves have several characteristics that set them apart
from any other kind of graphic figure.

a They lack the actually plotted points found in empirical figures.
b They may actually consist of curved (rather than straight) lines.
c The curves are generally simple in shape.
d The coordinates tend to be labeled qualitatively rather than quantitatively.

Theoretical curves are thus easy to recognize.

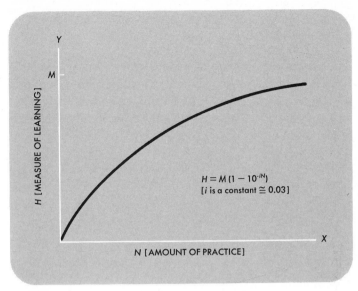

$$H = M(1 - 10^{-iN})$$
$$[i \text{ is a constant} \cong 0.03]$$

Figure 3.7
Example of a theoretical curve, in this case a *learning curve*. Note its smoothness as compared with the empirical curves above and its lack of plotted points. (Adapted from C. L. Hull, *Principles of behavior*. New York: Appleton, 1943.)

frequency distributions

Possibly the most pertinent graph in statistical analysis (as distinguished from data display) is the frequency distribution. It is rather like a data curve, with actual scores supplying values of the independent variable (on the X axis), but the *frequency* of each score is the dependent variable (on the Y axis). The more common scores will correspond to the higher values of the dependent variable (frequency), the rarer scores to the lower values. For example, in a frequency distribution of the height of adult human males, a low frequency of 3- and 8-foot heights and a high frequency of 5 to 6-foot heights would be expected: The frequency of a score depends on its abscissa value.

This is typical of distributions based on actual scores: The most frequently found scores are usually toward the middle of the range, so it is the middle scores that will correspond to the highest values on the Y axis, while the extreme (highest and lowest) scores become progressively rarer and correspond to low values on the Y axis. Figure 3-8 shows what frequency distributions often look like. Where the curve meets the X axis (below 50 and above 90), scores have 0 frequency; this means simply that no males measured were less than 50 inches or more than 90 inches tall.

Frequency distributions in which the most common scores are in the middle of the score range and the more extreme scores are rarer and rarer usually occur when only various **random influences** cause scores to differ from the mean. Random influences (influences that are just as likely to make a score larger than the mean as to make it smaller) will tend to balance each other out in most cases, so most scores will then be close to the mean. Only rarely will all the random influences push a score in one direction, toward one extreme or the other. Thus, extreme scores will be rare. In our example of the height of human males, it is very rare for all the factors that influence human height to push in one direction, toward a height over 7 feet or under 4 feet.

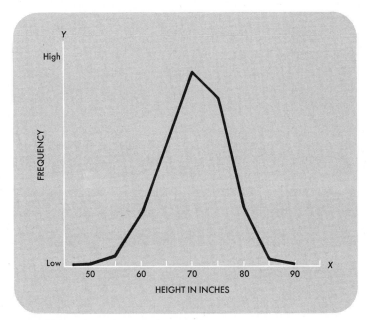

Figure 3.8
Example of a frequency distribution: the heights of all the men in a hypothetical town. Note the relatively symmetrical shape of the distribution. Extremely small and large scores are rare, so the distribution is "humped" in the center. Size of score (height) is the independent variable, and frequency of score is the dependent variable. In all frequency polygons as in this one, both ends of the graph must touch the X axis. This will happen if the two most extreme scores marked on the X axis do not actually occur among the data.

Another example of a distribution in which only random factors cause scores to differ from the expected mean is the proportion of heads obtained when a set of 10 coins is flipped. You would expect about half the coins, 5 out of 10, to be heads. Suppose you flip the set of coins 100 times, though; you almost certainly will not get 5 heads each time. Sometimes 6 heads will appear, less often there will be 7, and still more rarely 8, 9, or 10 heads. A total of 10 heads would be very rare, but not impossible. Similarly, a total of 4 heads would be fairly common, 3 heads less so, and 2, 1, or 0 heads very unusual. The extreme scores are much less common than the scores in the middle of the range. Measurements taken from nature tend to work the same way, but they are more complex.

Factors that influence the **shape** of frequency distributions (where they are high and where they are low), can be divided into two categories: *random* (those that influence the variance of scores evenly toward both ends of the distribution) and *biased* or *biasing* (those that tend to push scores one way more than the other). Figure 3.9 shows examples of **skewed frequency distributions**, in which scores fall more often toward one end of the abscissa than toward the other. Reaction time gives a skewed distribution: Most reaction times are quite short, not much greater than the minimum possible, but for some people they are unusually long, causing the upper tail of the distribution to be longer. A distribution that is not skewed is a **symmetrical distribution**.

A word of warning at this point. You have probably heard about so-called "normal distributions" or "normal curves" in previous exposure to statistics. Don't concern yourself with them in this chapter. They will be discussed in Chapter 7, when the groundwork has been set. None of the distributions discussed in the present chapter are "normal," so clear your mind.

grouping data

Whenever you have fewer than several hundred or thousand scores (with luck, most of the time), you are likely to find them clustering in the middle of the distribution that you are constructing, but probably very few that are *exactly* the same. Your distribution will look something like Figure 3.10, which shows the heights of 25 men. No two of them have exactly the same height, so the frequency count on the Y axis shows just one occurrence of each height. But there *are* more oc-

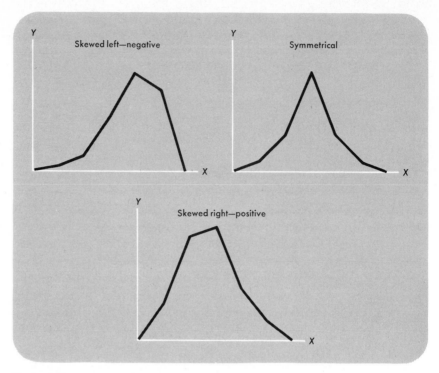

Figure 3.9
Skewed and symmetrical frequency distributions. Note that the direction of skewness is named after the direction in which the longer "tail" is pointing.

currences of men having scores toward the middle of the distribution, as expected. How can the graph be drawn so as to indicate this fact?

The answer lies in **grouping** the scores into a number of intervals along the X axis. The intervals can be chosen in many ways, but it is usually simplest to choose them so that their **midpoints** are integers chosen so as to be multiples of the integer width i. A common interval is five scores wide (thus extending 2.5 units to the left and right of the midpoint), centered perhaps on a multiple of 5. The data of Figure 3.10 are shown grouped in Figure 3.11, with intervals five units wide, centered on multiples of 5. As in this figure, it is usual to plot data in frequency distributions by entering a cross for each score; each cross is entered above the midpoint of its interval, one unit above the last cross that was entered in that interval. When the scores thus marked are stacked in each interval, the heights of the stacks represent the number of scores that fall within each interval.

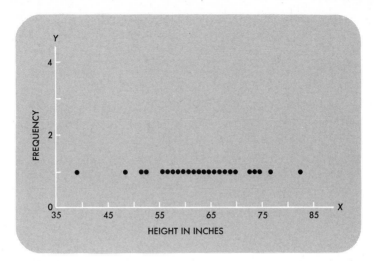

Figure 3.10
A frequency distribution of the actual heights of a small group of hypothetical men.

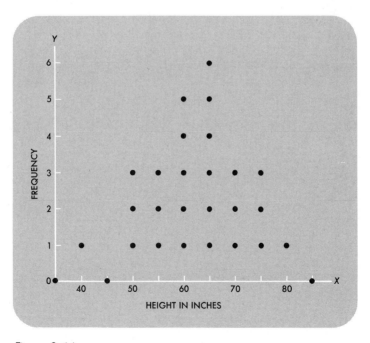

Figure 3.11
The data from Figure 3.10 grouped in intervals. Now it looks more like a graphic figure.

Guard against the troubles that result from setting the *limits*, rather than the *midpoints*, of the intervals at convenient numbers. If your limits are 0, 5, 10, 15, . . ., then the midpoints will be 2.5, 7.5, 12.5, . . ., and you will have two kinds of trouble. Where, for one, would a score of 5 belong on such a graph? *Between* two intervals, which is not allowed. Moreover, your midpoints (assuming that you are measuring scores in whole numbers) are not possible score values.

Problems such as these can be avoided by following a simple rule: Use only an odd value for interval width, and center each interval on a multiple of this width. So if $i = 5$, midpoints would be 5, 10, 15, etc.

If the relationship between Figures 3.10 and 3.11 is not clear, study Figure 3.12, which shows in detail how a graph like Figure 3.10 becomes one like Figure 3.11.

Notice that when you group data into intervals, you lose (for the purposes of the graph) subtle differences among data within the intervals; the graph cannot reflect, for example, the difference between a height of 52.8 inches and one of 53.1 inches. There is a remedy for this: increase the value of **n** (remember, **n** is the number of cases in your sample). With more cases you can make the intervals narrower and still have a fair-sized stack of crosses in each.

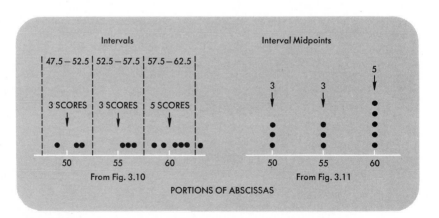

Figure 3.12
How the ungrouped scores of Figure 3.10 are rearranged in Figure 3.11. Any scores falling within an interval are plotted above the midpoint of that interval.

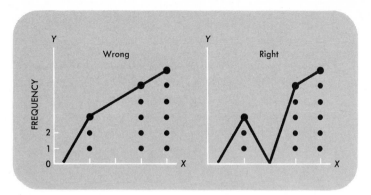

Figure 3.13
The left-hand panel illustrates a common mistake in handling an empty interval. A frequency of 0 is still a value on the Y axis, so it should be plotted like any other frequency. The right-hand panel is drawn properly.

Once you have your stacks of crosses nicely grouped, as in Figure 3.11, you can apply the final polish. Normally the finished figure is either a **frequency polygon** (as described earlier) or a **frequency histogram** (as discussed below).

The **polygon** is drawn by simply connecting every topmost cross by a straight line to the topmost cross in the adjacent intervals. The stacks of crosses are then erased, leaving only the polygon. Great care must be taken to *show* a 0 frequency in intervals that are in fact empty; see Figure 3.13.

A **frequency histogram** is made by erecting flat-topped vertical bars in place of the stacks of crosses: each bar is centered on the midpoint of the interval and drawn to the width of the interval; thus no spaces are left between them. Figure 3.14 shows a histogram of the same distribution as the one plotted in Figure 3.11.

How to choose between polygon and histogram? The choice is really arbitrary; in psychology the polygon is strongly favored. The polygon is probably best for data that vary continuously along the abscissa, because the relative smoothness of the polygon would suggest the potential continuity of the scores. Similarly, the histogram might be favored if the values along the abscissa were whole numbers or categories, because their discreteness would be symbolized by the separate bars of the histogram.

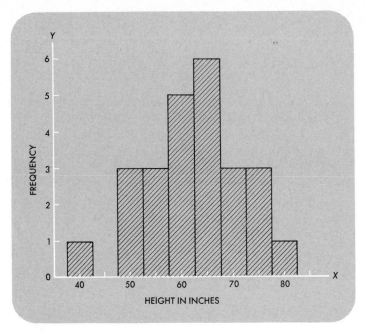

Figure 3.14
The data from Figures 3.10 and 3.11 graphed in a frequency histogram. Each vertical bar has the same width as the score interval, and the height of each bar corresponds to the number of scores within the interval.

graphing nominal data

So far we have not discussed the problem of graphing data from nominally scaled independent variables. If there is no natural *order* for the levels of the independent variable, they may be arranged on the abscissa in any way at all; thus the graphic figure could have almost any shape. There is a special type of graph reserved for such a situation: the **bar graph**.

In a bar graph, although the various levels of the independent variables may be arranged in any order (spaced well along the X axis), it is usually best to have some reason, (however arbitrary), for your arrangement. Each point along the X axis is labeled with the nominal value of that point.

All right, so you have forgotten what nominal scaling is. An example may help you to recall. Suppose you want to draw a frequency distribution of the relative frequencies of various occupations, such as dentist, clerk, and professor, in a particular town. The occupations may be

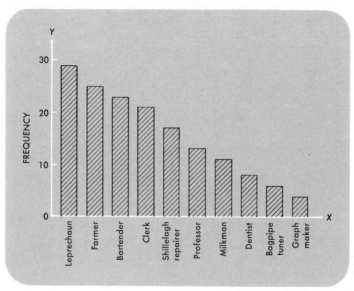

Figure 3.15
A bar graph showing the frequency of various occupations within a
mythical Irish village. The separateness of the bars symbolizes the
quantitative unrelatedness of the different occupations. Because the
bars can be arranged in any order, the bar graph has no fixed shape; a
common convention is to place them in order of decreasing frequency.

arranged in any order along the abscissa, so long as they are labeled
clearly. The values on the ordinate, as is usual with frequency distributions, are the frequencies of each occupation. See Figure 3.15. The
X axis is arranged in order of decreasing number: an arbitrary but
reasonable scheme, since frequency can then be rank ordered by a
glance along the X axis.

Obviously, the bar graph is similar to the histogram. The only difference is that the bars are separated, to symbolize the separate (discrete)
nature of nominal classifications.

cumulative frequency distributions

Sometimes it happens that we wish a graph to tell us not how many
scores fall *into* each interval, but instead how many scores fall *into and
below* each interval. Figure 3.16(a) shows a simple frequency polygon
with the score frequencies shown above each plot (so there are *five*

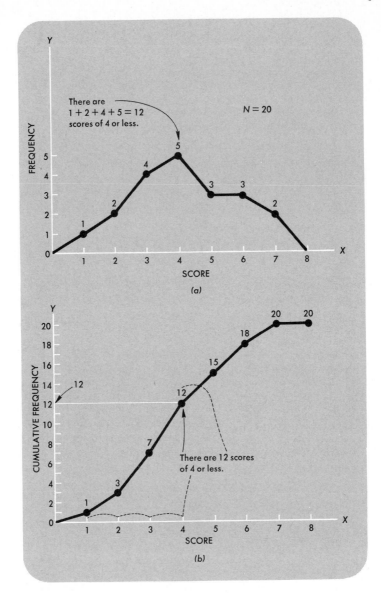

Figure 3.16

scores of 4, *two* scores of 7, and so forth). Figure 3.16(b) shows the same data in a **cumulative frequency distribution**, or **ogive**. Instead of showing the number of scores *in* each interval, Figure 3.16(b) shows the number of scores *in or below* each interval. Each interval along the abscissa in an ogive contains the number of scores within that interval *plus* the number of scores in all lower intervals.

In an ogive, when intervals rather than actual raw scores are being used, the values laid out along the abscissa should *not* be the midpoints of the intervals, but rather the upper real limits of the intervals. This is sometimes awkward. But ogives can be drawn just as easily from ungrouped data (so long as the number of individual scores is not so large as to be unmanageable); thus grouping into intervals is not necessary in the first place (although it may still be convenient).

The ordinate of an ogive can be rescaled from the actual frequencies (up to 20 in Figure 3.16) into percentages. In our example, 20 is 100% of the scores and 12 is 60% of the scores. After rescaling the ordinate you could then see that 60% of the scores were valued at 4 or less, that 15% of the scores are 3 or less, and so on. (What percentage of the scores in Figure 3.16(b) are 18 or less?) This feature of ogives will make them useful when we consider percentiles in a later chapter.

And that's all there is to graphs—for now, anyway.

active mental summary

Review the following terms until you understand them thoroughly.

COORDINATES	EMPIRICAL
X AXIS	LINEAR INTERPOLATION
Y AXIS	POLYGON
ABSCISSA	FREQUENCY DISTRIBUTION
ORDINATE	HISTOGRAM
RULE OF RIGHT ANGLES	BAR GRAPH
SCATTERGRAM	CURVE
INTERVAL WIDTH	OGIVE
INTERVAL LIMITS	CUMULATIVE FREQUENCY DISTRIBUTION
INTERVAL MIDPOINT	

self-test on chapter 3

1 Identifying parts of a graphic figure. Match the letters on the various parts of Figure 3.17 with the verbal list on p. 83. Just use light pencil. Score one point for each correct answer.

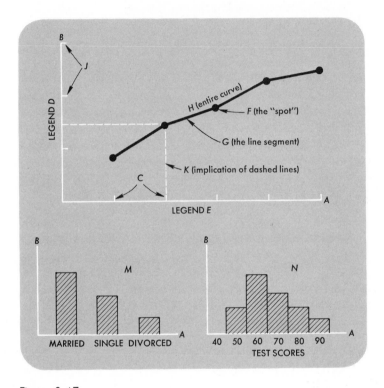

Figure 3.17

( ) Data curve
( ) Levels of independent variable
( ) Y axis
( ) Histogram
( ) Interpolation line
( ) Abscissa
( ) Description of dependent variable measure
( ) Rule of right angles
( ) Mean score
( ) Label for independent variable
( ) Ordinate
( ) Possible values of dependent variable
( ) X axis
( ) Bar graph
( ) Best for continuous data
( ) Best for nominally scaled independent variable
( ) Skewed distribution

SCORING of items on the other side of this leaf. Align the left-hand side of this scoring side with the completed answer blanks on the other side, and evaluate your memory for the anatomy of graphic figures. Review in the text as required.

EXERCISE IN CONSTRUCTION OF GRAPHIC FIGURES

Before starting, be sure to complete any necessary remedial study for Question 1. Some scratch paper will be useful at this point.

This exercise concerns a hypothetical experiment to determine whether accuracy of dart throwing was affected by various amounts of intake of the potent brew served in English pubs. The target consisted of a series of concentric rings, all of the same width, with the central ring (the target numbered 0 (zero error if you hit it) and with successive peripheral rings numbered 1, 2, 3, . . ., 14 (providing units of distance from the central aiming point, and thus a measure of error in throwing the darts). The subjects (abbreviated S) were 15 customers of the pub where the experiment was conducted. Five had received *no* experimental treatment; five had received 2 pints of experimental treatment; and five had received 4 pints of treatment. In each case, the total amount of treatment had been consumed within the 10 minutes prior to testing. Each *S* was given just one trial at throwing the dart from a distance of 10 feet. The data came out as follows.

H
C
B
N
G
A
D
K
F
E
B
J
A
M
H
M
N

0 pints	2 pints	4 pints
2	4	8
4	6	10
0	6	6
1	5	12
3	4	14

Your eventual object is to draw a graph that summarizes the data in an informative way. Completing the following multiple-choice questions should help you do so.

In the following questions, any number of answers may be correct. Circle the letter next to each choice that you consider correct. Score $+3$ for each correctly circled answer, -2 for each incorrectly circled one, and -1 for each correct one not circled.

2 In this experiment the independent variable
    a is the number of subjects
    b should be scaled along the X axis
    c is the error score
    d is the post imbibing period
    e is number of pints imbibed

3 For the purpose of graphing, the dependent variable is
    a number of pints
    b concentric rings of the target
    c each subject's individual score
    d all 15 scores averaged
    e the average of the 5 scores within each treatment group (the 0-pint, 2-pint, and 4-pint groups)
    f should be scaled along the X axis
    g should be scaled along the Y axis

4 If a 1-pint group had been included, the best estimate of its mean score would be
    a 1.0
    b 2.0
    c 3.5
    d 5.0
    e 7.5

5 Since the independent variable represents points along a continuous scale, the type of figure should be a
    a curve
    b polygon
    c bar graph
    d histogram

SCORING AND ANALYSIS OF QUESTIONS 2 THROUGH 6

Answers to Question 2

a F   The independent variable is always what is varied by the experi-
b T   menter to see the effect of the variation on the dependent variable.
c F   Remember that the independent variable is scaled along the abscissa
d F   or X axis. Amount of treatment (intake of brew) is the independent
e T   variable here, so (b) and (e) are correct. Number of subjects per group
and postimbibing period are both constant and therefore cannot exert
differential effects on the three groups.

Answers to Question 3

a F   For graphing purposes, the dependent variable is the average score
b F   within each treatment group. If you average all 15, you lose the differ-
c F   ences between groups, but plotting all individual scores would be too
d F   confusing. The dependent variable is plotted along the ordinate or
e T Y axis, so (e) and (g) are the correct answers.
f F
g T

Answers to Question 4

a F   The mean for the 0-pint group is 2.0, and for the 2-pint group it is 5.0.
b F   Thus the number of pints is 1 pint, and linear interpolation then
c T   requires that we take as our estimate of 1-pint performance the mid-
d F   point between 2.0 and 5.0, which is 3.5. Thus (c) is the only correct
e F   answer.

Answers to Question 5

a T   Choice (a), curve, is the only correct one. Polygons are reserved for
b F   frequency distributions that touch the X axis at both ends, and the
c F   present data do not form a frequency distribution at all. Bar graphs are
d F   used for nominal data. Histograms are best for discrete data, but
amount of beer imbibed is a *continuous* variable.

6 Each plotted point on the figure should lie on a vertical line rising from the proper point along the _____ axis, and along a horizontal line running from the proper point along the _____ axis (thus at the _____ of the two lines).

a X, Y, origin
b X, Y, intersection
c Y, X, interpolation
d Y, X, intersection
e ordinate, abscissa, intersection

If you answer all these questions correctly, you should now be able to graph the data. Draw a frequency polygon from the data from this experiment.

EVALUATION

Determine your score. The highest possible is 38 for Questions 1–6. Grade yourself out of 5 on your graphic effort, thus making the maximum possible score 43.

35 to 43	Excellent
20 to 34	OK
0 to 19	Review carefully where needed.
negative score	Seek help! Either review thoroughly on your own, or talk with somebody who seems to grasp all this clearly. Also, see comments on remedial study, page 58.

a F   **Answers to Question 6**
b T   This question involves the rule of right angles. The only correct answer
c F   is (b); for clarification, if needed, review the first section of this chapter.
d F   You can check your graph against the author's version below.
e F

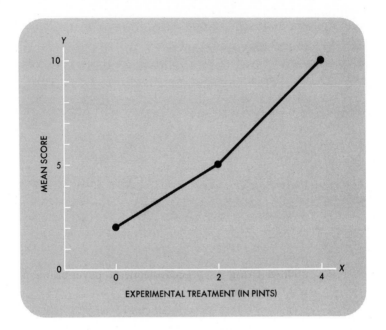

ﾉome ordinary problemﾉ

1 What would be the most appropriate type of graphic figure for each
   of the following sets of data?
   a frequency distribution of IQ in 1000 children
   b blood alcohol as a function of the number of beers consumed
   c crime rate comparison among eight cities
   d blood alcohol as a function of the weighed amount of alcohol
      consumed
   e expected population distribution of IQ

2 In each of the above examples, what is the independent variable,
   and where would it be represented on the graph?

3 What is the other variable in each case, and where would it be
   represented?

4 From the depths of your own random knowledge (or imagination), sketch graphic figures that would probably arise from the above examples.

5 In *any* frequency distribution, what type of variable is plotted on the X axis?

6 Below are some raw scores. Plot them in a frequency polygon having some rationally chosen number of intervals, interval width, and interval midpoints.

41 20 31 53 37 24 39 55 40 24 37 42
57 23 41 46 36 27 54 43 60 25 39 49

7 Take a pair of dice and roll them repeatedly, making a note of the result of each throw. After you have collected about 72 scores, draw a histogram of your results.

8 Take just one of the dice used in the previous problem, and hold it in your hand ready to roll. Before you roll it, jot down the value you *think* the die will show; then roll the die and copy what it shows. Do this 25 times or more. (But not too much more.) Make a scattergram of your results, with *predicted score* on the abscissa and *obtained score* on the ordinate.

9 If such a thing as precognition, or psychokinesis, or some other possible extrasensory thing were operating in Question 8, what shape would you expect your scattergram to have? Did it?

10 Why are frequency distributions usually humped in the middle and tapered toward the sides?

11 Draw cumulative frequency distributions of the data in Questions 6 and 7 (use a separate ogive for each). What is the typical shape of an ogive obtained from scores that are symmetrically arranged in an ordinary frequency polygon?

chapter 4
averages: measuring togetherness

In Chapter 3 we mentioned the common tendency of scores to cluster near the middle of a frequency distribution, as though they differed from some central value only by various combinations of accident. This raises the question of what this central value might actually be. Another question, a more practical one, is how to find some reasonably central single score value that in some way represents all the values present in the distribution. This, of course, is the familiar problem of **averaging**.

Averaging techniques are rather straightforward; of the three kinds most often used, you already know how to calculate the hardest one. The formula for the **arithmetic mean** $\bar{X}$

$$\bar{X} = \frac{\Sigma X}{n}$$

was introduced in Chapter 1 and used in Chapter 3. This is the type of average most often used in statistics; luckily, it is also the one most familiar to all of us, so it represents nothing new to learn.

A second type of average is called the **median**. As you probably know, the median of a divided highway is the strip down the middle that divides it in half. Similarly, the median of a group of scores is the score value in the middle once they are rank ordered. As an example, look at the scores 75, 35, 62, 23, 100, 40, and 16. Rearranging them in order of size, we have 16, 23, 35, 40, 62, 75, 100. Counting from either end, the middle value, the median, is 40, because there are three values on each side of 40. Easy, isn't it?

Sometimes it is harder than that to calculate a median. When there are so many scores that rank ordering them is difficult, or when you have an even number of scores so that the median must be between, rather than at, two of them, the process becomes more complicated. It is especially difficult when there are tied scores at or around the location of the median. Fortunately, we don't use medians all that often in psychology, but you should know how to compute one if necessary.

A third type of average is the **mode**. No calculation at all is needed here, especially when you have a graphic figure of your data. The mode is the **mo**st frequently **de**noted value, the one that occurs most frequently in the distribution. Look back at Figure 3.11; you can easily recognize the mode as 65 inches, the highest point of the distribution.

So we have met three types of average. Must they all have the same value? No. For example, the mean of the distribution in Figure 3.11 is 62.4, not 65. Then which is the "correct" one? Any or all may be "correct" depending on the circumstances; they are suited to different purposes. Incorrect choice of averaging method is one common way of making statistics "lie," or at least give the wrong impression. For each measure of central tendency, we will examine instances of improper use, the characteristics of the measurement, and the situations in which it is most appropriate.

arithmetic mean

Imagine that the small village of Paradise has a total of ten resident families. One family living in Paradise owns a factory in the village and has a yearly income of $91,000. Each of the other nine families, whose working members are all employed in the factory, has an annual income of only $1,000. One worker complains to the government that their incomes are well below the poverty level. The government investigates (i.e., sends a letter of inquiry) and is in turn informed that the average income in Paradise is really very high: $10,000, to be exact. And this is true: the *total* village income is $100,000 among ten families, so the mean income per family is $10,000. Obviously, though, this figure tells the government nothing at all about the way the working families in this village actually live. What is wrong with the statistic?

If a distribution is sharply skewed, or asymmetrical, use of the mean as a *typical* score is often deceptive. A single extreme score can affect the value of the mean so much that it is not even near the value of most scores in the distribution. In such cases, the assumption that the mean represents a typical score is false and may be pernicious. Instead of the mean income of Paradise, the median income would be more informative about general living conditions. In examples as extreme as this one, you might even prefer the mode, $1,000.

The advantages of the mean are:

1 It has mathematical properties that will be very important to us in future work. In particular, if a group of scores is divided into smaller subgroups of equal size, the average of the means of all subgroups is equal to the mean of the original group.

2 It can be calculated directly from the raw data, arranged in any order.

3 It is reasonably stable in value; a few changes in a set of data will not usually affect the value of the mean much.

4 Most people understand it: the mean is *the* average for most people.

the median

The shortcomings of the median are very simple. First, it involves merely ordinal scaling, a step down from cardinal scaling. When you find the median of a set of cardinal data, you discard information that could be important to your experiment. Second, the median can be very time consuming; you may have to rank order several hundred scores. Third, few people understand it. Fourth, it is somewhat sensitive to small changes in the data; if a score moves only a short distance from just below the median to just above the median, the median will thereby be raised. The mean is not as sensitive to very small changes.

The main virtue of the median is its ordinal method of derivation. Remember: it is the score value that is exceeded (by no matter how much) by half the scores. It has nothing to do with the cardinal values of the scores, once the scores are ranked by those cardinal values. For example, the distributions

1 2 3 3 3 5 9 11 13 13 13 14 40

and

1 2 3 3 3 5 9 11 13 13 13 14 14

are identical except for the last score in each; they have the same median, 9, but their means, 10 and 8, differ appreciably. If the extreme score in the first distribution, 40, is doubtful or unreliable (and such extreme scores are often considered unreliable), it is better to use the median, which remains the same whether the largest score is 14 or 40. Conversely, a major misuse of the median is the use of it when a very extreme score (like the owner's income in Paradise) *is* valid. A median income of $1,000 in that village would possibly suggest, wrongly, that the *total* income in the village was only $10 \times 1,000 = \$10,000$. When the score distribution is symmetrical, the median has a value equal to the mean. If the distribution is skewed, the median usually lies closer to the bulk of the scores than the mean does.

The choice between the mean and the median is really a matter of your purpose and convenience. If you have only a few scores, not terribly skewed, the median can be found very quickly, either as the middle score of an odd number of scores, or as the mean of the two middle scores when the total number of scores is even. The mean has the advantage of utilizing all cardinal values of the data, but its sensitivity to those cardinal values is sometimes its downfall. The median is the better choice when you suspect that the extreme scores in the distribution are unreliable. It is also preferable if your distribution is skewed and you fear that the mean is unduly affected by the extended tail of the distribution; in a skewed distribution, the median usually lies closer to the bulk of the scores than the mean does. Sometimes mean and median are combined: If a rat is being run in a maze for five trials per day, the median of the five trial scores is often adopted as that day's score. Then the latter may be averaged, using the mean, with the medians of the other rats in the experiment. In short, either mean or median may serve as data, depending upon your purposes, experience, data characteristics, and good judgment.

the mode

The mode is the poor relation among measures of central tendency. Its use in the social sciences, by comparison with the mean and the median, is rare. This is unfortunate, because it is the easiest of the three to calculate—if calculate is the right word!

The mode is not very stable, as we can see in Figure 4.1, in which the addition of very few scores can accomplish an enormous change in the

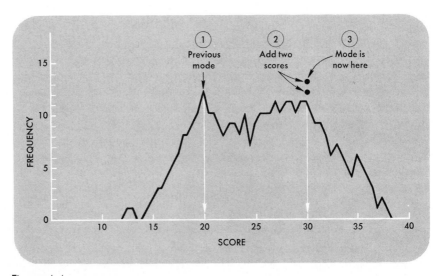

Figure 4.1
The mode is very fickle. In this distribution, with 354 scores, the mode is 20. If only two more scores happen to be included in the distribution and both are 30, the mode changes to 30. A 0.006 or 0.6% change in *n* thus causes a tremendous change in the value of the mode. This example is extreme, but it illustrates how easily the mode can change.

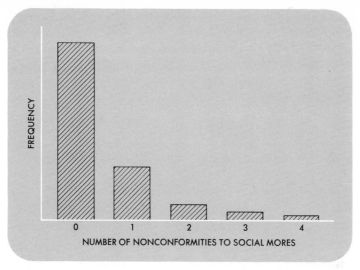

Figure 4.2
Example of the most skewed type of frequency distribution: the J-shaped curve of social conformity. Adapted from F. H. Allport, The J-curve hypothesis of conforming behavior. *J. Soc. Psychol.*, 1934, **5**, 141–183.

modal value. A distribution may possess more than one mode, with consequent problems of ambiguity. In symmetrical "humped" distributions the mode corresponds closely with the mean and median; in skewed distributions the mode falls at the "hump," while the mean falls in the longer tail and the median falls between the mode and the mean.

The most important use of the mode is for sharply skewed distributions. For example, a distribution may be J-shaped, as in Figure 4.2. This graph indicates that most people will conform to important social mores, while a few will do something contrary to these mores once, even fewer twice, and very few three times or more (in a situation that allows data to be gathered). The mean of such a distribution would give a highly incorrect average for social nonconformity, and the median would be better but still bad. The mode may be the best average to choose for J-shaped or U-shaped (sharply bimodal) distributions (those with two clearly distinguished modes). A sharply bimodal distribution is found, for example, if we search for male, as contrasted with female, body characteristics in grasshoppers. "Male" and "female" occupy the extreme ends of a physical continuum, as seen in Figure 4.3.

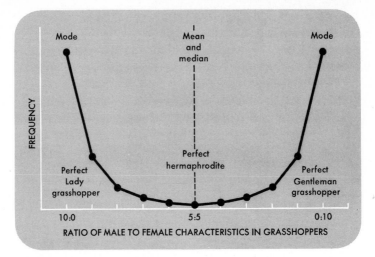

Figure 4.3
Example of a U-shaped frequency distribution: the relative frequencies
of individuals having various combinations of male and female body
characteristics among grasshoppers. Neither the mean nor the median
is the best choice under these circumstances; the mode is the best.

This is an instance of a "U-shaped" distribution; the mean and median
would fall somewhere in the middle, suggesting that the "average
grasshopper" is a perfect hermaphrodite, a rather unreasonable suggestion, to say the least. Instead, using the two modes, we can conclude that the "average grasshopper" is *either* predominantly male *or*
predominantly female.

averages sullied—and sillied

Now that we know that any one type of average need not have the
same value as any other average computed on the same data, we can
recognize that noxious habit of some people, many newspapers, and
practically all politicians, who persist in referring to "*the* average
income of folks in this country," "*the* average number of kids in American families," and so on. There are two sins (or two sillies) being com-
mitted by talking like this. First, such statements are far less precise
than they should be. They are imprecise not only because *at least five*
types of averages exist (of which we have looked at only three), but also
because the data going into such averages can be transformed in a

multitude of ways, and the resulting average then untransformed. Some half-dozen transformations are commonly used and *may* be combined with any choice of average. Theoretically, therefore, one skewed distribution *could* yield $6^5 = 7776$ *different* averages! Clearly, talking about *the* average is useless. The second silly sin is that any average chosen is usually chosen for some good reason—good from the standpoint of the user. Considering that statistics are usually trotted out to convince people of something, it is likely that the type of average chosen is the one most likely to support the user's claim, rather than the one that best typifies the data. Before you draw general conclusions from an average (such as that "the average family" in the village of Paradise has an annual income of $10,000 and thus is far from poverty-stricken), insist on knowing *which* average is meant, *why* it was chosen, and *what* different types of averages would have shown.

Another important point: the concept of "average" *is applicable only to scores*, not to the people who produce those scores. Being of average *height* doesn't make you an average *person*. Indeed, our previous dis-

Figure 4.3a
React with hostility to those who try to talk of "the average man in the street." (*Reprinted through the courtesy of Howard Post and United Feature Syndicate. Copyright United Feature Syndicate.*)

cussion about the essential compatibility of human uniqueness with a large number of psychological dimensions should tell you that a person whose score was exactly the mean (or median, or mode) of *all* physical dimensions (let alone all economic, political, psychological, etc., ones) would be utterly rare. So for practical purposes, there are *no* "average people."

on calculating averages

Now that you have the basic conceptions of the various types of average well in mind, we can concentrate on practical matters of calculation.

mode

When you have only a moderate number of raw scores, the mode can be seen clearly on the frequency distribution. Even with a large number of scores, counting is the worst difficulty involved in finding the mode when raw data are used.

As seen in Figures 3.10 and 3.11, though, raw data are sometimes assembled into intervals for the purpose of graphing. One interval might be five scores wide, e.g., from 2.5 to 7.5, with 5 as the midpoint, which is so labeled. Assuming that all scores are integers, we would place into the interval labeled 5 all scores whose values were 3, 4, 5, 6, or 7, and tally them as though they all had the midpoint value of 5. As we saw in Chapter 3, the shape of a frequency distribution may change greatly when this sort of thing is done. And sometimes, because it is so labile, the value of the mode can change quite drastically, across the width of several intervals.

There is not much you can do about this modal unfaithfulness except possibly avoid using the mode with even greater determination than before. If necessary, it is possible to refer to the **modal interval** of the distribution (as distinct from the **modal score** of the raw data) and hope that they are not too dissimilar.

median

While the basic idea of the median is simple enough, its calculation is usually not simple at all, for two reasons.

1 Frequently there are huge numbers of scores, some of them tied with other scores.

2 Data are often unranked, with the task of ranking them a truly formidable one.

Under these conditions, finding the true median is almost impossible.

However, an *estimate* of the median value can be obtained quite easily if we have the data arranged in intervals in a frequency distribution. It helps if the distribution is a histogram like the one shown in Figure 4.4, but a frequency polygon will do as well. In Figure 4.4, for example, there are 200 scores in the distribution, so the median must be the abscissa value below which 100 scores lie. Of those 100 scores, 85 are in the leftmost three bars, labeled 15, 20, and 25; since there are 40 scores in the fourth bar from the left (labeled 30), the median must be somewhere in that interval. With an even number of scores (200) the median must be *between* the 100th and 101st scores, hence between the 15th and 16th scores in the fourth interval. If we assume (as we must, not having any other information) that scores are equally distributed

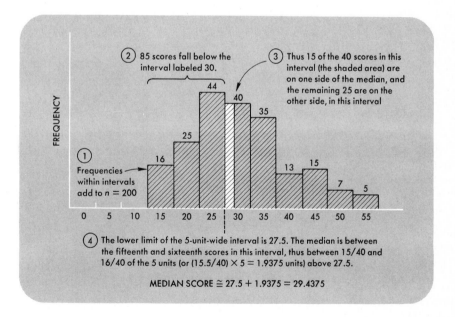

Figure 4.4

Estimating the median of data in a frequency distribution. The necessary assumption is that the scores are evenly distributed within the interval containing the median.

within the interval, then the median must be 15.5/40 of the way through the fourth bar from its left-hand margin, and 15.5/40 of the interval width is

$$\frac{15.5}{40} \cdot \frac{5}{1} = 1.9375$$

Since the center of the interval is 30 and the interval is 5 units wide, its left-hand edge is at 27.5; thus, 27.5 + 1.9375 = 29.4375 is our estimate of the median. That is, 1.9375 units of the bar (shaded area in Figure 4.4) make up the remaining $15\frac{1}{2}$ scores that must be added to the 85 in the three bars to the left to total 100. Note, however, that this value will probably *not* be identical with the median obtained by counting all the ranked scores; it is a good working approximation, though.

In case you are still slightly uncertain about the median, it may help to realize that there are really two kinds. The point that lies between the upper and lower halves of the distribution is one kind. It follows from the definition of the median; it is easy to calculate *when* the distribution has relatively few cases and *if* there are no tied values exactly in the zone where the median is located.

The other kind of median is the approximation illustrated in Figure 4.4, which stems from the basic definition of the median *plus* the assumption that the scores are evenly distributed within the median's interval. This assumption *must* be wrong much of the time, so the approximation of the median may not match the exact median considered above. But the approximation avoids the embarrassments of the exact median: it is easily found for any number of scores, and its value is not affected by the presence of tied scores within the median's interval. It would be easy to decide that the approximation is inferior to the exact median, but such a decision would be wrong. The two types of median were equally chosen and defined quite arbitrarily by humans to serve human purposes. Both of them are abstractions of human thought and therefore should have equal credence in our eyes. Neither was originally handed down from Mount Olympus engraved on slabs of stone.

arithmetic mean

Like the median, the mean is best calculated from the raw scores; once that is done, you have the mean itself, and there is no need to worry about how good your approximation is. These days, when machines are usually available to do the computations, the raw data are usually used in direct calculation of the mean. Sometimes, however, we want a quick approximation of the mean, and if the scores are already grouped into a frequency distribution such as the one in Figure 4.4, we can easily find such an approximation.

Look, for example, at the leftmost bar; it is evident that 16 scores occur in the interval 12.5–17.5, presumably (again, because we have no evidence to the contrary) centered on the midpoint value of 15. As an estimation, we could say that there were 16 scores of 15. Now if we were calculating the mean directly, we would sum those 16 identical scores along with all the remaining scores, and then divide by 200 (the total number of scores). But adding 16 scores whose values are identical (15 in this instance) is the same as multiplying the common score value by the number of scores having that value ($15 \times 16 = 240$). From the second interval we get (still as an approximation) 25 scores of 20, whose sum is $20 \times 25 = 500$. We can do the same thing for each score interval. Thus the value ΣX can be approximated by adding these products (midpoint times frequency) for all the bars of the histogram.

Frequency	Midpoint	Product
16	15	240
25	20	500
44	25	1100
40	30	1200
35	35	1225
13	40	520
15	45	675
7	50	350
5	55	275

$$\Sigma X \cong \Sigma (f \times \text{midpoint}) = 6085$$

Our quick estimate of the mean is therefore

$$\frac{6085}{n} = \frac{6085}{200} = 30.425$$

This estimate should be within half an interval width of the actual mean, and most of the time it will be much closer than that.

TO BE ORDERLY AND
SYSTEMATIC...............

........WEAR GLASSES, TIE
AND SHIRT THE
RIGHT WAY ROUND

.........HAVE CHAIR
RIGHT WAY UP

............USE OTHER
END OF PENCIL

.........USE PAPER
RATHER THAN OLD
HANDKERCHIEF

clean living with calculations

Of course, it's one thing to know how to do something and another thing actually to do it. Here are some suggestions for making life with calculations easier.

When you are making manual calculations (as with relatively few scores or on exams), the secret is to be orderly and systematic. Don't use obscure scribbles on old napkins. Don't lay the figures out like a secret code that nobody (including you) can decipher. Instead, use decipherable scribbles; make sure they're decipherable by labeling what they are. Whatever you are doing your calculations on, start at the top of the page and write small, straight, and clearly. These precautions will actually help you remember what you are doing; they will circumvent that old familiar "midquestion stab of panic" that occurs when you notice that you have lost all track of what all the calculations you've been doing are *for*.

First, write down the best formula for your purposes. It could be for a measure of central tendency, or for another statistical quantity not yet discussed; but your general approach should always be the same. Two things are typically wanted: portrayal of the scores, and various operations to be performed on them. List the scores in a column on the left of the page, and head the column with a descriptive symbol X (or whatever letter you are using to denote the scores). If you want to find ΣX, it is easy to sum them to the bottom. If you want to rank order the scores, create a second column (with some heading like "rank X") and strike off the scores in the first column as you enter them in the second. Or, if you wish to square the scores (and later you will!), create a second column headed "X^2", and enter the square of each score next to the original score. Thus, the various operations called for in your formula can be done by setting up appropriate columns and doing the appropriate arithmetic on the sums at the bottom. Be sure that all sums are properly labeled; this will save time in the end.

Example: Find the mean of these scores and the value of ΣX^2.

5 8 2 9 8 5 6

X	X^2
5	25
8	64
2	4
9	81
8	64
5	25
6	36
43	299
ΣX	ΣX^2

$$\bar{X} = \frac{\Sigma X}{n} = \frac{43}{7} = 6.14$$

$$n = 7$$

$$\Sigma X^2 = 299$$

As a last step, check again to be sure that you have achieved what the question called for. For example, you were asked to find ΣX^2, and the answer found was 299. But, if you had misread the problem as calling for $(\Sigma X)^2$, your answer would have been $43 \times 43 = 1849$, not 299. Laying the problem out as shown helps you to be sure that you squared the scores first and then added, rather than adding them first and then squaring the sum.

Squaring scores in the fashion shown is often necessary in statistics and will be a common feature in later chapters.

When you perform calculations on machines, a different set of problems arises. The major one is to remember what you are doing! If you are adding a large number of scores, set yourself a rhythm for entering them and resist all interruptions until you have finished *and copied and labeled the sum* on a clean piece of paper with the formula for what you are doing at the top. If interrupted or distracted, you can easily omit a score or enter it twice. Your answer may then be "only slightly incorrect, I knew the *principle* all right," but your instructor will probably give you 0 anyway. And even if you enter all scores correctly and the machine gives you the correct sum, if you forget to copy it before turning the machine off, you may lose the answer and your time will have been wasted.

Be sure that you understand how to use your calculator. There are several types, and they do not all work the same way. Check yourself in advance by giving the machine a few simple problems whose answers you already know; if you can't get it to add 2 and 2 correctly, *something* is wrong!

Some calculators have memory registers that do not clear (return to 0) when the machine is switched off. If you come along and work on a machine that has not been cleared, any data already there will mix with your own, with disastrous results. Always clear your machine (unless it is one that you *know* clears automatically when turned off) as soon as you turn it on.

Even when your machine has a memory system, take the precaution of writing down answers as they occur. You will be glad you did!

And keep your wits sharp: remember that the machine is stupid. If you get nonsensical results, it won't help you to blame them on the

Figure 4.4a
. . . And keep your wits sharp!

machine. It *will* help you to recognize that the results are nonsense and to start over again, possibly even with that satisfied feeling that comes from having figured out what went wrong in the first place. The machine only does what people make it do. Garbage in, garbage out.

More words to nibble on:

ARITHMETIC MEAN
MEDIAN
MODE
J-SHAPED DISTRIBUTION
U-SHAPED DISTRIBUTION

exerciser for a very rainy evening

Find 16 checkers, or poker chips or sugar cubes if they are more your style; just be sure they are all the same size and weight. Then find a very thin, lightweight, stiff, plastic or metal ruler, at least 12 inches long (preferably 18 inches long). Finally, find a short piece of dowel, or any reasonably hard cylindrical object, perhaps two or more inches long and $\frac{1}{2}$ inch to $\frac{3}{4}$ inch in diameter. (If you are unusually dexterous and patient, you could manage with a *cylindrical* pen, but something thicker and less smooth would be better.)

These are the raw materials with which the properties of the mean, median, and mode can be demonstrated.

Set up the ruler and dowel on a flat surface so as to form a rough balance scale, as shown in Figure 4.5. It will help if you secure the dowel to the table with a bit of tape. The checkers (or whatever), all having the same weight, can be used to represent the various scores

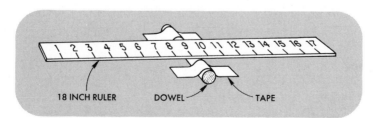

Figure 4.5

occurring in a particular distribution. The range of score values is represented by the ruler scale. The height of any stack of disks on the ruler represents the frequency of a score on the abscissa, just as the height of a column of cross-marks in a frequency distribution (for example, Figure 3.11) represents the frequency of a score.

1 *Demonstration of the mean as a weighted balance.* Place the ruler on the dowel at its 9.5-inch mark (assuming an 18-inch ruler; if you are using a 12-inch ruler, place it on the dowel at the 6.5-inch mark). Does the ruler balance? Now put counters on the ruler to represent the scores that follow; center each counter in the 1-inch space to the *left* of the score you are entering into the "graph."

With an 18-inch ruler, use this distribution.

```
 1  2  3  4  5  6  7  8  9  10  11  12  13  14  15  16  17  18
          X  X  X  X  X  X  X   X   X   X
                   X  X  X  X
                      X  X
```

With a 12-inch ruler, use this distribution.

```
 1  2  3  4  5  6  7  8  9  10  11  12
    X  X  X  X  X  X  X  X  X   X
             X  X  X  X
                X  X
```

Now, without letting the counters move, adjust the ruler on the dowel so that it balances perfectly. It will balance either on the point labeled 9.5 (if you are using the 18-inch ruler) or on the point labeled 6.5 (with the 12-inch ruler). Calculate the mean of your distribution arithmetically.

$$\frac{5 + 6 + 7 + 8 + 8 + 9 + \ldots + 13 + 14}{16} = 9.5 \text{ or}$$

$$\frac{2 + 3 + 4 + 5 + 5 + \ldots + 10 + 11}{16} = 6.5$$

The ruler balances on the mean of the distribution; this is always true. What is the median value? Eight scores (half of $n = 16$) lie below the value of 9.5 (or 6.5) and half lie above it. Thus the median equals the mean. The mode: two scores (9 and 10 or 6 and 7) occur three times each; midway between them is 9.5 (or 6.5). Mean, median, and mode are identical. Now move your two outermost counters one space further out. Does this have any effect on mean, median, and mode?

As a general rule, what must be true about a frequency distribution if its mean, median, and mode are to be identical? If necessary, experiment to find out.

2 *Demonstration of the mean as affected by skewness of distribution.* Set up the counters as in Question 1. Adjust to the point of balance at the mean, and then move only the rightmost counter to the last interval on the right. The ruler will no longer balance. Readjust it to its new point of balance, and note as precisely as possible the scale value of the new point. (Go ahead, write it down: what if the whole contraption suffers a fatal accident while you putter about?) Work out the mean arithmetically; again, the ruler balanced on the mean of the distribution.

Was the median affected? The mode? When a distribution is skewed, what happens to them?

Repeat the demonstration more dramatically by placing the counters according to one of these distributions.

```
1  2  3  4  5  6  7  8  9 10 11 12 13 14 15 16 17 18
            X  X  X  X  X                      X  X
            X  X  X
            X  X  X
            X  X  X
```

or

```
1  2  3  4  5  6  7  8  9 10 11 12
         X  X  X  X  X     X  X
         X  X  X
         X  X  X
         X  X  X
```

Find the point of balance; it will be well to the right of the median and the mode. Check the mean arithmetically.

3 *Demonstration of score freedom and loss of freedom, once the mean is set.* Set 15 of the counters fairly regularly along the length of the ruler, leaving one counter over for later use. Now, balance the ruler on the 9.5 point; this is to be the mean value for all 16 scores. Actually, *does* it balance on 9.5? Probably not. Whether it does or not, place the remaining counter on the ruler in such a way as to make it balance perfectly at 9.5.

Could this last counter have been placed just anywhere? Or was it obliged to go in just *one* particular location?

As a general principle, if $n - 1$ of the scores and the mean of all n scores are given, the last score is not free; its value is dictated by the other information you have. This principle will return in a later chapter.

*r*ome ordinary problem*r*

Here is a huge bunch of scores, to which the following ten problems refer. They are already arranged in ascending order.

4	17	23	27	32
8	17	23	27	32
11	18	24	28	32
12	20	25	28	33
13	21	26	28	34
15	22	27	29	34
15	22	27	29	37
16	22	27	30	37

1 Using the raw scores, find the value of $\bar{X}$.

2 Group the data into intervals and draw up a frequency histogram.

3 What characteristic is possessed by the frequency distribution? Would this characteristic have any bearing on your choice of average?

4 Since the data are ranked, it is easy to determine the median. Is it identical to the mean?

5 What is the modal score?

6 Now determine the mean using the frequency × midpoint method. Is it identical with that found in Question 1?

7 Using the grouped data from Question 2, approximate the median. Compare it with the median found from the raw data. Are they identical?

8 How does the modal interval compare with the modal score?

9 For the above questions, the interval width was left unspecified. Now, however, repeat the process using an interval *one score wider* than the one you chose. Repeat the calculations for mean and median using the new intervals. Are the results the same?

10 Based on the result of Question 9, can you state a general principle that greatly eases the problem of setting up score intervals for purposes of graphing and approximation?

11 A student was running an experiment in which reaction time (RT) was measured many times for every subject. A measure of central tendency had to be taken for each subject's scores. However, our student noted with dismay that the apparatus was unreliable in measuring the very lowest and very highest values of RT (each comprising about 20% of the total), although the middle 60% were measured accurately. Which measure (or measures) would be most appropriate for getting indications of central tendency? Is any measure particularly inappropriate? Explain. (Remember that the RT distribution is quite skewed.)

12 From the following data, sketch the shapes of the distributions which they represent, assuming reasonably smooth general contours. If necessary, refer back to the general conclusions you came to in earlier problems.

Distribution	Mean	Median	Mode(s)
1	45	45	45
2	45	25	—
3	25	45	—
4	45	45	15, 75
5	45	55	10, 75

13 In a unimodal distribution (one with only one mode) that is skewed, one tail will be steeper than the other. There will be a "sharp end" and a "blunt end." To which tail will the median be closest? Toward which end does the mean lie? Where is the mode?

14 Make a frequency distribution of the following data.

2 6 3 3 8 4 3 1 2 2 5 4 10 4 3 6 5 2
3 3 4 9 3 3 5 2 9 3 4 7 2 1 4 2 4

a What is the shape of the distribution?

b Give its modal value(s).

c Calculate the mean and the median; is their difference characteristic of the type of distribution?

d Perform a transformation on the data by dividing each score into 1 (leave as a fraction) and then multiplying each reciprocal by 10. Make a frequency distribution of the new scores.

e Add 1 to each score, and repeat the transformation in (d).

f Add 10 to each score, and repeat the transformation in (d).

g From what you have learned in the preceding questions, can you make any general statement concerning the effect of these transformations on distributions?

h Are the relationships within the data disturbed or invalidated by any of the above transformations?

15 Find the mean $\bar{X}$ of the scores

$$2 \quad 3 \quad 3 \quad 4 \quad 4 \quad 4 \quad 5 \quad 5 \quad 6$$

Now, find the difference $X - \bar{X}$, between each score X and the mean. (If a score X falls below the mean, note that $X - \bar{X}$ is negative.) What is the sum of all the differences, taking all repetitions of scores into account? Does any general rule seem to be operating here? If in doubt, try it again with the scores

$$5 \quad 5 \quad 7 \quad 9 \quad 9 \quad 9 \quad 12$$

chapter 5
variability: measuring apartness

In Chapter 4 we considered the fact that scores usually cluster around some central value, which can be defined and measured in various ways. But many of the scores in a distribution will differ from the central value. There is *variability* in any distribution. Now we shall consider the *amount* of clustering tendency around any such central value.

In Figure 5.1 the graphs of two distributions of scores, both having a mean of 5, are shown. Although the distributions have the same mean, one is more widely spread out than the other. If you wanted to describe the two distributions as thoroughly as possible using only *two* quantities for each of them, it would certainly be a good idea to start with their means. But then the other quantity ought to reflect the difference between the two distributions. The one in (a) is closely clustered with little spread, while the one in (b) is much more spread out. Graph (a) shows *small variability*, while graph (b) shows *large variability* of scores, but with the same mean and number of scores n. So the problem is how to express variability as a single numeric quantity. There are *five* measures of variability that we shall consider here. They are called range, quartile deviation, mean deviation, standard deviation, and variance. As was the case with the three measures of central tendency we considered in Chapter 4, each measure of variability has its uses and its weaknesses, but each one does express the "spread-outness" of scores as a single numeric value.

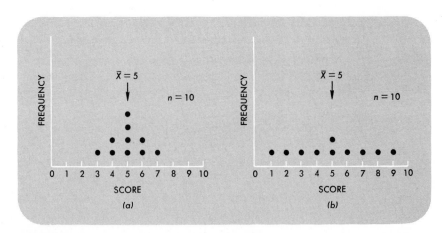

Figure 5.1
Two distributions of scores having the same means but with different variabilities. How can such different degrees of variability be expressed numerically?

We shall be using the example in Figure 5.1, so perhaps you should return to it; compare the two panels and be sure you understand the situation. Note that in each case $n = 10$ and $\bar{X} = 5$. Compute the means with pencil and paper if necessary; however, this example is so simple that n and $\bar{X}$ should be obvious at a glance.

range

Look at the difference between the time it snowed the most compared to the least

The **range** as a measure of variability is common in everyday life. The ages of the children in a family might *range*, for example, from 3 to 9. Then the difference between the largest and smallest scores (ages) would be $9 - 3 = 6$. However, between 3 and 9 *inclusive* there are 7 ages, not 6 (if in doubt, count them). Thus, to include both extreme scores, it is necessary to add 1 to the difference between the largest and smallest scores. When you have done so, you have calculated the **range** as statisticians define it.

$$\text{Range} = \text{Largest} - \text{Smallest} + 1$$

Let's apply this formula to the data of Figure 5.1. In graph (a), the smallest score is 3 and the largest is 7, so

$$\text{Range} = 7 - 3 + 1$$
$$= 5$$

Inspection of the graph shows that there are 5 columns of crosses, to match a range of 5. In graph (b), the smallest score is 1 and the largest is 9, so

$$\text{Range} = 9 - 1 + 1$$
$$= 9$$

(The range is not always equal to the number of such columns, but it's nice when it works out that way.) So it seems that the range works as a measure of the spread of scores. And nothing could be easier to compute.

So what's wrong with the range? There must be something wrong with anything so simple! Unfortunately, there is.

Basically, the range is undemocratic. Its value depends only on the two most nontypical scores in the distribution, those at the extreme ends. Change one of them, and the range changes too; this could be quite a problem if the extreme scores are unreliable. Like the mode, the range is too changeable because it neglects too many of the scores. It would be better to find a measure of variability that uses more than just two score values; ideally, *all* the score values should have some say in the matter.

The range is useful with a distribution of very few scores that are reasonably closely grouped anyway. One doesn't meet such distributions very often, however.

quartile deviation

Sometimes called the "semi-interquartile range," this is a step up from the range. The **quartile deviation** is **half the distance between the 25th and 75th percentiles**. This makes it necessary to introduce percentiles.[1]

A given percentile, say the 25th, is the point on the X axis *below which* just that percentage (in this case 25%) of the scores in the distribution lie. As is the case with the distributions in Figure 5.1, the 25th and 75th percentile points may not be precisely apparent. Mathematical rules for locating the **quartile points** (the 25th, 50th, and 75th percentiles) of a distribution will be discussed in the next chapter, but we do not really need them now.

The basic concept of the quartile deviation is really very simple, as seen in Figure 5.2. Graph your frequency distribution as usual, without grouping the scores into intervals; then lop off 25% of the scores from each end, noting the abscissa values that mark the cutoff points. Then the quartile deviation is half the difference between these cutoff values.

Calculation of the quartile deviation is no more difficult than finding the quartile points of the distribution. If no scores are tied, this task is quite manageable. Because the quartile deviation is rather less important than other indices of variability, we will confine our attention to this simple situation.

First, arrange the scores in order of increasing size; a small collection of scores might then look like this:

9   14   15   18   25   26   27   29   31   35   39   45

---

[1] Percentiles will be considered much more extensively in Chapter 6. For now, the general idea will do.

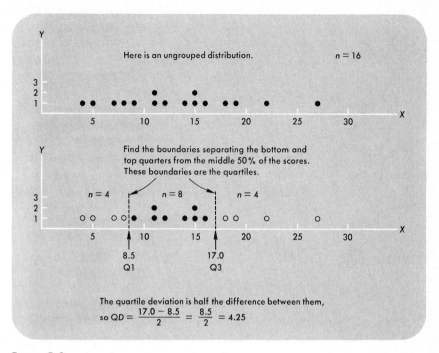

Figure 5.2
The calculation of the quartile deviation. Part (a) shows the distribution of ungrouped scores. Part (b) shows the bottom one-fourth (4 of the 16 scores) cut off at a certain abscissa value, and the top one-fourth is cut off similarly. The distance between cutoff points (the *quartile points*) is the *interquartile range*; half of the interquartile range is the quartile deviation.

Now, find the point at which 25% of the distribution (3 scores, in this case) is exceeded, and the remaining 75% is still to come. Clearly, this point is between 15 and 18, but exactly where? For simplicity, we place it midway between them, at 16.5. This is the *first quartile point,* QP_1. The third quartile point, QP_3, is found similarly, at 33 (halfway between 31 and 35). The difference $QP_3 - QP_1$ is $33 - 16.5 = 16.5$. Half of 16.5 is 8.25, so this is the quartile deviation. In general, letting QD represent the quartile deviation,

$$QD = \frac{QP_3 - QP_1}{2}$$

You may have noticed a point of similarity between the quartile deviation and the median: both are based on ordinal scaling and have the strengths and weaknesses that ordinal scaling implies. Quartile deviations are useful when, for any reason, you distrust the scores in the extreme tails of a distribution but are confident about the more central ones. This benefit is shared by the median. The major difficulty of the quartile deviation, apart from its retreat from cardinal to ordinal scaling, is that it does not reflect how much the extreme scores in a distribution deviate from the middle ones. The quartile points enclose the middle 50% of scores, so we know that the remainder of the scores fall outside these points but not how far outside. As we've seen, this can be a blessing if those extreme scores are dubious, but it is equally a drawback when they are reliable. We still need a measure that involves *all* the scores in the distribution.

mean deviation

Here is an obvious approach to the measure of variability: Since we are interested in how much scores tend to deviate from the mean of the distribution, why not calculate the *average distance* between the scores and the mean? This quantity is called the **mean deviation** of scores from the mean.

Look at Figure 5.1(a) again. Four scores are equal to the mean (5), four scores (the scores of 4 and 6) deviate from the mean by 1, and two scores (3 and 7) deviate from the mean by 2. Why not add up these amounts of deviation and then divide by n? This would give us

$$\frac{2 + 1 + 1 + 0 + 0 + 0 + 0 + 1 + 1 + 2}{10} = 0.8$$

Go on, work it out yourself. Don't skip this. The mean deviation for Figure 5.1(a) is 0.8.

Work out the mean deviation, MD, for Figure 5.1(b). Your answer should be 2.0. Remember that deviations of 0 still contribute to the value of n in the denominator. The formula for the mean deviation is thus

$$MD = \frac{\sum |X - \bar{X}|}{n}$$

Note that in the numerator, the difference between each score X and the mean $\bar{X}$ is an *absolute* difference; the sign of each difference, whether plus or minus, is ignored. Now you may recall from Chapter 1 that absolute values can be tricky. In this instance, they would have undesirable future payoffs in some of the usages for which variability measures are required. Therefore, the mean deviation is not usually found except as a conceptual device. Since the mean deviation, unlike the range and the quartile deviation, is calculated from the actual values of *all* scores in a distribution, it is sensitive to how much the extreme scores differ from the mean.

We have found that the mean deviation for the distribution in Figure 5.1(a) is 0.8, and for the one in Figure 5.1(b) it is 2.0. The greater spread of the scores in (b) yields a larger value of MD. Conceptually it should hang in there pretty well. But the mean deviation gets to be too much trouble if it's taken too far, and we need an improvement on it.[2] Fortunately, the needed improvements are conceptually akin to it.

variance

The **variance** of a frequency distribution is defined as

$$s^2 = \frac{\sum (X - \bar{X})^2}{n}$$

in which s^2 (s squared) is the symbol for variance, $\bar{X}$ is the mean, and X stands for the varying score values. This formula is very like the formula for the mean deviation. The only difference is that in calculating s^2 each difference $X - \bar{X}$ is *squared* before being summed. This has some obvious and some subtle advantages. Obviously, it removes the whole issue of absolute values, because the act of squaring any number renders it positive. Less obvious is that the variance has desirable mathematical properties that we will meet later in the book.

Let us systematically calculate the variance of the scores in Figure 5.1(a). Arrange the scores vertically in a column labeled X, and then

---

[2] The difficulties with the mean deviation involve the fact that such deviations have a *direction*, with X being *lower* or *higher* than $\bar{X}$. When arithmetic operations are performed on the deviations of scores, the direction of difference will have a sign, which would interfere adversely with the other calculations that we wish to undertake.

find the difference between each score and $\bar{X}$, listing these differences in a second column; then square each value $(X - \bar{X})$, listing the results in a third column:

X	$X - \bar{X}$	$(X - \bar{X})^2$
3	-2	4
4	-1	1
4	-1	1
5	0	0
5	0	0
5	0	0
5	0	0
6	$+1$	1
6	$+1$	1
7	$+2$	4

$$\Sigma\,(X - \bar{X})^2 = 12$$

The third column is then summed, giving $\Sigma\,(X - \bar{X})^2$, and the final step is to divide this sum by n, which is 10 in this example. The variance is thus $\frac{12}{10} = 1.2$.

The expression $X - \bar{X}$ (pronounced "X minus X-bar") is unduly noisy and lengthy for frequent use, so we rename it x (little x). Lower-case letters, as contrasted with capitals, denote **deviation scores**.

$$x = X - \bar{X} \quad \text{and} \quad x^2 = (X - \bar{X})^2$$

If we are using Y to name the scores, we can say $y = Y - \bar{Y}$, etc. It is best to call x "little x" to distinguish it from X (big X).

A conceptual crisis may be nibbling at your mind with regard to the variance. We started with a set of scores and transformed them to their x values (their distance from the mean), but then we squared the x values, making them larger. To counteract the effect of the squaring, we can take the square root of the variance. This square root is called the **standard deviation**, s, of the distribution.

standard deviation

The standard deviation is the most widely used index of variability. It has the same conceptual features as the mean deviation without the disadvantages of the latter. It is easy to convert back and forth between the variance and standard deviation, and the variance is a widely used and often necessary quantity. The relationships among the formulas show interesting parallels:

Mean $(\bar{X})$	Mean Deviation (MD)	Variance (s^2)	Standard Deviation (s)
$\dfrac{\sum X}{n}$	$\dfrac{\sum \lvert X - \bar{X} \rvert}{n}$	$\dfrac{\sum x^2}{n}$	$\sqrt{\dfrac{\sum x^2}{n}}$

It must be stressed that the variance and standard deviation formulas shown here merely describe the variability within a group of *known* scores. As we will find later, we sometimes wish to *estimate* what the variance or standard deviation *would be* if all possible scores from the same source were included, that is, if the sample of scores were larger. The variance and standard deviation formulas must be modified slightly in order to give such an estimate fairly. The rationale for the modifications will be considered in a later chapter, but to avoid possible confusion, we note now that the version of the variance used for estimation has the formula

$$\tilde{s}^2 \text{ or } \sigma^2 = \frac{\sum x^2}{n - 1}$$

instead of

$$s^2 = \frac{\sum x^2}{n}$$

The symbol σ is the Greek lower-case *sigma*, or *s*. For now you can file this version of the variance away. The two kinds of variance are so close conceptually that you don't have to worry about the new one. Not yet, anyway.

A word of caution: You may well be dismayed at the grinding, machine-like boredom of calculating s^2 and s for a large number of scores. Maybe, you wonder hopefully, there is a quick way of getting these values directly from the mean deviation? Unfortunately, there isn't. There is no fixed relationship between MD and s^2. It is necessary to go through the calculations as shown above, although there are ways of reducing the agony somewhat, as we will soon see.

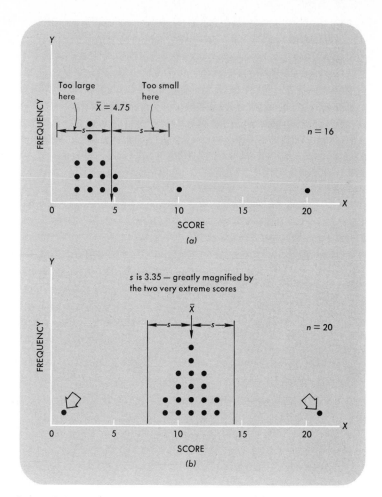

Figure 5.3

Two situations in which the standard deviation might give a misleading picture. In part (a) the distribution is tremendously skewed, so that the standard deviation is both too large to describe accurately the variability to the left of the mean and too small to describe accurately the variability to the right of the mean. Part (b) shows how every extreme score can magnify the size of the standard deviation so that it does not accurately represent the variability of the majority of scores. Here, two scores keep the standard deviation from being valid for the remaining 18.

disadvantages of the standard deviation

Even the standard deviation sometimes has problems: nothing can do everything! There are times when s fails to convey the information you want. Two such cases are shown in Figure 5.3. In Figure 5.3(a), the distribution is sharply skewed, with extreme bunching of scores just to the left of the mean and an extreme spread of the scores to the right of the mean. In this case s is both too large to reflect the actual variability to the left of the mean and too small to reflect the actual variability to its right. In Figure 5-3(b) a few extreme (large and small) scores have enlarged the value of s beyond any good description of the remaining scores, which are clustered in the middle. In both of these situations the quartile deviation might be a better choice for describing variabilities.

error and standard deviations

If you return to the cartoon of the leprechaun in Chapter 3, you will see the effect of his progressive intake of Irish Dew upon his dart-throwing performance; his error increases with his intake. Moreover, his "average amount of error" can be measured by taking the distance between each dart and the center of the bull's-eye and finding the mean of all such distances. This process would be analogous to finding the mean deviation and, by extension, to finding the standard deviation of a distribution. In general, the value of s can be thought of as a measure of some error that makes scores differ from their mean. Factors (such as Irish Dew) that increase error also increase s. Factors that decrease error decrease s. This idea will be important later in this book.

constant error and variable error

Suppose our leprechaun of the previous paragraph had been cold sober —and also a dead shot. But suppose too that he was cross-eyed. Aiming at the center of the target, he indeed gets all his darts clustered in a tight little group, but away from the center. Obviously, there is *error* involved here, but there is very little *variability* in his performance. The standard deviation of his performance will be small. When there is little variability *within* a set of scores (or within a cluster of leprechaun-tossed darts), there is still room for error *between* the overall performance and some other standard (the score that *should* have been obtained, or the bull's-eye). Such error is termed *constant error* because it affects all scores in a similar way, though not necessarily to the same extent. The standard deviation is *not* a measure of constant error. Why does such constant error occur? Because some

An illustration of constant error.

An illustration of variable error.
If a leprechaun were extremely well oiled, how easy would it
be to decide from his dart-throwing performance whether he
was cross-eyed as well?

constant factor (such as being cross-eyed) is influencing the scores
(or the fall of the darts), always in a particular direction.

Variable error, on the other hand, may be thought of as causing
scores to deviate in all directions from some expected mean value. As
implied in our example, the two types of error are independent of one
another; either kind may occur irrespective of the other. (Of course,
both kinds may occur together, as would happen if our leprechaun were
drunk *and* cross-eyed.) The effect of an independent variable in an
experiment should be roughly the same for all scores that it affects, so
an experimental effect is something like a constant error. (It ideally
changes scores without affecting the amount of variability among
them.) So are the effects of uncontrolled and unsuspected variables
that have not been designed out of the experiment (eliminated by the

way the experiment is planned and executed). Problems may occur when a well-controlled experiment yields data in which the constant factor, the experimental effect, is mixed with variable error.

Think of a ratio between CE (amount of constant error) and VE (amount of variable error)

$$\frac{CE}{VE}$$

If the ratio is large, it would mean that something is affecting the data besides random error. If the ratio is 0 or near 0, it would mean that the data were affected only by random forces. If the ratio is between the two extremes, you would wonder. The purpose of statistical analysis of experimental data is first, to separate CE from VE; second, to form a ratio as shown[3]; and third, to decide what the size of ratio implies about the experiment's problem.

Albert Einstein said, "God is subtle but he is not dishonest." The subtlety consists in facts being served up to us in small doses, surrounded by clouds of variable error. Considerable skill, straight thinking, and experience are often required to find the particles of truth and separate them from the clouds of error. The role of statistical inference, discussed later in this book, is to provide the basic methodology for doing so. It should be an exciting prospect.

the calculation of s

You probably still feel uneasy about the ritual involved in calculating s. In particular, it is very time-consuming to lay out your columns of X, x, and x^2, particularly when n is very large or when $\bar{X}$ or the values of X are not easy to work with. Luckily, alternative, time-saving versions of the formula for s exist. The formulas you have seen already are simple and *conceptual*, but conceptual formulas are rarely suitable for calculations; for the latter you need a **working formula**. Here are several working formulas for s; they all use the raw score values of X, rather than x or x^2. The first one is very useful if you already know the value of the mean.

---

[3] The various *statistical tests*—the t test, F test, etc.—to be found later in the book all involve such ratios, made in different ways.

$$s = \sqrt{\frac{\sum X^2}{n} - \bar{X}^2} \qquad s = \sqrt{\frac{\sum X^2}{n} - \left(\frac{\sum X}{n}\right)^2}$$

$$s = \sqrt{\frac{n(\sum X^2) - (\sum X)^2}{n^2}}$$

Lest these formulas frighten you, recall once again that they do not have to be *solved*. They are merely symbolic recipes for doing simple arithmetic in a proper sequence. Squaring these gives working formulas for the variance, s^2.

$$s^2 = \frac{\sum X^2}{n} - \bar{X}^2 \qquad s^2 = \frac{\sum X^2}{n} - \left(\frac{\sum X}{n}\right)^2$$

$$s^2 = \frac{n(\sum X^2) - \sum X^2}{n^2}$$

derivation of the working formula for the variance

This algebraic derivation is strictly optional. You can skip it and lose nothing. But since you may be determined to master every step of the argument, here is how the conceptual formula for s^2 becomes the working formulas above.

Initially, we have

$$s^2 = \frac{\sum (X - \bar{X})^2}{n}$$

which can be expanded as

$$\frac{\sum (X^2 - 2X\bar{X} + \bar{X}^2)}{n} = \frac{\sum X^2 - 2(\sum X)\bar{X} + \sum X^2}{n}$$

Since

$$\sum X = n\bar{X} \quad \text{and} \quad 2(\sum X)\bar{X} = 2n\bar{X}^2$$

we can rewrite the second line as

$$\frac{\sum X^2 - 2n\bar{X}^2 + n\bar{X}^2}{n}$$

which simplifies to

$$\frac{\sum X^2 - n\bar{X}^2}{n}$$

Rewriting $\bar{X}$ as

$$\frac{\sum X}{n},$$

we get

$$\frac{\sum X^2 - n(\sum X^2/n)}{n}$$

which simplifies to

$$\frac{\sum X^2 - (\sum X)^2/n}{n}$$

and then to

$$\frac{\sum X^2}{n} - \frac{(\sum X)^2/n}{n}$$

and to

$$\frac{\sum X^2}{n} - \left(\frac{\sum X}{n}\right)^2$$

This is the same as

$$\frac{\sum X^2}{n} = \bar{X}^2$$

which is one of our working formulas. It is also the same as

$$\frac{n\sum X^2 - (\sum X)^2}{n^2}$$

which is another of our working formulas. By taking the square root, we get the value of s as well.

mathematical independence
between mean and standard deviation

When you calculate the value of the standard deviation, you use the value of the mean. It therefore might seem that the value of $\bar{X}$ must affect the value of s in some way, so that if the mean were to be different, the standard deviation would need to be different too.

But in actual fact, the values of $\bar{X}$ and s are mathematically quite independent of each other. As we saw in Figure 5.1, a given mean value can be associated with two (or in fact any number of) different standard deviations. The purpose of this section is to show that a given standard deviation can be associated with any number of different mean values—in other words, the two statistical concepts are mathematically independent of each other.

1. the paper doll method

Draw the coordinate axes of a graph; label the ordinate *Frequency*, and label the abscissa in arbitrary score units, say from 0 to 10. Make the graph fairly big, but don't draw any frequency distribution within the coordinates.

On a second piece of paper draw a symmetrical curve like the one in Figure 3-9 that would fit very comfortably within the coordinates on your first sheet. Cut out the bell-shaped frequency distribution with scissors. This curve represents a distribution whose degree of variability is fixed: the paper cutout is not made of rubber. It also has a mean value, of course, which you can mark in with pencil.

Now place the cutout anywhere within the coordinates of the first sheet, as if they were parts of one graphic figure. The mean value now has a number attached: the value written beneath the abscissa on the first sheet. The standard deviation of the cutout distribution is still unchanged; it refers to deviations of hypothetical scores away from the newly labeled mean.

Now slide our cutout to the left or right along the abscissa. You have changed the value of the mean (and of all the hypothetical scores) by some constant amount. But the standard deviation did not change; the cutout has not altered its shape. The values of mean and standard deviation therefore do not affect each other.

2. the inductive example method

Determine the mean and standard deviation of the following sets of data.

A	B	C
1	8	6
2	9	8
3	10	10
4	11	12
5	12	14

It is best to use

$$s = \sqrt{\frac{\sum (X - \bar{X})^2}{n}}$$

You will note that sets A and B have different means and equal standard deviations, but B and C have equal means and different standard deviations. So the mean does not determine the standard deviation, nor does the standard deviation determine the mean.

estimating the standard deviation from grouped data

Just as a close estimate of $\bar{X}$ may be achieved from data grouped into intervals, s may be estimated by squaring the midpoint value of every interval, multiplying by frequency of scores within the interval, and then summing the products. This gives a value closely similar to $\sum X^2$. Substituting it for $\sum X^2$ in the working formula

$$\sqrt{\frac{\sum X^2}{n} - \bar{X}^2}$$

gives instead

$$\sqrt{\frac{\sum (\text{frequency} \times \text{midpoint}^2)}{n} - \bar{X}^2}$$

Since midpoints are usually integers, easily squared, and frequency is always an integer, the calculation is easy. But you must remember that it merely yields an approximation of s. Values of s calculated in this fashion will not be identical (except by luck) with s calculated from raw scores; for large masses of data the approximation may be a little large, but almost never more than a quarter of an interval width larger than the raw score s.

using calculating machines

All colleges and universities and a large number of high schools now provide calculating machines for student use. There are three types of machine.

electro-mechanical or rotary calculators

These work by means of rotating interlocked gears, which are connected to dials that show numbers through little windows. Such machines are now outdated, slow, noisy, fatiguing, and exasperating. They are, however, cheap (on the second-hand market) and better than nothing. Advanced institutions have gotten rid of them, and therefore we won't consider them further here.

electronic calculators

These are small, silent, very fast, and easy to use. On most machines the various stages of calculation are displayed as lighted numbers in a window. Some machines print them on paper tape. Either way, you can use the machine in a conversational manner (getting its answer to each intermediate problem before giving it the next calculation to do) through its keyboard, which contains two sorts of keys (or buttons):

1 Numerical entry keys, usually in a 4×3 array consisting of the numerals 0 through 9, a decimal point, and a negative sign. The digits of the number to be entered are pressed in "reading" sequence, from left to right (4 followed by 0 followed by 6 for 406), and you

must enter the decimal point unless the number being entered is an integer.

2 Operation keys, such as add ($+$), subtract ($-$), multiply ($\times$), divide ($\div$), and square root ($\sqrt{}$). Some machines have other operating buttons, such as ENTER, CUMULATE, CLEAR, and CLEAR ENTRY, which also give instructions to the machine. The instruction booklet that accompanies the machine will explain how these keys are to be used.

Electronic calculators are manual devices; you must enter your numbers and instructions manually, through the keyboard. In the interests of compactness and low price, the calculator is usually designed to handle only a very limited amount of data without *overflowing*, which means halting operation or giving an incorrect answer because it has been asked to do something it cannot do. (For example, a calculator may have a button that raises a number to any power you wish, but no small calculator can hold the number 1000^{1000}.) There is always a visual indicator (usually a signal light) of overflow. Another important common control is a *decimal place setting*. This allows you to increase the number of places to the right of the decimal point at the expense of places to the left, or vice versa.

The basic electronic calculator may be very small, literally pocket-size. Its compactness arises from use of *integrated circuits*, in which many electronic components are incredibly miniaturized and stuck to a tiny "chip" of plastic. The main parts of the machine are:

1 A source of power (batteries or a wall plug)
2 The keyboard, mentioned above, which controls
3 Various *registers*, which are storage places for the numbers entered through the keyboard, as well as for instructions symbolized by the $+$, $-$, $\times$, and $\div$ buttons and any others the machine may have.

Small calculators usually have only two or three registers for data, so that only simple calculations may be done; e.g.,

$$3 \quad + \quad 2 \quad = \quad 5$$

(in register A)	(entered into register B)	(new value in register A, replacing old value, 3)

Larger electronic calculators have more registers, permitting storage of certain numbers in "memory" until you are ready to use them again. Special functions such as $\sqrt{}$ require quite a bit more sophistication.

buying your own calculator?

Electronic calculators are getting quite inexpensive; as little as $5.00 might buy you one that will add, subtract, multiply, and divide. These are the basic functions of any machine. Should you spend the money for a shirt-pocket device? If so, should you spend a little for extra operations? And if so, which are most useful?

1 A "floating" decimal point, one that will move automatically to the best location for portraying your data as completely as possible.

2 Rechargeable batteries, preferably with a recharger. This feature will save you enough money in an average year to pay for this textbook.

3 At least one memory register.

4 A square root button. This is now found on only a few pocket-sized machines, but it is worthwhile if you like finding standard deviations.

5 A button, often marked X^2, that will square a number. This is very convenient, in conjunction with a memory, for calculating $\Sigma \; X^2$ quickly.

The first four features listed are the most useful in statistical pursuits. If your only use for a calculator is for statistics, don't seek or pay extra for any other special functions unless they come as part of the deal anyway. Remember that the various gadgets found on the cheaper machines are designed for ordinary consumers, not statisticians.

the electronic computer

This is the third type of calculating device. You probably won't have access to one unless you are an advanced student. Basically, it is a very large calculator, containing thousands of registers (instead of only two or three) in which *instructions* as well as *data* may be stored. Some registers are set aside to control the *sequence* with which the scores and instructions are handled. Devising such instructions is called **programming**; the person who does it is a **programmer**; the set of instructions, once perfected, is called a **program**. Programs exist for all common statistical chores, such as finding standard deviations.

The computer is a very fast-acting device, not easy to keep busy, so in the interests of having it earn its keep, data are fed into it very fast. Both data and program are put into the computer by means of punched cards (on which scores and instructions are coded by means of a "language" that the computer can "understand") or magnetic or paper tape. The computer gives its answer via a *teleprinter*, basically a very fast typewriter controlled by the machine, or by punching it out on cards. Both reading in (via cards or tape) and reading out (via printer or cards) are tremendously fast—several hundred times faster than would be possible manually.

Computer technology is moving ahead very quickly. A recent innovation is *time sharing*, which allows several users to work simultaneously on a large computer, usually by means of separate *remote terminals*, which look like electric typewriters with television screens and can be situated practically anywhere that is convenient, sometimes in a different city than the computer itself. The user communicates with the machine via the typewriter and the machine replies on the screen or the typewriter. Such keyboard *input–output* (as computer scientists call it) is much more convenient than punched cards or tape for most relatively small tasks.

To talk to the machine (give it instructions, data, etc.), the user must employ a "language" which the computer understands, which the user can learn, and which is oriented to the kind of problem the user wants to tackle. There are several such "languages," such as FORTRAN (an acronym for *for*mula *tran*slation), FOCAL, BASIC, and COBOL, each for a somewhat different purpose or make of machine. The symbols or "words" of such languages are those characters ordinarily available anyway on a typewriter keyboard, such as parentheses, asterisks, and various letters. But luckily, it is possible more and more often to use something like ordinary language in interactions with the computer.

What can be done by computers is extremely impressive and useful, but never forget that the computer is a big, fast, reliable idiot; it can do only what it is told, very repetitively and very quickly. It is not a god, and so far as intelligence and creativity are concerned, it will never replace people.

words, more words

Warning: Not knowing what these words mean can be dangerous to your academic health!

VARIABILITY	QUARTILE POINTS
RANGE	DEVIATION SCORE
QUARTILE DEVIATION	CONSTANT ERROR
MEAN DEVIATION	VARIABLE ERROR
STANDARD DEVIATION	THEORETICAL FORMULA
VARIANCE	WORKING FORMULA

exercises for another rainy weekend

1 Find the standard deviation of the data shown on page 103. For some feedback on procedures and setup, see page 133 *after* you have tried your best with this material.

2 Obtain or construct a target like the one shown in Figure 5.4. Secure it at eye level on a cork or similar backstop. Now obtain 10 or 12 darts. From a distance of **4 feet**, throw the dats one at a time, aiming at the vertical line that is solid rather than dashed. (Should you miss the target entirely with any shot, ignore the result; keep trying until all darts hit the target.) When all darts are thrown, record your scores. The dashed vertical lines have the score values printed below them, and you can estimate the intervening values (e.g., a hit midway between 18 and 19 scores 18.5). Now determine your mean score and standard deviation.

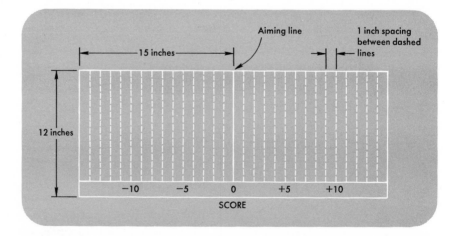

Figure 5.4
Dart-throwing target. Try to hit the center aiming line, and use the dashed vertical lines as an easy measure of your errors.

Now remove the darts, stand **12 feet** from the target, and repeat the procedure. Leave the darts in the board so you can distinguish the new scores from the previous ones. Find the mean and standard deviation of the new distribution of scores.

Questions (a) to (g) pertain to this exercise. Jot down your thoughts on a piece of paper. Discussion of these questions appears below.

a Which score distribution had the greater variability?

b Can you explain the difference in variability?

c Could the dart-throwing exercise have been an actual data-gathering experiment? What concrete aspect of the dart-throwing would have been analogous to n?

d Were the means of both score distributions the same?

e If not, which mean lies closest to the target line, whose value is 0?

f To what can we attribute the fact that the other mean is further from the aiming line?

g Is there any relationship between variability of scores and their mean value relative to the aiming line?

COMMENTS ON EXERCISES 1 AND 2

Exercise 1

X	X^2
6	36
3	9
8	64
2	4
5	25
5	25
9	81
38	244

$\sum X$ $\sum X^2$
$n = 7$

(1) Set up column of X values; find $\sum X$
(2) Set up column of X^2 values; find $\sum X^2$
(3) Count the number of scores, n.
(4) Write down your choice of working formula, e.g.,

$$s = \sqrt{\frac{\sum X^2}{n} - \left(\frac{\sum X}{n}\right)^2}$$

(5) Substitute obtained values into formula, and work it out arithmetically.

$$s = \sqrt{\frac{\sum X^2}{n} - \left(\frac{\sum X}{n}\right)^2} = \sqrt{\frac{244}{7} - \left(\frac{38}{7}\right)^2}$$

$$= \sqrt{34.857 - 29.469} = \sqrt{5.388} = 2.321$$

(6) Congratulate yourself.

Exercise 2

a The odds are overwhelming that there was far more scattering of the darts from 12 feet. Thus the *variance* and its square root, the *standard deviation*, will almost certainly be greater from 12 feet.

b Such difference in variability occurred simply because from 12 feet there is literally more room for errors to occur than from 4 feet.

c *n* refers to the *n*umber of scores in any distribution. In this exercise the number of darts that hit the target is *n*.

d The odds are overwhelming that the two means are *somewhat* different. Are they in fact?

e Again, there is a tremendous probability that the mean closest to the aiming line was that obtained from 4 feet. You were closer to the target, and if that won't make you more accurate, nothing will!

f The mean score from 12 feet is further from the aiming line because being farther away, you were less accurate.

g Yes. The *reduced accuracy* of performance from 12 feet is also reflected in the *greater variability* of performance under that condition. In other words, the amount of variability (the standard deviation) is a measure of the amount of error that has crept into experimental data (possibly measurement error; possibly other kinds).

further problem/

A certain amount of good old *drill* is decidedly useful for establishing friendly relations with means and standard deviations. Try the short exercises below; the arithmetic is still simple.

Calculate at least $\bar{X}$ and s^2, and also s if possible.

1   5, 5, 5, 5, 5, 5
2   1, 2, 3, 3, 5, 7, 7, 8, 9
3   0, 0, 4, 4, 5, 6, 6, 10, 10
4   0, 0, 4, 4, 5, 6, 6, 10, 30

People feel much better after a bit of exercise.

self-test on chapters 4 and 5

Directions

By this time you should know the directions. Happy thinking, good friends.

1 Here are some numbers:

$$4 \quad 2 \quad 3 \quad 6 \quad 3 \quad 3 \quad 3 \quad 0 \quad 0 \quad 6$$

    a The value of ΣX^2 is 900.
    b The mean, median, and mode are identical.
    c The median value is 3.
    d The value of s must necessarily be an integer.
    e The variance equals 12.8.
    f The variance equals 90.
    g The variance equals 3.8.
    h The standard deviation is approximately 1.94.
    i The range is 7.

2 The mean of a group of five scores is 2.80. You have lost one of the scores, but the others are 2, 3, 4, and 3. The lost score
    a is still lost and cannot be determined.
    b is 2.
    c could be any value between 0 and about 5.
    d has a fixed, known value because we know $\bar{X}$ and the remaining scores.
    e could not be known if another one were also missing.
    f is 2.4.

3 Whereas the median may be considered an expression of numerical equality (of the number of scores on each side), the mean may be considered
    a an expression of equality of total "leverage" on each side of itself.
    b affected more by some scores than others.
    c identical to the median.
    d preferable to the median.
    e none of these.

ANSWERS AND DISCUSSION OF SELF-TEST

Answers to Question 1

a F
b T
c T
d F
e F
f F
g T
h T
i T

For this series of numbers, the total, ΣX, is 30, and the sum of the square values, ΣX^2, is 128. There are ten values, so $n = 10$. The mean therefore is $30/10 = 3.0$; the distribution is symmetrical, so the median should also be 3, but check to see. The score 3 occurs more often than any other, so the mode is 3. The value of s cannot be an integer, since you do more than add to obtain it. The variance is 3.8; s, the square root of 3.8, is indeed about 1.94. The range is $L - S + 1$, which is $6 - 0 + 1 = 7$.

Answers to Question 2

a F
b T
c F
d T
e T
f F

If $\bar{X} = 2.80$ and $n = 5$, then ΣX must be $2.8 \times 5 = 14$. The scores given add up to 12, so the lost score must be the difference, 2. Therefore, (b) and (d) are correct. However, if *two* scores were missing, we could not know what their values were.

Remember that with the mean known, all scores except one *could* have any value. The last one, however, would have to have the specific value necessary to make the mean work out to what we know it is.

Answers to Question 3

a T
b T
c F
d F
e F

Recall the ruler-and-checkers experiment of Figure 4.5, which showed that the scores on one side of the mean exert the same leverage as those on the other side and that extreme scores have the most leverage. Thus (a) and (b) are true. The last three choices are wrong, particularly (d); there is *no* general reason for assuming that one measure of central tendency is preferable to another.

4 The mean and standard deviation, when compared with the median, mode, range, and quartile deviation,

a are more accurate.

b make use of all the data, while some of the others do not.

c should not be used with ordinally scaled scores, that is, ranks.

d are more *reliable* because they are insensitive to slight shifts of certain scores.

e are free from important problems in usage.

5 The median and quartile deviation are preferable to $\bar{X}$ and s

a when very large amounts of data must be processed.

b when fairly small numbers of scores fall into a somewhat skewed distribution.

c when scores fall into a J-shaped curve, or "hill-slope," distribution.

d when the mean or standard deviation cannot be used for some good reason.

e when the data are cardinally scaled.

f when the data are ordinally scaled.

6 You are told that *one* score in a distribution has the value of 8.0 and that $s = 0.00$. From this information, you know

a that the center point of the distribution is *approximately* 8.

b that the mean cannot be ascertained.

c that the other scores are ascertainable.

d that the range equals 1.0.

e nothing at all about the value of n.

f that the mean is *exactly* 8.0.

g that all the scores are identical, with value 8.0.

Answers to Question 4

a F The word "accurate" is misleading; accuracy depends upon care in calculation rather than choice of type of calculation. Whether $\bar{X}$ and s give an accurate *picture* of the data depends upon the data, not the technique.

b T They *do* make use of all the data, whereas the range uses only two scores and the mode uses only the most frequently occurring scores.

c T The mean and s employ cardinal arithmetic techniques, not justified when the data are merely ordinal.

d T They certainly are more reliable, in the sense of being quite insensitive to minor score variations (the mass of scores toward the center of the distribution gives great "inertia" to the values of the mean and s).

e F They are not free of problems, though; highly skewed data distributions, which are frequently found, cannot (without transformation) be summarized fairly by $\bar{X}$ and s.

Answers to Question 5

a F The median and the quartile deviation come from strictly ordinal pro-
b T cedures. With large amounts of data, tedious rank ordering is neces-
c F sary first, and bad unless required anyway because of the mean and
d T standard deviation being inappropriate. The median and the quartile
e T deviation are easiest to calculate on fairly small amounts of data; they
f T are appropriate for moderately skewed data, but not when the data are extremely skewed. In general, use the median and the quartile deviation when they are not inappropriate to the data and when there is a good reason for avoiding the mean and the standard deviation. They can be used for both cardinal and ordinal data, but some information is lost when cardinal scores are converted to ranks.

Answers to Question 6

a F The most important thing to note is that if $s = 0.00$, then *there is no*
b F *variance*, so every score must be identical to the one you know about,
c T 8.0. Clearly, the mean must also be 8.0; n is unspecified, and the range
d T is $8 - 8 + 1 = 1$. This is a good one to try on your friends who don't
e T have this book.
f T
g T

EVALUATION

Score three points for each correct choice, subtract two points for each wrong one you thought correct, and subtract one point for each correct one you thought wrong. Maximum possible is 66.

50 to 66	You're swimming!
33 to 49	Afloat, anyway.
10 to 32	Review where needed.
0 to 9	Oops.
Negative score	Seek help!

This is the time to review your performance on all previous problems and self-tests. The term is getting on, and if you have troubles now, they will get worse before they get better. There are several things you can try.

a Reread the first few chapters. This time, don't try to memorize them. Just try to understand them.

b Try going to another statistics text. Another may suit your style better than this one.

c Acquire a programmed text if you have Skinnerian leanings. A proper teaching machine system is even better.

d Reevaluate your attitudes, motives, ambitions, and hang-ups.

e Fall in love with somebody who understands statistics with amazing clarity.

f Confer with your instructor.

g Write to the author. I probably can't solve your problems in any way that's useful to you now, but your help can improve future editions of this book.

ſome ordinary problemſ

1 A quality-control inspector in the Chug-a-lug Motor Company is examining the work of two machinists whose job is to produce Gizzmoes machined to a length as close as possible to 1.5000 inches but no larger. Machinist A's Gizzmoes have mean measure $\bar{X} = 1.4998$ with standard deviation $s = 0.0015$; machinist B's Gizzmoes have mean measure $\bar{X} = 1.4997$ with standard deviation $s = 0.0003$. Which machinist is the better worker? Why? Use the concepts of constant error and variable error.

2 On a piece of graph paper, sketch two frequency distributions having the same mean but different standard deviations. Then sketch two more distributions having different means and the same standard deviations.

3 Find the standard deviation and quartile deviation of the data given on page 108 using the raw-score method.

4 Estimate the standard deviation of the data on page 108, using the frequency × midpoint method. Are the two values of s identical? If not, can you suggest why not?

5 The standard deviation is commonly thought of as a measure of variability. But sometimes, as when distances are being measured by surveyors, the same stretch of ground will give a different result each time the measurement is repeated. *The actual distance does not change* (as far as we know!), but the measures of that distance do form a variable distribution. What would the standard deviation of that distribution reflect?

6 Calculate the standard deviation of these scores.

$$8 \quad 10 \quad 11 \quad 12 \quad 12 \quad 14 \quad 16 \quad 18$$

Use each of the following formulas for s

a
$$s = \sqrt{\frac{\sum (X - \bar{X})^2}{n}}$$

b
$$s = \sqrt{\frac{\sum X^2}{n} - \bar{X}^2}$$

c
$$s = \sqrt{\frac{n \sum X^2 - (\sum X)^2}{n^2}}$$

Which formula is the most difficult as a working formula?

7 Subtract 3 from each of the scores in the preceding question and recalculate s, using the formula of your own choice. Has the value of s been changed? Has the value of the mean been changed? Can you draw any conclusion about the interdependence of the mean and the standard deviation?

8 What would happen to the values of the mean and standard deviation in a distribution whose scores

a were increased by adding a constant value to each?
b were doubled in value?
c were multiplied by 10?
d were multiplied by -1?

9 A nearly symmetrical distribution of 100 scores falls between 25 and 980; however, 95 of those scores fall between 250 and 700. The mean is 590.

a Sketch the distribution as well as you can based on this information.

b Assuming that the distribution is unimodal, between what values is the mode certain to be found?

c Which measure of variability would be the most representative? Explain your answer.

10 This may sound silly, but go through all the motions of calculating the standard deviation of the following small distributions.

a 7
b 7, 7, 7, 7, 7
c 5, 6, 7, 8, 9
d 2, 5, 7, 9, 12

11 Using the data of Question 10, compare the values of ΣX^2 and $(\Sigma X)^2$. Are these values the same within each set of data? Try to understand, or at least explain, your answers.

12 Add 5 to each of the scores in 10(c) and recalculate s. How does its value compare with the previous one?

13 Multiply each of the scores in 10(d) by 2 and recalculate s. How does its value compare with the previous one?

14 Consider the results of Questions 12 and 13, and draw some general conclusion.

15 Calculate the value of s, the range, and the quartile deviation on the following data.

1 4 5 6 9 12 15 16 17 18 21 24 27 28 30 33

Now repeat the calculations on the data as slightly changed.

1 4 5 6 9 12 15 16 17 18 21 24 27 **48 79 90**

Draw a conclusion from a comparison of the quantities calculated on these sets of data.

16 Here are some frequency distributions. They are so simple that you should be able to tell what the mean of each is simply by looking at the graphs. State what you think each mean is *without* computing it, and then check your answer by doing the arithmetic.

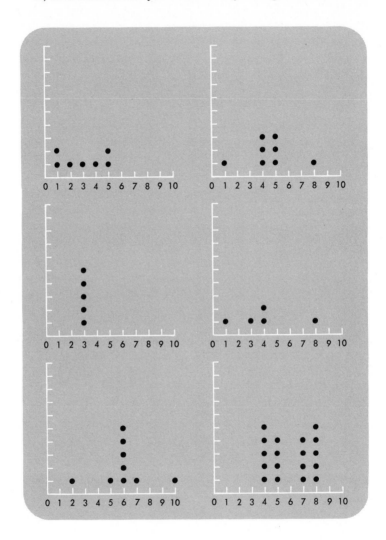

17 Some statisticians and engineers use the term **root mean square** in place of the term **standard deviation.** Explain why.

chapter 6
percentiles
and
standard
scores

Let us return to a problem first introduced in Chapter 1. Recall the dilemma of the educator who had to follow the progress of many high-risk children of different ages, studying different subjects in many classrooms with diverse teachers, and subjected to various grading standards and methods. The problem was to detect and measure, quantitatively and precisely, the childrens' progress relative to their peers within each class. In this chapter we consider this problem, its solution, and its surprisingly important impact upon statistics generally.

A score may be interpreted in two ways: relative to some fixed *external standard,* or relative to the group of scores of which it is a member. As an example of an external standard, you probably know that formerly school pupils had to get 50% on a test to pass it. Such pupils were competing against this arbitrary standard, not against one another: all might pass, or all might fail. Whether the test was unreasonably difficult or ridiculously easy, the passing score would be 50. Students might perform really well, but fail because the test was difficult and graded too stringently; or the opposite could occur.

As an example of an *internal comparison,* consider the more modern education policies, under which a grade of A might mean merely that the actual score was superior to four-fifths of the other scores on the examination, regardless of what percentage of questions was answered correctly. On a really difficult examination, a score of 40% might get an A; on a ridiculously easy examination, the same score might well be an F. Figure 6.1 shows how a distribution of scores on an easy examination, as contrasted with those on a hard examination, might look. An *external standard* would then be an arbitrary cut-off point on the abscissa, whereas an *internal standard* would be some indicator of the overall average performance of each group (such as the mean score). Let's suppose that the same pupils took each examination and that one pupil, Bruce, scored 40% and 95%, as indicated in Figure 6.1. The two scores look vastly different at first glance, but *in comparison with his classmates,* Bruce did about equally well on both examinations, scoring close to the top on each.

Nowadays we use letter grades to avoid such problems, because they can be assigned relative to an internal standard

They are often not fine-grained enough, however, because they allow only a few distinct scores (such as A, B, C, D, F), and they find acceptance only in connection with academic grades. In the behavioral sciences we need a generally useful internal-comparison device that permits great precision, because there are few adequate external standards for measurements of human characteristics. Certain approaches

144

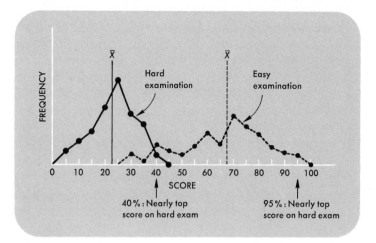

Figure 6.1
Frequency distributions of scores received on an easy examination and a difficult examination. Note the absence of any necessary relationship between *percent correct* and *relative quality* of performance.

to the problem are obvious and possibly acceptable, but many have disadvantages. For example, you could state the raw value of a score and then quickly follow it with further information to put the score into context. Bruce's teacher might say to the parents, "Bruce got 40%, but it was an unfair exam and he actually got the second-highest mark in the class. Most kids got about 25%, so Bruce did amazingly well." This approach is rather windy, and would assume hurricane proportions if applied to a whole class of 40 children. It would be preferable to build the same information into the score itself. Fortunately, there are two easy ways to do this.

The first employs a rank-ordering technique and thus is appropriate for ordinal as well as cardinal data. This is the **percentile** approach. The second requires cardinal scaling; it relates each score to the mean of its distribution, in units based on the size of the standard deviation. This approach is called transforming raw scores to **standard scores.**

percentiles revisited

We touched upon percentiles previously, in connection with the median and the quartile deviation. Now we must become more involved.

Let's set the stage by imagining any collection of scores. Now let's arrange them in ascending order of size, from smallest to largest. For simplicity, suppose there are just ten scores with no duplication of score values. For example,

Original scores	23	69	47	102	13	25	78	92	31	17	(Line 1)
Rearranged in order	13	17	23	25	31	47	69	78	92	102	(Line 2)
Rank value of each	1	2	3	4	5	6	7	8	9	10	(Line 3)

In the second line, note that the *differences* between adjacent ranked scores can be (and are) quite variable. However, when the cardinal scores are replaced by their ranks, as in the third line, these variable differences are no longer expressed. (This is the essential difference between ordinal and cardinal scaling.)

Since there are ten scores in our collection, each score, or rank, is one tenth (or 10% of the total number. So we add another line to show this.

Rank value of each	1	2	3	4	5	6	7	8	9	10	(Line 3)
Percentage of total number of scores	10	10	10	10	10	10	10	10	10	10	(Line 4)

Now we can examine the notion of **percentiles**. A percentile is the *point* on the scale of rank-ordered scores that *exceeds a* given percentage of those scores and is *less than the remainder*. The percentile point may, but need not, correspond with an actual score. In our example, note (line 2) that somewhere between 13 and 17 (the two smallest scores) is the hypothetical point to the right of the lowest 10% of the distribution and to the left of the highest 90% of the scores; if we choose that hypothetical point midway between 13 and 17, the tenth percentile is at 15, as shown by the dashed vertical bar. It does not matter that there is no actual score whose value is 15. Pursuing our example, we have

Original scores in order	13	17	23	25	31	47	69	78	92	102	(Line 2)
Rank value of each score	1	2	3	4	5	6	7	8	9	10	(Line 3)
Percentage up to and including each score	10	20	30	40	50	60	70	80	90	100	(Line 5)

24 = 30th percentile

To take another example, the score value that exceeds 30% of the scores and is less than the remaining 70% is midway between 23 and 25, at 24 as shown by the solid vertical bar.

Percentile points are locations on the *scale* of original scores; **percentile ranks** are the percentile points at which the actual scores fall. Percentile points and percentile ranks are closely related, but do not confuse them. If you start with an actual score and find what percentage of the scores it exceeds, you are finding a *percentile rank*; if you want to know what point on the scale is to the right of a particular percentage of the scores and to the left of the remaining scores, you are finding a *percentile point*.

Calculating a **percentile rank**, *PR*, verges upon being quite easy, at least in principle. We'll tackle it with the same example. What is the *PR* of the score of 25, fourth from the bottom in line 2?

The score 25 is *greater than* the three scores to its left and *smaller than* the six scores to its right. So far so good. But the three below it and the six above it make up only nine of the ten scores in the distribution. The difficulty is that we have not counted the score of 25 itself. We must count it somehow. It would seem that 25 is neither above itself nor below itself, but in fact we can regard 25 as lying *half below itself and half above itself*. No, don't laugh—this idea is not illogical. Remember, percentiles are *points* on a scale, whereas a score like 25 really represents a *range* on that scale, between 24.5 and 25.5 (remember about significant figures?). Adding 0.5 (half of the score range represented by 25) to the three scores below 25 and again to the six scores above 25 accounts for all ten scores: 3.5 + 6.5 = 10. So the raw score 25 exceeds 3.5 of the 10 scores, or 35% of them. Stating it officiously, 25 has a *PR* of 35.

The formula for working out percentile ranks is a formalization of the preceding discussion.

$$PR \text{ of score } X = \frac{(\text{Total number of scores exceeded by } X) + 0.5}{\text{Total number of scores in the distribution}} \times 100$$

percentiles have their problems

Yes, indeed! A *minor* one is the tiny chore of rank ordering perhaps several hundred scores. And how do you find the percentile rank of a score that occurs more than once? Another thing that makes percentiles and percentile ranks less fun than they might otherwise be is the fact that they are appropriate only when there are quite a few scores— 100 at the very least, and preferably several hundred or more. You can go through the motions (as we just did) with fewer scores, because the formula will grind out an answer; but the percentile ranks will be only approximate, and percentile points, most of them falling *between* actual scores, will be rather meaningless. After all, with fewer than 100 scores, the idea that there actually are 100 percentile points is a bit grandiose; it implies more precision than is actually justified by so few scores.

There is another problem, a philosophical one. Recall that in our example with ten scores, we assumed that the percentile points should be interpolated when necessary, not only *somewhere* between the actual scores that they fell between, but *halfway* between them. Why *halfway*? This smacks of an interval-scaling assumption, and we're supposed to be working on a merely ordinal level. Illogical. Possibly there is no better choice, but still illogical. At least we can say that the problem gets smaller as the number of scores increases, because with more scores the intervals between scores will tend to be smaller.

Still another difficulty with percentiles is that people sometimes misinterpret them. They may fail to realize that even such an impressive looking percentile rank as 95 may not correspond to much objective merit (assuming that the scores in question measure something that allows us to think in terms of objective merit); if all scores in a distribution are poor, the score with percentile rank 95 will also be poor. Imagine what would happen if an airline pilot's skill were evaluated by strictly internal comparison. In a roomful of chimpanzees, some would *rank* quite high in flying skills (*anything* can be rank ordered), but you would scarcely feel safe flying via *Pan Troglodytes Airlines*. Some things require absolute adherence to an external, objective standard. University teachers have been concerned that academic standards would invisibly deteriorate if percentile scores were used exclusively; student groups might slack off uniformly, whether deliberately or not.

Nonetheless, percentiles are used a lot, seemingly by mere custom, in much educational and psychological testing, despite certain grave disadvantages. As we have seen, distributions *usually* contain many closely similar scores in the middle; since the scores are often not perfectly reliable, two scores that differ only because of error in measurement could be given quite different percentile ranks. To use percentiles is thus to force rank-ordering on many scores that may well be essentially the same, those that cluster in the middle of the distribution. In a class of 100, even the 40th- and 60th-ranked students often have closely similar scores. Sometimes an alternative approach is desirable. Standard scores, which will be introduced in the next section, have the advantage of using the actual values of the scores, thus retaining the cardinal information lost by converting to ranks. The percentile approach is, of course, the only choice for purely ordinal data.

an easy way to find percentile ranks with large amounts of data

Sometimes it happens. You can't talk your way out of transforming 906 scores into their percentile ranks. Here is a relatively easy way to do it, using a **cumulative frequency distribution**.

First, set up an abscissa on a large sheet of graph paper. Scale the abscissa with the scale used for measuring your scores. Then, without grouping the scores, draw a frequency polygon, placing a cross on the graph above the appropriate abscissa value of each score. (The scores do not have to be ordered first.) Cross each one off your list as you put it into the growing polygon so you don't enter the same score twice—before *and* after the telephone rings. When you finish, you'll have the world's largest and messiest frequency polygon!

Now you can begin building a cumulative frequency distribution. Get another large sheet of graph paper. At the extreme bottom, duplicate the abscissa of your frequency polygon; then place an ordinate on the left, scaled in such a way that its top end represents a frequency of 906, your total n. The ogive, built as described in Chapter 3, is shown in Figure 6.2. It is easy (if not very quick) to do, working from your first graph to the new one. When finished, join the topmost crosses to give an ogive running from the bottom left to the upper right. (This has not yet been done in Figure 6.2.)

The final step is to rescale the ordinate in *percentages*. This *can* be the trickiest part, but it is simple once you understand how to do it. Since 906 scores are 100% of them, mark 100% at the top of the ordinate, exactly on a level with the uppermost point on the ogive. Then mark the abscissa with 0%. To get the intermediate percentage points, just subdivide the ordinate into 100 equal parts between these two extremes. These parts will *not* usually coincide with the convenient

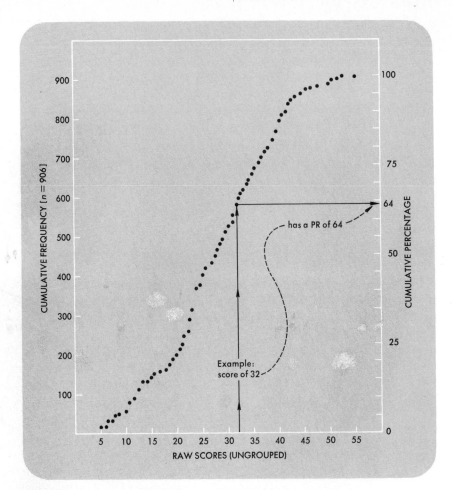

Figure 6.2
Cumulative frequency distribution of 906 scores. The scores are ungrouped and thus are located directly above their score values on the X axis. The ordinate on the right makes a cumulative percentage distribution, allowing easy determinion of percentile ranks and percentile points.

lines already drawn on your paper. To find the percentile rank of any score, merely use the rule of right angles. Locate the score on the abscissa, erect a perpendicular from it to the ogive, and then draw a horizontal line across to the ordinate. Where it strikes the latter is the percentile rank of the score (within a margin of error too small to cause any difficulties).[1] This procedure is shown in Figure 6.2 for a raw score of 32.

[1] With over 900 scores, you don't really need to worry about any given one being half under, half above itself.

and an easy way to find percentile points with large amounts of data

The cumulative percentage distribution you just made can also be used to find percentile points. Simply reverse the steps used in finding percentile ranks. If you want the 40th percentile point, start at 40% on the ordinate, draw a horizontal to the right as far as the ogive, and then drop a perpendicular to the abscissa. The score value where it meets the abscissa is the 40th percentile. A cumulative frequency distribution (once you have it!) makes percentiles easy.

standard scores

Let's visualize a distribution of scores whose mean and standard deviation are known. Figure 6.3 shows such a distribution, in which the mean is shown as a vertical line intersecting the abscissa at its value, 4.8. The value of s is 1.8, and other vertical lines have been placed at intervals of 1.8 to the left and right of the mean. Now let us look at the positions of scores M and N in Figure 6.3. In terms of the distances marked off by our *new* lines, how far are they from the mean? We see easily that score M is half a standard deviation to the left of the mean and that score N is $2\frac{1}{2}$ standard deviations to the right of the mean. Suppose we think of $\bar{X}$ as the origin of our graph and of s as our unit of measure. Then we could say that M is $-0.5\ s$ and N has the value $+2.5\ s$. Bearing in mind that scores tend to cluster around the mean,[2] and to a degree already measured by the standard deviation, we can see that a score that is very much higher (or lower) than $\bar{X}$ in s-units is very high (or low) in relation to the entire distribution (i.e., is an *extreme* score). Similarly, a score that is near the mean in terms of s-units is obviously close to the average for the entire group (i.e., is a fairly typical score). Indeed, once the raw data have been reexpressed in terms of the s-units, they can be replaced entirely by these s-units. Figure 6.4 shows the distribution of Figure 6.3 with the abscissa relabeled this way.

If we extend the foregoing line of thought, we can now think of every score in the distribution in terms of s-units, rather than the original unit of measurement. A score (in s-units) of -0.75 would thus be a distance of $\frac{3}{4}s$ to the left of the mean, $\bar{X}$; a score of 1.1 would be a distance of $1.1\ s$ to the right of $\bar{X}$. To transform any raw score X in this way, we

---

[2] This is the case in most distributions that are of interest to us in social science, anyway.

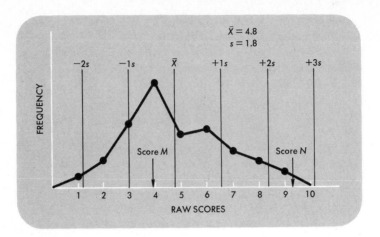

Figure 6.3
Frequency distribution showing the abscissa scaled in the original raw-score units. Starting at the mean, whose value is 4.8, we can count outwards in units corresponding to the standard deviation of 1.8. Thus score M is half an s unit below the mean, and score N is $2\frac{1}{2}s$ units above the mean.

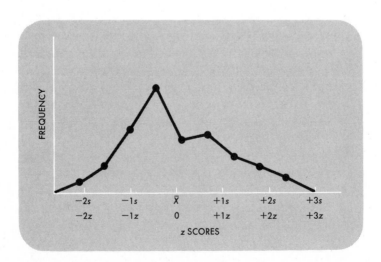

Figure 6.4
Figure 6.3 redrawn with s units and corresponding z scores on the abscissa. Note that the raw-score labeling can be recalculated if necessary, as long as the numerical value of s is known.

find its distance from the mean, $X - \bar{X}$, and divide that distance by our new unit of measure, s. So

$$z = \frac{X - \bar{X}}{s}$$

where z is the transformed score. Scores transformed this way are called z scores.

Because we are using the score values themselves for arithmetic computations, the raw data must be cardinally scaled. You should recognize that since $\bar{X}$ and s are both constant in any distribution, the raw scores are not at all rearranged by the z transformation. The ordering of scores remains exactly the same; moreover, if X_1 is a raw score twice as far from the mean as raw score X_2, and if we transform them into z_1 and z_2, then z_1 will be twice as far from the mean as z_2. So, the z scores represented in Figure 6.4 are exactly equivalent to the raw scores in Figure 6.3 except for the scale of numbering on the abscissa.

From Figure 6.4 you should already have noted that z scores are negative to the left of the mean value, while they are positive to the right of the mean. The z score of the mean $\bar{X}$ is 0.

Sometimes it is awkward to work with both positive and negative scores that also involve decimals. (Many people are very uneasy in company with decimals.) Hence, z scores themselves can be transformed again to eliminate both these problems. To make all the scores positive, we can add a fixed value to each z score, a value larger than the most extreme negative z value likely to occur. In the usual symmetrical distribution, very few z scores will be less than -5.0 (and then only when the number of scores is very large), so we can add a **constant** such as 5.0 (or any other larger constant we might desire) to each z score. The effect would be something like this:

z score:	-3.0	-2.0	-1.0	0	$+1.0$	$+2.0$	$+3.0$
z score $+ 5$:	$+2.0$	$+3.0$	$+4.0$	$+5.0$	$+6.0$	$+7.0$	$+8.0$

We now have scores between $+2$ and $+8$. There will still be decimal fractions between these new values, however, and they can be eliminated by multiplying them by 10. The resulting scores are termed **T scores**.

z score:	-3.0	-2.0	-1.0	0	$+1.0$	$+2.0$	$+3.0$
T score $= 10(z + 5)$:	20	30	40	50	60	70	80

The formula for T is thus $T = 10\,(z + 5) = 10z + 50$.

The mean of a T distribution is 50, and its standard deviation is 10. You will note that a T transformation gets rid of only one decimal place, which is fine if z is expressed to just one decimal place. If greater precision is wanted, you must multiply by 100, or 1000, to eliminate the decimal point. But expressing a T transformation too precisely may well not be justified by the accuracy or precision of the raw scores. Observe that multiplying $5z$ by a constant value other than 10 will give a T whose distribution will *not* have a mean of 50 or a standard deviation of 10. You should be able to figure out these values for yourself. A graph can be made with raw scores along the X axis and equivalent z or T scores along the Y axis, and all the points will fall exactly along a straight diagonal line, as in Figure 6.5. X, z, and T have a *linear*, or straight-line, relationship. For this reason z and T transformations are often termed **linear transformations**.

caution . . .

It is clear by now that percentiles and standard scores offer **internal standards** of comparison for scores within a distribution. However, it does not follow that one distribution is similar to any other. A percentile rank of 90 (or a z score of $+3.0$) from one distribution may or may not be comparable with a similar percentile rank or z from another distribution. Only if the two distributions are *known* to have the same mean, standard deviation, and shape can percentiles and standard scores be used for comparing scores between as well as within the distributions.

It should be apparent that standard scores are similar to percentiles as far as the selection of chimpanzees as pilots goes. The chimpanzee with a z score of $+3.5$ is a top chimp, but probably still should not be entrusted with a jumbo jet.

There is room for discussion of whether human values conflict with the use of internal standards of evaluation. It does not often happen that every student in a course displays exquisite mastery of the course, but it does happen. There will still be a distribution of scores, however. Is it good when the bottom-ranking (but excellent) student necessarily gets an F, or, in a "Mickey Mouse" course, when the top (but incompetent) student gets an A? The answer is, of course, that judgment and discernment must be (and typically are) used to combine external standards of quality with internal standards of evaluation.

You may recognize that the procedure of grading all students in a course on a common scale is useful and legitimate, but that averaging grades from many classes to find a grade-point average is perhaps less

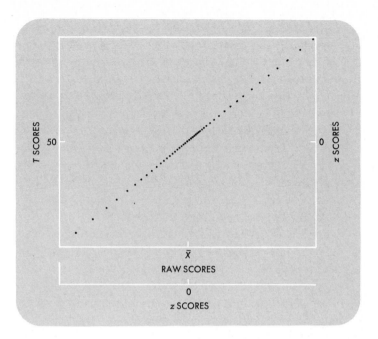

Figure 6.5
Relationship between raw scores and z scores or T scores. Each score
is shown as a dot. Why are the dots more numerous toward the center?
Why is the conversion of raw scores to standard scores called a linear
transformation?

legitimate. After all, some courses are easier than others, and bright
students are more likely to seek out certain courses. If you want a high
GPA above all else, it follows that you should seek easy courses populated by dull students, preferably at a mediocre college.

For the same reason, it is not really legitimate to average percentile
ranks, although it is sometimes done. If we do it, we must remember
that any percentile rank is relatively unreliable near the center of a
humped distribution and relatively reliable in the tails.

Be sure to note the effect of percentile and standard score transformations upon the information carried by the data. Percentiles reduce cardinal scaling to ordinal, thus discarding some information.[3]
Standard scores retain cardinal scaling and lose no information, since
the relationship among the raw scores is perfectly preserved in the
spacing of the z and T scores. However, ratio-scaled raw data will be
converted to an interval scale by a z or T transformation.

[3] This *can* be a good thing, depending on your purposes.

the *z* score, statistics, and research

We have seen that *z* scores offer the handiest method for discerning and expressing relationships between scores and the distributions to which they belong. Let's look briefly at an interesting application of *z* scores. This is merely an example to introduce the topic of statistical analysis, which will be discussed later in this book.

A nutritional researcher has discovered a food element that may be related to longevity. The stuff is very hard to purify and since only a few grams of it are available, a wide-scale study of its properties is impossible. So the researcher breeds 10 female mice; after the 10 litters (of 6 baby mice each) are born and weaned, the 60 baby mice are divided into two equal-sized groups, including half the mice from each litter in each group. One group gets the rare food element in its diet; the other group is given the same diet but without the food element. Except for the difference in diet, the two sets of mice are treated identically, living out their natural life spans in a land (for research mice, anyway) of wine and feasting. At the end of their sybaritic existences, their life spans are measured, and the two groups are compared. It turns out that the mice that did not get the food additive (the *control group*) had a mean life span almost identical with the norm for laboratory mice. While the mice that were given the additive (the *experimental group*) had a considerably longer mean life span. There was, however, some slight overlap of life spans between the two groups. Naturally, the researcher would like to know whether the longevity difference was due to the food element or to mere accident. After all, none of the experimental mice lived to an unheard-of age; they just tended to live longer than the control mice. Now obviously both sets of scores were affected by variable influences to some degree, which might have had the effect of creating an apparent difference between means where none should have occurred. So let us look at the *means*. There are two possible reasons for the difference between them.

1 The mean life spans may be different because of the nutrient that was given to only one group; that is, the nutrient may systematically increase age.

2 They may be different merely because there is variability in the life spans of mice; the mice in the experimental group might have shown unusual longevity even if they had not been given the food additive. (Not very likely, you say? But is it safe to ignore the possibility?) By

chance alone, such an event *might* occur. By chance alone, a series of ten coin-tosses *might* show all ten heads; at least, they probably would *not* show exactly five heads and five tails. Strange outcomes can arise from pure chance.

The rationale for statistical analysis of empirical data is to permit a rational decision between these two possible answers, each of which is dignified by a formal name:

1 The **experimental hypothesis** says that the difference between means is caused by experimental treatment (the nutrient in this case), which consistently affected the experimental group's scores (but not the control group's). Built into this hypothesis is the assumption that a similar difference would *necessarily* be found if other mice were similarly tested; in other words, that the difference extends in principle to the entire possible population of mice from which the researcher happened to choose this sample.

2 The **null hypothesis** says that the difference in means is merely accidental, that if the experiment was repeated on the whole population of mice, no nutrient-related difference in longevity would be found. *Null* signifies *nothing*—the null hypothesis is the "nothing hypothesis" that says that the experimental treatment has *no effect* on the groups' mean scores.

So what does this have to do with *z* scores??

Well, suppose we could magically take a huge number of different samples of mice, find the mean longevity in each sample, and plot all these means in a **frequency distribution of means**. The distribution would reflect the effects of mere chance on the natural life span of mice. Figure 6.6 shows how the distribution would look. Note that the raw scores (age at time of natural death) have been transformed into *z* scores. The mean *z* score is, as always, 0, and the distribution is bell-shaped; in fact, it is a very particular type of bell-shaped curve, called **standard normal distribution** or **Gaussian distribution**. Why, you may ask, must the distribution be shaped this way? And how, without actually finding the distribution of the means from many samples, do we know the standard deviation of that bell-shaped distribution? (We would need to know the standard deviation in order to calculate the *z* scores.) The next two chapters will answer that. Taking it on faith for the present, let us glance at Figure 6.6 and note the *very* low frequency of *z* scores above +4.00. The odds are strongly against finding a *group* of

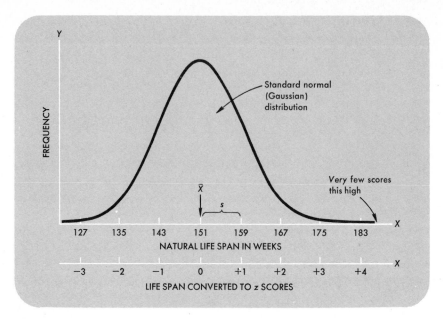

Figure 6.6
The standard normal distribution has a particular shape, as shown here. If very many groups of laboratory mice are measured for mean life span, the mean life spans for all the groups will fall into a distribution of this shape. The abscissa can be scaled in z units, making the distribution generally applicable to any kind of measure. For further details, see Chapter 7.

normal mice whose life spans average four or more standard deviations above the mean for all mice. Yet suppose our nutritionally supplemented experimental mice *did* have such an average, relative to the mean of the control group and population. A *possible* finding, but hardly *probable*—if we tentatively favor the null hypothesis. Well, we could decide to favor the experimental hypothesis instead. But then we would feel very foolish if it turned out that we'd jumped to a wrong conclusion. Which is the correct conclusion to draw from the experiment? We would like to come to *some* conclusion.

But this is an uneasy situation, and it is impossible to decide with certainty. A mean z score of $+4$ *could* have occurred by accident (the null hypothesis being true), or it *could* have happened because the nutrient actually increases life span (the experimental hypothesis being true). It is a question of probability. How improbable must the data be, assuming the truth of the null hypothesis, before we decide to adopt the experimental hypothesis? *That* is the question. Before we can approach it, we need to discuss probability.

important look backward and forward

Briefly review mentally the contents of this book thus far. We have been dealing with the concepts of central tendency and variability, as embodied primarily in the mean and standard deviation. In this chapter these have been combined conceptually to produce the z score. Because no score can be understood apart from $\bar{X}$ and s, the z score is a most important concept. Indeed, it is a central organizing concept in statistical analysis. Very possibly you should review and redigest now, until you are thoroughly comfortable with *everything* up to this point.

As a review, ponder the following questions.

1 What is meant, conceptually, by the mean? Explain in words rather than with a numerical formula.

2 Explain in words what is meant by the standard deviation?

3 Based on the concepts you have just explained, now try to write the formulas for $\bar{X}$ and s, without recourse to mere memory or peeking.

4 What information, if any, is lost when a raw score is transformed to a standard score? What information, if any, is gained by this procedure?

Contemplate these, in as much depth as possible.

EXTERNAL STANDARD OF MEASUREMENT
INTERNAL STANDARD OF MEASUREMENT
RANK ORDERING
PERCENTILE POINT
PERCENTILE RANK
STANDARD SCORE
z-UNIT or s-UNIT
z SCORE
T SCORE
FREQUENCY DISTRIBUTION OF MEANS
NULL HYPOTHESIS
EXPERIMENTAL HYPOTHESIS

exercise for a rainy coffeebreak

Find a fairly wide elastic band, the bigger the better, and cut it open so you can lay it flat on a table. Tack it down at both ends, if necessary, to make it lie still (*not* into the genuine Empire rosewood table!). You now have made a potential *abscissa*.

Now take a felt-tip marker and dab it here and there along the band. You are creating scores along the abscissa. If you are knowledgeable and artistic, there will be more dots—scores—toward the middle.

Next, estimate where the mean score of your distribution lies, and mark this location with a stripe running across the band. Then estimate the standard deviation, and put stripes one *s* unit to the left and the right of the mean. (They should be about 30% of the way from the mean to the extreme score at either end.) Add stripes corresponding to the remaining *s* units, as in Figure 6.3.

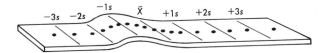

Now the demonstration can proceed. Stretch the abscissa; by doing this you separate the scores more, thus increasing their variability. If you now let the band return slowly to its normal size, you decrease the variability again. So you have seen the value of *s* increase and decrease in a way that depended entirely on how you changed the score values. But what happened to the *z* scores of this distribution? The value of each score *X* changed, *s* changed, but the *z* scores all remained the same! When the band is stretched, each dot is still the same number of *s* units from the mean that it was before the stretching, because the *s* units move *with* the dots.

You can probably understand this by merely *imagining* the demonstration. However, should you actually be doing it, be sure *not* to aim the rubber band in the direction of your professor, who may be sipping hot coffee. The exercise can be repeated to demonstrate the equivalent properties of percentiles. Mark ten scores along the band, mark off every tenth percentile point, and stretch the band. The percentile ranks of your scores will not be affected. Of course, if you didn't handle your rubber band with care the first time, you may have to do this with a different professor, maybe even at a different college.

self-test on chapter 6

1 The z score of 0.00 and the 50th percentile are identical
   a always.
   b occasionally but unpredictably.
   c conceptually in the sense that they are both averages (of the mean and median, respectively).
   d when the distribution is symmetrical.
   e as in (d), but with the qualification that percentile points are sometimes rather arbitrarily chosen, as when they fall among many tied scores.
   f never.

2 When is a z-score transformation preferable to a percentile transformation?
   a always
   b when the data have cardinal scaling
   c when the data are only ordinal
   d when the distribution is very strongly skewed
   e when the n is rather small

3 100 integer scores are grouped into intervals five units wide, and you want to find the 30th percentile point. The three leftmost intervals in the distribution contain 26 scores; the next one (known to contain the 30th percentile) contains eight scores, some of them tied. The 30th percentile point
   a can be found only by assuming an even distribution of scores within the interval.
   b would in this case lie exactly halfway through the interval.
   c would in any event be found in a manner equivalent to the method for estimating the median for grouped data.
   d would be equivalent to a z score of -2.00.
   e would be estimated by use of *strictly* ordinal assumptions.

SCORING AND ANALYSIS of Self-Test

Answers to Question 1

a F **b** T   If you are alert, you will have realized that this could have read, "the
c T **d** T   mean and the median are identical. . . ." They are identical numerically when the distribution is symmetrical, but rarely identical otherwise.

e T   This qualification was newly introduced in this chapter.

f F

Answers to Question 2

a F   *z* scores are not *always* preferable to percentiles. Sometimes the cardinal assumptions required are not met in the data, or the actual need
may be to express a ranking concept (as when scholarships are to be
awarded to the top 20% of those graduating from Kwogmire Kollege).

b T   Cardinal data can always be transformed into *z* scores, and since *z*
scores discard less information than percentiles, the former are preferable if they meet your needs in other ways.

c F   With merely ordinal data, you have no choice: percentiles.

d F   With a skewed distribution, *z* scores may not be so useful, based as they
are on the mean and standard deviation, which do not reflect skewness. Percentiles might well be preferable.

e T   With *n* smaller than about 100, you shouldn't use percentiles; *z* scores
are best *if* they are not inappropriate for other reasons.

Answers to Question 3

a T   When, as here, any percentile is 4 scores into an interval containing
b T   8 scores, we must assume *something* about the distribution of scores
within that interval. The most harmless assumption is that all 8 scores
are spaced evenly throughout the interval, even though this is sometimes known to be impossible. (Eight integers cannot be evenly spaced
across an interval five units wide!) At any rate, 4 scores out of 8 takes
us halfway across the interval that contains the 8 scores.

c T   The procedure for finding any percentile point is identical with that for
finding the median from grouped scores, as given in Chapter 4.

d F   There is no way of telling what the equivalent *z* score should be, because the cardinal scaling of *z* scores has been abandoned; percentiles
are merely ordinal.

e F   The concept of *even* spacing across the interval is not ordinal; it is
interval. With percentiles, cardinal assumptions tend to slip in
through the back door anyway.

4 Among the 100 scores of the preceding question, the raw score having the lowest value

 a would have a $PR = 0$.

 b would have a $PR = 0.5$.

 c would have a $PR = 1$.

 d would have a z score very close to 0.

 e would have a relatively large z score.

5 The mean value of those 100 scores is 47.56, with $s = 3.45$. A given score has a value of 45.835. Its z score is

 a likely (but not certain) to have a PR of less than 50.

 b 0.05

 c essentially 0

 d -0.5

 e $+0.5$

6 IQ is measured as a variety of T score, with $\bar{X} = 100$ and $s = 15$. Therefore, we *know* that

 a mean population IQ may increase if intelligence levels do.

 b the mean population IQ is *approximately* 100.

 c the mean population IQ is theoretically *exactly* 100.

 d inasmuch as the number of mentally retarded people is slightly greater than the number of geniuses, the median IQ must be less than 100.

 e population variability in IQ is constant, in spite of the successful efforts of nutritionists and educators to raise intelligence levels.

7 Suppose you have done a bit of research, gathered your data, and made a frequency distribution of the scores. Your abscissa is scaled in z scores. Which of the following are true?

 a If you want to find percentile ranks, you can construct an ogive as easily with z scores as with raw scores, but then you must "untransform" the abscissa scores again.

 b It would be relatively easy to find percentile ranks corresponding to your z scores using a cumulative frequency distribution.

 c If your distribution is a humped one, you probably have few, if any, z values as small as -4 or -5.

 d You will surely find that the 50th percentile has the same raw score value as the z score 0.

 e If you lost the original raw scores but still had the complete set of z scores and the values of $\bar{X}$ and s, you could reconstruct the raw scores.

 f If you lost everything except the complete set of percentile ranks, you could reconstruct the raw scores.

Answers to Question 4

a F   There are 100 scores and 100 percentile ranks, so each score must
b T   correspond with a rank. Since the first score is not entirely higher or
c F   lower than itself, its rank cannot be 1 or 0. Considering it half above
      and half below itself, its rank is 0.5.
d F   A z score of 0 belongs to the mean, which is a long way from the lowest
e T   score; we would expect the latter to have a large negative z score.

Answers to Question 5

a T   If the distribution is skewed, the median (50th percentile) *might* be
      much less than the mean, so 45.835 *could* be above the median.
b F   If you said "essentially zero," you were being lazy.
c F
d T $z = \dfrac{X - \bar{X}}{s} = \dfrac{45.835 - 47.56}{3.45} = \dfrac{-1.725}{3.45} = -0.5$
e F

Answers to Question 6

a F z scores, T scores, and IQ scores have fixed means of 0, 50, and 100,
b F   respectively, regardless of how the underlying raw scores as a whole
c T   may change.
d T   Remember that IQ is not itself intelligence; it is merely a standard
e T   score *measure* of intelligence.

Answers to Question 7

a T   True, if the ranks are to be given as values on the raw-score scale.
      Remember that z scores are a linear transformation of raw scores, so
      what is possible with one set is possible with the other.
b T   The abscissa would be marked off in z units, the ordinate would be
      marked off in percentages, and each z score would have a percentile
      rank on the ordinate.
c T   True too. See Figures 6.4 and 6.6, for example.
d F   Not necessarily. In a skewed distribution, the median and mean prob-
      ably do not coincide.
e T   Yes, you can get raw scores back from z scores if you know the values
      of $\bar{X}$ and s. Since

$$z = \frac{X - \bar{X}}{s}$$

then

$$X = \bar{X} + sz$$

f F   No, you can't do this. Percentiles no longer contain the information
      needed to transform them back into raw scores.

8 An experiment is done to compare an experimental group with a control group. The control group has a mean score similar to that of the untreated population from which it was sampled. The mean score of the experimental group

   a would, if very close to the control group's mean score, favor the experimental hypothesis.

   b would, if very close to the control group's mean score, favor the null hypothesis.

   c would *not* tend to favor the null hypothesis if it has the value of $+5z$ relative to the standard normal distribution of untreated population mean values.

   d can be interpreted only in the light of knowledge about the untreated population mean score and its standard normal distribution.

   e allows a choice between the experimental and null hypotheses, but only with some knowledge about probability.

EVALUATION

$+3$ for each correctly circled choice; -2 for each incorrect but encircled one; -1 for each correct one you missed. It's a cruel, cold world. Maximum possible score is 72.

45 to 72	Exquisite
35 to 44	Less exquisite
5 to 34	Not exquisite at all
less than 4	Do something about any problems you feel about this chapter. You must have *some*! Now ! Don't wait!

Answers to Question 8

a F If the experimental group has a mean value very close to the control

b T group's mean value, the chances are good that the difference is accidental, as it probably would be if you tossed ten coins and got six heads
and four tails, rather than five heads and five tails. However, if you got
ten heads and no tails, you would wonder about the coins!

c T Look at Figure 6.6. A z score of $+3$ or greater has very little chance of
occurring! Getting an experimental group mean of $+5z$ is rather like
tossing ten heads in a row: You wonder about the fairness of the coins
in one case, and you wonder about the null hypothesis in the other!

d T Yes, or you have no basis for making up the z scores that we've been so
glibly using here.

e T True! That's the domain of the next chapter. *And* the one after that.

> Questions 7 and 8 are *full* of some new concepts. If you made *any*
> correct choices, you are doing beautifully. If you got them all
> wrong but had a "feel" for the issue, you are still doing as well as
> most students. So cheer up, warm up, and onward to Chapter 7!

ſome ordinary problemſ

1 Find the 75th percentile point of the data below.

Interval midpoint	Frequency
25	2
50	13
75	48
100	86
125	140
150	231
175	237
200	169
225	47
250	27

This question seems to throw a curve at you; its n is not the convenient value of 100 to which previous examples have accustomed
you, and the interval width is not the familiar old 5 units. And the

interval *midpoint* is not the *upper limit*! Use your head, work in a logical sequence, and you'll find that nothing is different. (An ogive from grouped data could be used, although it won't be as exact as one made from raw scores.)

2 Calculate the 75th percentile again, using the instructions given in the following formula. Is the formula a correct recipe for percentiles? Collect your wits on this one, and work it out step by step.

$$\text{Value of given percentile} = X_{LL} + IW \frac{(P - f_{LL})}{f_I}$$

where X_{LL} is the score value of the lower limit of the interval containing the desired percentile

IW is the width of the interval

P is the desired percentile

f_{LL} is the accumulated frequency of scores up to but not including the lower limit of the interval containing the desired percentile

f_I is the frequency, or number, of scores in the interval containing the desired percentile.

3 Find the z scores of the following distribution.

2 3 4 5 5 6 6 6 7 7 8 9 10

4 Create a graph with the raw scores of Question 3 on the abscissa and the z scores on the ordinate. What is the characteristic of this graphic relationship?

5 Change the last score in Question 3 from 10 to 15. Does this change affect the z scores of any but the changed score itself?

6 A distribution has mean 48.0 and standard deviation 6.0. What are the raw scores corresponding to

$$z = -2.5$$
$$z = +2.33$$
$$T = 75$$
$$T = 30$$

7 Cumulative frequency distributions based upon data from humped frequency polygons are often said to have an S shape. What is meant here? Is it true? What gives an ogive such an S shape?

8 This one will challenge you. A distribution of scores is humped in the center, and the score with an equivalent z score of 0 is at the 60th percentile. Sketch what the distribution might look like. Which way is it skewed?

9 A humped distribution of scores is converted to percentile ranks; none of the raw scores are tied. What would be the shape of a frequency distribution of those percentile ranks? Provided that there are no tied raw scores, would the shape of the raw-score distribution affect the shape of the distribution of percentile ranks?

HOW TO TRANSFORM A CUMULATIVE FREQUENCY ORDINATE TO A CUMULATIVE PERCENTAGE ORDINATE

Now that you have tried it, you realize that correct and exact percentage scaling of an ogive ordinate (as in Figure 6.2) can be difficult. Here is a really good method.

Take an $8\frac{1}{2} \times 11$ inch sheet of paper and draw 21 radiating lines, as shown in Figure 6.7. The common starting point for each line is at the bottom of the left-hand long edge of the paper, and the destination of each line is a point on the right-hand long edge. These 21 points are half an inch apart. The lowest of the 21 lines is horizontal.

Now, measure your cumulative *frequency* ordinate from the abscissa to the value of n, the highest point reached by the ogive. Suppose it is 5.56 inches long. We want to divide this distance into 100 equal parts. We can start with 20 equal parts to give every fifth percentage point, and our "fan diagram" will do that for us. The vertical distance between the two outermost lines in the fan diagram (the first and twenty-first) varies as you move from left to right. Find the point at which the two are exactly 5.56 inches apart (vertically); then the intervening lines cut this 5.56-inch distance into 20 equal segments. The length of each segment is equal to 5 percentage points on the cumulative percentage ordinate you want to construct. You can either place your ogive graph onto the fan diagram and trace the percentage points or measure the distance corresponding to 5% with a ruler. Within these 5% intervals, the nearest percentage point can be estimated by eye.

In using the fan diagram, be sure to keep your ogive abscissa exactly superimposed on the bottom line of the fan, or your intervals will *not* be constant in size.

Figure 6.7
The fan diagram for dividing any length of cumulative frequency ordinate into a cumulative percentage ordinate. Regardless of how far apart the outermost lines are, the intervening ones divide the total distance into 20 equal parts, corresponding to every fifth percentage point on the cumulative percentage ordinate.

chapter 7
probability: measuring likelihood

Both classical logic and classical science have an important limitation: they look on any given statement as either true or false and leave it at that. This may not seem to be much of a limitation at first glance. After all, either a tossed coin will land heads up or it will not. Either LSD causes chromosomes to break or it does not. "Obviously," you may say, "all we need to do is find out, and then we'll know."

Unfortunately, contemporary scientific questions can rarely be answered so simply. Consider the well-known claim that frequent use of LSD damages chromosomes. (This example is not from the behavioral sciences, but all areas of science have to deal with problems that have no clear-cut, yes-or-no answers.) Biologists find a certain "normal" incidence of chromosome damage in people who have never used LSD. Some investigators have noted, however, that the incidence of damage is somewhat higher in the LSD users they have tested. Suppose that the normal incidence of breakage is 10 in 100,000 chromosomes, but the users of LSD tested are found to have a breakage incidence of 20 in 100,000. (These are purely hypothetical data.) What may be concluded from this evidence? The 20 in 100,000 incidence might have nothing to do with LSD; perhaps the relatively small number of users tested just happened to be a group of people with unusually high incidence of chromosome breakage. It is, after all, *possible* to find a small sample of people who have never used LSD and who show a mean chromosome breakage incidence higher than the population mean, just as it is possible for a flipped coin to show heads four times in succession even though it is a fair coin. But perhaps a doubled incidence (from 10 to 20 per 100,000) of breakage is too great to be dismissed as accidental; then we would have to conclude that use of LSD (the only consistent difference between the high-incidence group and the population in general) caused the change. If a flipped coin showed heads 20 times in succession, we would be reluctant to believe that it was a fair coin, although a fair coin almost certainly *will* show 20 heads in a row once in a very long while. Similarly, if a group of LSD users shows double the normal incidence of chromosome breakage, perhaps we should be reluctant to dismiss this as an accident.

Few, if any, experiments yield conclusive proof or disproof of a hypothesis, although those not well acquainted with experimental practice may not realize this. Then how can we draw a conclusion from an experiment like the LSD one? If we cannot, we have wasted our time in performing (or imagining) it. Fortunately, we are able to avoid a standoff by resorting to **probability theory**, which is a seductive (but not immoral) alternative to that true-or-false (also called two-valued) logic. The argument goes like this: Given what we know about the natural incidence of broken chromosomes, which *on the average* is 10 per 100,000, what is the *probability* that 20 per 100,000 might merely be a fairly extreme (say, two or three standard deviations from the mean) case of normal incidence? Suppose this probability is relatively high; then we would feel justified in retaining the conservative position (the null hypothesis) that LSD does not affect chromosomes. However, if the probability of 20 being found in a normal population is very low, we would feel more justified in rejecting the null hypothesis and adopting the alternative one, namely, that LSD *does* have something to do with chromosome damage.

How low is *very low*? The answer is really arbitrary, depending on how much of a chance you want to take on being wrong, and on the way in which you would prefer to be wrong if you must be. (For example, how much of a chance will you take on concluding incorrectly that LSD is harmless, perhaps tacitly encouraging its use, and thus letting many people damage their health with it?) Quite often we find the values 0.01 and 0.05 being used; that is, the experimenter would not conclude that LSD damages chromosomes unless the figure of 20 per 100,000 was less than 1% (or 5%) likely to be found when a sample is chosen from a population that has a mean of 10 per 100,000. But how do we determine how likely it is? To do so we must use some probability theory. By now you should be convinced of its necessity.

various concepts of probability

Probability is generally best defined as *the likelihood of something happening*. This likelihood is usually symbolized by p. It can be thought of as the percentage of occurrences of a particular event (Such as rolling a 4 with a die) to be expected in the long run. In the long run, for example, you can expect a fair die to show a 4 on $\frac{1}{6}$, or $16\frac{2}{3}\%$, of all rolls, because each of the six sides of the die is equally likely to come up on each roll. There is a second likelihood, q, of the thing *not* happening, and $p + q$ must equal 100%, because it is *always* true that a thing either does or does not happen. Except for this constraint, p or q may have any value between 0% and 100%. Usually, probabilities are expressed in decimal form, for example, 1.00 instead of 100% and 0.50 instead of 50%. Henceforth we will use decimal notation.

It may be difficult to define an event unambiguously, to say definitely whether it has occurred. Before you can talk about the incidence of broken chromosomes, you must know precisely when a chromosome is to be classified as broken: what about bent or partly broken chromosomes? You cannot assign probabilities to events unless you clearly define the events whose probabilities you are considering.

In general, once all events in question are clearly defined, the probability of a particular event, p_{event}, is the number of ways in which the event can happen divided by the total number of possible events. For example, when you roll a die, there are six possible outcomes, and there are three ways in which you can roll an even number (2, 4, or 6). So the probability of rolling an even number is $\frac{3}{6} = \frac{1}{2} = 0.5$.

Real agony can arise when probability theory is applied too literally to individual cases. Suppose you know that with your grade-point average, you have an 80% chance of being admitted to medical school, but you are not admitted. Yet your friend, who has a lower GPA, and thus only a 40% chance of being admitted, is accepted. It seems unfair. But your 80% probability of being accepted means that 80%, *not* 100%, of students with your average are accepted (so some are not), and your friend's 40% probability of being accepted *will* pay off in some cases, as it did in this one. Probability guarantees absolutely nothing in an individual case.

Probability is never retroactive. It projects forward into the future, never backward into the past. What knowledge is for past events, probability is for future (or unknown) ones.

Actual values may be given to *p* in several ways.

theoretical probability

Suppose you roll a single die; since it has six sides all equally likely to come up, there is a $\frac{1}{6}$ or 0.166 probability of rolling a 6 (or of rolling any particular number you have in mind). This is a *theoretical* probability, calculated on the *theoretical assumption* that all sides of the die are equally likely to come up. Most betting games that involve chance (such as poker, roulette, and craps) consist in trying to use the theoretical (or *formal*) probabilities involved to your own best advantage.

empirical probability

Empirical probability is based not on a theoretical assumption about relative likelihoods, but on past experience. A baseball player's batting average can be interpreted as an empirical probability. If a player has a batting average of 0.200, for example, he has been getting a hit once in every five times at bat (on the average), and the team manager will usually use that figure as an indication of the player's chance of getting a hit next time at bat: one chance in five, or a probability of $\frac{1}{5}$ of getting a hit. Theoretical and empirical probabilities often match each other, but sometimes they do not. When you roll an honest die, your formal chance of getting a 4 is $\frac{1}{6}$, 0.166. Over a run of actual rolls, you may find that you do indeed get 4 $\frac{1}{6}$ of the time; then theoretical and empirical probabilities do match. If the die were loaded, however, you would doubtless find that 4 occurred less often (or more often) than formally predicted, and in this case, theoretical and empirical probabilities would not match.

subjective probability

This is the term used to describe the likelihood that we all assign to possible events in everyday life based on our own unique reasons. Such probabilities may be actual numbers, but often they are not. We will not consider subjective probability, except to point out that it is based partly on theoretical and empirical probability and partly on attitude, motivation, and experience. The Russian Roulette player doesn't *really* believe in that $\frac{1}{5}$ or $\frac{1}{6}$ probability of *his* head being blown off; the downtown jaywalker doesn't *really* expect to be run over.

There is, however, another use of the term "subjective probability," in a relatively new field called **Bayesian statistics**. In Bayesian statistics, subjective probability can be fixed upon in advance (for any reason) and then compared with empirical outcome. It is systematic, legitimate, and useful as well, but as an approach it's beyond the scope of a book that is trying to stay slim. A book on the subject is available for interested people.[1]

combining separate probabilities

A probability is assigned only to a single event, but of course we are very often more interested in complex events, whose net outcome depends upon a large number of events combining in a certain way (or in any number out of all possible ways). There are two basic rules that (like all Great Truths) are very simple to remember.

[1] Novick, M. R., and Jackson, P. H., *Statistical Methods for Educational and Psychological Research,* New York: McGraw-Hill, 1974.

the multiplication rule

The multiplication rule is used to obtain the probability that *all* of two or more separate and unrelated events will occur. Simply multiply all their probabilities. Suppose you have three ordinary coins. Each one, if tossed, has a probability of coming up heads of $p_{heads} = 0.5$. What is the probability of tossing all three coins and getting *all* heads? The multiplication rule tells us instantly.

$$p_{all\ heads} = 0.5 \times 0.5 \times 0.5 = 0.125$$

Writing this with fractions instead of decimals, we get

$$p_{all\ heads} = \tfrac{1}{2} \times \tfrac{1}{2} \times \tfrac{1}{2} = \tfrac{1}{8}$$

or one chance in eight. Normally, we use the decimal version as shown first; of course, $\tfrac{1}{8} = 0.125$, so the answers are the same.

The most important thing to bear in mind when using this rule is that the separate probabilities (0.5 in the case of each coin) must not affect each other or be mutually related in any way. The assumption must be made that a head coming up on coin 1 will not affect the probability of a head coming up on coin 2, and so on. With coins and dice, this assumption is easy to justify, but with events in the practical world, it sometimes isn't. Suppose you are attending the Great Annual Statistics Picnic (GASP) and are drawing partners for the three-legged race. A quick check has shown that 100 people are there besides yourself, half of them female. So there seems to be a 0.5 probability of getting a female as a partner, which you happen to prefer. You would, however, also like to get a good conversationalist as a partner, and you know that 20% of the people there have this characteristic. It would *seem* that you have a probability of $0.5 \times 0.2 = 0.10$ of drawing a female partner who is a good talker. *But* this may not be true; suppose, for example, that all 20 good conversationalists are female. Then the 0.10 probability is invalid, because you have a 0.5 chance of drawing a female partner, times a 20 in 50 chance of getting one who is a good talker, which works out to $0.5 \times 0.4 = 0.20$, a 20% chance. If the good talkers had all been male, you obviously would have no chance (probability 0) of getting a good conversationalist who is female. The whole problem would be, either way, that sex and conversational ability were *related* in this group of people; the probability of choosing a good conversationalist *depends* on whether you choose a male or a female partner. If half the good talkers were male and the other half female, sex and conversational ability would be independent of each other in this group.

The assumption of mutual independence or unrelatedness is an important one, as we will see in later sections of this chapter and in future chapters. In the next chapter, we shall see that the concept of a random sample rests upon drawing each member of the sample quite independently of every other member.

When considering the joint occurrence of several related events, you must keep in mind the possibility that the number of *possible* events may change after the first one occurs. For example, suppose you are being dealt a Bridge hand (13 cards) from a deck of 52 cards. Since one-fourth of the cards are hearts, you have a 0.25 chance of getting a heart as your first card. But what is the probability of getting a heart as your second card? There are 12 hearts remaining, but among only 51 cards, so you have a $\frac{12}{51}$ chance of being dealt a second heart. So the probability of your first two cards both being hearts is

$$\frac{13}{52} \times \frac{12}{51} = 0.0588$$

which is just less than a 6% chance. (This assumes, of course, that the cards are fair.) What is the chance of being dealt *all* hearts, one of those weird once-in-a-lifetime events? We merely extend the argument.

$p_{\text{all hearts}}$

$$= \frac{13}{52} \times \frac{12}{51} \times \frac{11}{50} \times \frac{10}{49} \times \frac{9}{48} \times \frac{8}{47} \times \frac{7}{46} \times \frac{6}{45} \times \frac{5}{44} \times \frac{4}{43} \times \frac{3}{42} \times \frac{2}{41} \times \frac{1}{40}$$

This is very cumbersome; it might preferably be written

$p_{\text{all hearts}}$

$$= \frac{13 \times 12 \times 11 \times 10 \times 9 \times 8 \times 7 \times 6 \times 5 \times 4 \times 3 \times 2 \times 1}{52 \times 51 \times 50 \times 49 \times 48 \times 47 \times 46 \times 45 \times 44 \times 43 \times 42 \times 41 \times 40}$$

but it is simpler to employ the factorial notation (!) of Chapter 1 instead.

$$p_{\text{all hearts}} = \frac{13! \times 39!}{52!}$$

Now the last formula looks very different, but it really is not. The denominator is the product of the numbers from 52 all the way down to 1, but the product should actually stop at 40. This is the reason for the

Table 7.A

FACTORIALS

N	$N!$
1	1
2	2
3	6
4	24
5	120
6	720
7	5040
8	40320
9	362880
10	3628800
11	39916800
12	479001600
13	6227020800
14	87178291200
15	1307674368000
16	20922789888000
17	355687428096000
18	6402373705728000
19	121645100408832000
20	2432902008176640000

A special case: 0! has the value 1. This table can be extended by multiplying the bottom value in the $N!$ column by the next integer in the N column. Obviously, though, the values of $N!$ will become even more unwieldly with larger N. It is better to use logarithms of factorials, which are available in various handbooks of statistical tables.

39 factorial in the numerator, which cancels the product of the numbers from 39 down to 1 out of the denominator. In general, the formula for the probability of drawing a particular combination of g things (without replacement) out of N distinct possibilities is

$$p_g = \frac{g!(N - g)!}{N!}$$

Things look even better once you know that there are tables of factorials that relieve you of all but the simplest arithmetic. Look at Table 7.A. You may be interested to know that you have only one chance in 596,410,911,600 to get all hearts in a bridge deal. This corresponds to a probability of approximately 0.0000000000016. You may have to wait awhile! (On the other hand, a rare event can occur at the next opportunity and not again for a long while. It *can* occur anytime, even conceivably on two deals in a row!)

As an example, suppose that you know that only 2% of the population have IQs of 135 or higher. In a sample of 100 children, if the sample selection process was random, you would thus expect about two to have IQs above 134. What would be the probability of having 3 (or 4, or 5, or 6) with IQs above 134? You would expect the probability to decrease sharply as the number of geniuses looked for in the sample increases. This is indeed what happens, and in each case, p could be specified exactly.

the addition rule

The **addition rule** is used to calculate the probability that at least one of several mutually exclusive events will occur. When you roll a single die, you will get 1 *or* 2 *or* 3 *or* 4 *or* 5 *or* 6, so one and only one of these events can happen. The probability of getting each of them is $\frac{1}{6}$ (because the die is six-sided), so we find the obvious: that the probability of getting 1 or 2 or 3 or 4 or 5 or 6 is

$$p_{1 \text{ or } 2 \text{ or } 3 \text{ or } 4 \text{ or } 5 \text{ or } 6} = \frac{1}{6} + \frac{1}{6} + \frac{1}{6} + \frac{1}{6} + \frac{1}{6} + \frac{1}{6} = 1.0$$

In other words, it is certain that you will throw *one* of these mutually exclusive alternatives. Note the term **mutually exclusive**. It's the key concept here.

The events just mentioned are also *exhaustive,* in the sense that all possible events have been included. The addition rule works equally well with several mutually exclusive but nonexhaustive events. The probability of throwing a 1 or 2 or 3 is $\frac{1}{6} + \frac{1}{6} + \frac{1}{6} = \frac{3}{6}$ or 0.5. That's the theoretical probability, and the empirical one too, unless you are unwise in your choice of craps partners. This example is nonexhaustive because a 1, a 2, and a 3 do not jointly exhaust all the possibilities. You could throw a 4, 5, or 6 instead. When all separate possibilities add up to less than 1.0 the events to which they refer are nonexhaustive.

In drawing a partner for the three-legged race at the GASP, you have a 0.5 chance of drawing a male, and a 0.5 chance of drawing a female, because there are equal numbers of each attending (besides yourself). The probability of drawing *either* a male *or* a female is, obviously, 1.00.

Previously, you were concerned with the probability of drawing a conversational female partner. Let's imagine now that you've altered your goals a bit and will be content with either a female *or* a conversationalist. This situation is a little more complicated, but still simple enough. You know that there is a 0.5 chance of drawing a female (because half of the group is female) and a 0.20 chance of drawing a conversationalist (because 20 out of the 100 picnickers are in this category). Since you want to know the probability of drawing one *or* the other (possibly both), we use the addition rule, so it looks as if

$$p_{\text{female and/or conversationalist}} = 0.5 + 0.2 = 0.7$$

But there is a complication! This computation is the same as adding the 50 women and 20 good conversationalists and dividing the sum by 100:

$$50 + 20 = 70 \qquad \frac{70}{100} = 0.7$$

Presumably, some of those 100 people are *both* female *and* chatty, and they are being counted twice by this use of the addition rule. (Suppose, in fact, that we wanted the probability of drawing a male and/or a nonconversationalist. If we simply added the probability of drawing a male, 0.5, to the probability of drawing a nonconversationalist, 0.8, we would get 1.3, which is nonsensical, since probabilities cannot be greater than 1.) A modification is necessary: we should subtract the probability of being *both* female *and* conversational; that way, we count such people only once. If we assume independence between these characteristics (as discussed above), we can use the multiplication rule to find that

$$p_{\text{female and conversational}} = 0.5 \times 0.2 = 0.1$$

as we have already seen. This latter joint probability is subtracted from the sum of the other two probabilities to give the correct *and/or* formula:

$$p_{\text{female and/or conversationalist}} = 0.5 + 0.2 - (0.5 \times 0.2)$$
$$= 0.6$$

Usually when considering *and/or* probabilities of this kind, we do not use the phrase *and/or*, but merely *or*. We have looked at the probability of drawing a female *or* a conversationalist, and the probability of drawing a male *or* a silent type. Usage of the *inclusive or* in this fashion presupposes the possibility of sometimes having *and* as well as *or*.

the Venn diagram

When considering complicated problems of probability, it often helps to use a *Venn diagram*. A simple one, pertaining to the proportions of males and females in our picnic of 100 people, is shown below. Note that the probability of any event (drawing a male or drawing a female) being *somewhere* in the Venn diagram is 1.0.

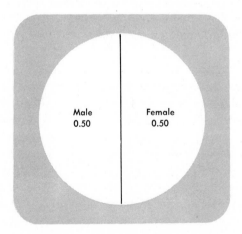

Another Venn diagram, for conversational ability among the same 100 people, would look like this.

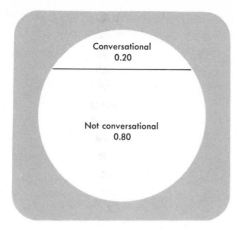

It is not necessary, but often useful, to let the area roughly represent the amount of probability.

The two problems that we've been considering via arithmetic can also be solved very quickly if we combine the two Venn diagrams just shown:

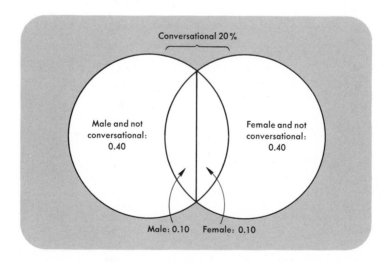

The probabilities shown in each of the four areas correspond to the four possible combinations of events. Again, the probabilities in the four areas total to 1.00, so we can merely add the probabilities of the areas in which we are interested. The probability of drawing a female *or* a

conversationalist is 0.6; the same logic shows the probability of drawing a male *or* a silent type to be 0.9.

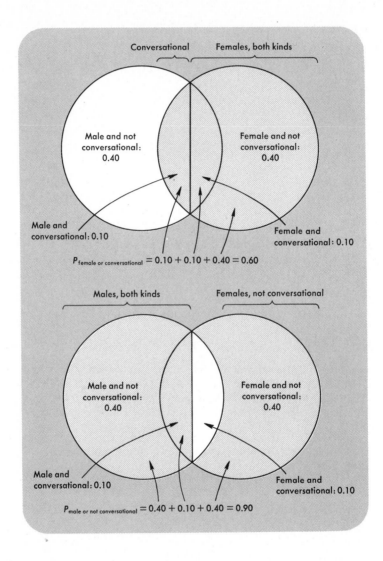

Venn diagrams are useful in many other ways, particularly in connection with *set theory* (a branch of mathematics that you may or may
not have met). But further elaboration is not necessary here.

It should be clear from our discussion thus far that probability is somewhat similar to percentage. Indeed, past percentages serve as the basis
for empirical probabilities. The simple probabilities with which we
started, such as the chance of being paired with a female in the three-

legged race, are based upon the percentage of ways in which this specified event can happen. Fifty females in 100 people means there are 50
ways of being paired with one of them out of 100 possible pairings, so
the probability of being paired with a female is 50% or 0.5 Of course,
you will not find such an easy conversion to percentage or probability
when the total number of possibilities is other than 100. If the picnic
had involved 150 other people, 50 of them female, your chances of
drawing one would have been $\frac{50}{150}$, or 0.33. Remember the basic definition of a probability whenever you want to calculate one:

$$\text{Probability of event } A = \frac{\text{Number of outcomes favoring event } A}{\text{Number of possible outcomes}}$$

As we have seen already, event A can be something as simple as rolling
a 6 with a die or tossing a head with a penny; it can also be a complex
end result that can happen in any number of complex ways, as we will
see later in this chapter.

Probability is a large area of mathematics that can only be touched
upon in this book. If you are interested in pursuing the topic, an excellent book by Warren Weaver (similar in tone to the one you are now
reading) is available in paperback.[2]

The assignment of a probability to an event, as we've seen here, depends upon the assumption that each of all possible outcomes is exactly
as probable as the rest. If partners for the three-legged race are deliberately being assigned so as to mix the sexes, your chance of pairing with
a female is about 1.0 if you are male, about 0 if you are female. Then
partners are not being assigned *at random*; the notion of randomness is
central to any consideration of probability.

randomness

The basic notion of randomness is simply stated. If there are N possible
events that are mutually exclusive and exhaustive, the probability of
any one of them occurring must be $1/N$. For example, among the 100
other people at our picnic, your likelihood of drawing any given person
as a partner must be $\frac{1}{100}$ or 0.01. Essentially, the term "random" refers
to a *procedure* rather than an end product; to talk of a random sample
is to talk about what procedure was used in getting the sample. The
end result is not necessarily determined. Suppose I were to show you
ten coins that had been flipped, all of them having landed heads. Is that

[2] Warren Weaver, *Lady Luck,* Doubleday, Anchor Books, Garden City, 1963.

a random result? Yes, if each was tossed fairly. But if I had arranged the coins so as to "look" random (say, alternating heads and tails), the result is emphatically *not* random! As we shall see in the next section, the assumption of randomness is a crucial part of the probability theory on which statistical inference is based.

There is a great deal of loose talk involving the word "random." Sometimes our popular usage of the word is close to being accurate, sometimes not. In the jargon of social science, especially its popularized forms, we hear frequent mention of "random samples" and "random sampling." The justification for such terminology is almost always faulty. For most purposes, researchers desire *representative* samples. While randomness is one route to representativeness, it is not the only one and often (as when doing research surveys) not the best one. The importance of *randomness* is concentrated in the statistical testing of experimental hypotheses.

probability distributions: combining the additive and multiplicative rules

The most important aspect (to us) of probability involves mutually-exclusive and independent events. This time we shall explore the example of IQ level.

Suppose that IQ (within our own culture) is determined by only ten factors, each of which may function to *increase* or *decrease* IQ by some fixed amount (the same for each factor). Let us assume further that each factor operates randomly, to increase IQ half of the time and to decrease it the other half of the time. Then we have ten independent possibilities, each of which can assume one of two mutually exclusive states. How would we expect IQ measures to be distributed in such a situation?

We can visualize this by means of a device called a *quincunx,* first devised by Sir Francis Galton. It is pictured in Figure 7.1. Look carefully at this figure.

This device consists of a sloping panel, with a hopper above loaded with balls, closely spaced parallel stalls below where the balls can collect, and several staggered rows (ten, in this case) of short wooden dowels in between. Each of the rows represents a factor determining IQ. As the ball representing any given individual rolls down the panel, it successively encounters each row (or factor) and is deflected either to the left or to the right, tracing the type of path shown in Figure 7.1.

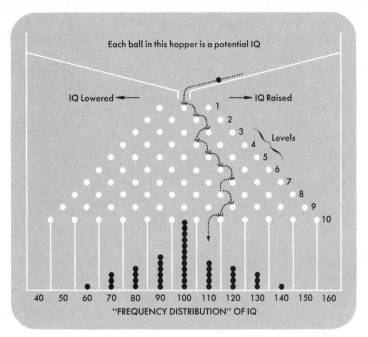

Figure 7.1

The effect of ten independent determiners of IQ as visualized in a quincunx. Each of the levels in the device is one determiner, serving to increase or decrease the score. The increases and decreases will usually cancel each other, making scores tend to cluster toward the center; however, a very high score can occur if the ball consistently bounces to the right.

From the common starting point, the hopper, there are many possible paths to the final destination, whose location is analogous to an IQ.[3]

What determines the final destination of any given ball? Chance, obviously. Upon hitting any dowel, the ball has a 0.5 chance of going left and a 0.5 chance of going right. Since it will encounter ten such choices on its way down the panel, the probability of any given pathway (including the one shown in the figure) being taken is

$$0.5 \times 0.5 \times 0.5 \times 0.5 \times 0.5 \times 0.5 \times 0.5 \times 0.5 \times 0.5 \times 0.5 = 0.5^{10}$$

which is about 0.0009 or 1 chance in 1024. An interesting phenomenon may now be observed if enough balls are run and allowed to rest where

[3] When (if?) you make your own quincunx, it is important to have the dowels spaced very evenly. Be sure too that the spaces between dowels in the rows are only slightly greater than the diameter of the balls (children's marbles are good). Quincunxes are also available commercially.

they land: As in Figure 7.1, they pile up toward the center and taper off toward each side. (The distribution shown in Figure 7.1 is not likely to occur with just 40 balls, but it could. Its actual shape would probably be even more piled up in the center.) Why should this situation occur?

To answer this question, let's consider a much smaller quincunx, having only two levels, as in Figure 7.2. As seen here, there is only one way of getting a ball into stall A, namely, by having it bounce left twice. Similarly, a ball can enter stall C only by bouncing right twice. However, there are two ways of getting into B: by going first right and then left, and by going first left and then right. So if four balls were run, the best bet is that one would land in A, two in B, and one in C.

A mathematical version of this interesting phenomenon was first developed by Blaise Pascal in 1665. His **arithmetic triangle** (pronounced arith*met*ic) can have as many levels as desired; they are analogous to the levels of a quincunx, and the numbers on each level refer to the number of different ways in which hypothetical balls could reach hypothetical stalls at that level. The Pascal triangle corresponding to Figure 7.2 is

$$
\begin{array}{ccccc}
 & & 1 & & \\
 & 1 & & 1 & \\
1 & & 2 & & 1
\end{array}
$$

The triangle can be extended to any number of rows by letting each numeral in the new row be the total of the two numbers to its right and left in the row above. Thus, the Pascal triangle for our ten-level quincunx would be as shown in Figure 7.3. Certain features of the Pascal triangle and quincunx are evident at a glance. The total number of ways that a ball can reach any row is double the total number for the row above; these totals can be expressed as successive powers of 2, from $2^0 = 1$ at the top to $2^{10} = 1024$ at the bottom. The distribution of cases within each level is called the **binomial distribution** for that level, and the entire Pascal triangle is called the binomial expansion. (*Bi*-means "two" and *nominus* means "name." This terminology harks back to certain hobgoblins called binomials that you probably met in high school. They were the two-term expressions like $x + y$ or $x^2 - 2x$. When you raise such creatures to a power, as in $(x + y)^{10}$, the final answer will be a fairly horrendous mess in which, amazingly, the numbers of a row of the Pascal triangle appear. Thus we talk of **binomial distributions** and **binomial expansions**.)

Let's return to our ten-factor IQ example and discover the *exact* probability distribution, expressed as the number of balls in each stall in Figure 7.1. We see that if 1024 balls were run, they should be arranged

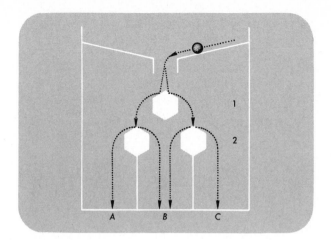

Figure 7.2

The principle of the quincunx, shown in a very simple one with only two levels. There is only one way in which a ball can reach A or C: by bouncing twice to the left or twice to the right, respectively. There are two ways, however, in which B can be reached: first left and then right, or first right and then left. Hence, we should expect twice as many balls to land in B as in either A or C.

Levels												Total N at Each Level:
1						1						1
2					1		1					2
3				1		2		1				4
4			1		3		3		1			8
5		1		4		6		4		1		16
6	1		5		10		10		5		1	32
7	1	6		15		20		15		6	1	64
8	1	7	21		35		35		21	7	1	128
9	1	8	28	56		70		56	28	8	1	256
10	1	9	36	84	126		126	84	36	9	1	512
11	1	10	45	120	210	252	210	120	45	10	1	1024

Figure 7.3

The Pascal Triangle for 11 levels. The N associated with each level is the number of different ways in which successive "either-or" events may occur.

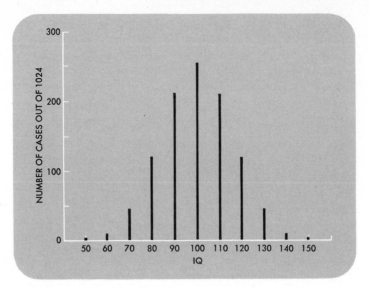

Figure 7.4
Frequency distribution (hypothetical) of IQs as obtained in the 11-level
quincunx of Figure 7.1. Over 1000 balls would need to be rolled (or 1000
cases collected) for the distribution to reach this shape.

in 11 stalls in quantities corresponding to the bottom row in Figure 7.3.
Perhaps we could visualize better with a new graph: Figure 7.4. Here
the vertical lines stand for stacks of very tiny balls; the alert reader
will note that this new version of the quincunx has the appearance of a
frequency histogram (see Chapter 3). If it were transformed into a
frequency polygon by connecting the tops of the vertical lines, it
would approximate a rather beautiful curve. But most important, it is
not only a frequency distribution of what *happened* to 1024 balls; it is
also a theoretical distribution of what *would happen* if 2048, or 4096,
or any large number of balls, were run. It is a *probability distribution*.
"A *what?*" you ask. Well, suppose our ordinate were scaled not by the
number of *cases* having each IQ but in *percentages* of the total number
of subjects having each IQ; and remember that percentages can easily
be made into probabilities. If we redraw Figure 7.4, making it a polygon and relabeling the ordinate with probabilities, we have Figure 7.5.
The surprising thing about Figure 7.5 is that even though we obtained
it (via Figure 7.4) from the fixed numbers in Figure 7.3, Figure 7.5
would have looked almost the same no matter what level of the Pascal
triangle we had started with.

In comparing Figures 7.4 and 7.5, an interesting and subtle (but, as

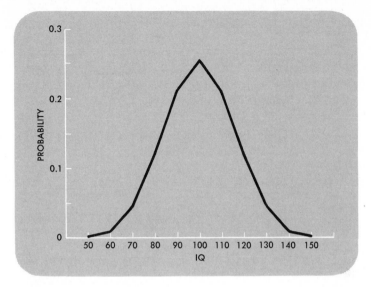

Figure 7.5
Figure 7.4 redrawn as a frequency polygon, with the ordinate redefined
as probability instead of number of cases out of 1024.

we shall see in future chapters, very important) fact emerges. Removing any percentage of the balls from Figure 7.4 will decrease the area under the polygon in Figure 7.5 by the same percentage. As an obvious example, suppose that exactly half the balls are removed from each stack in Figure 7.4. You would then have, in the remaining distribution, an area exactly 50% of the original. With the balls which were removed, you could construct an exact twin of the remaining polygon, as Figure 7.6 clearly shows. The shape of the binomial distribution does not depend upon how many scores go into making the curve, as long as a certain minimum number (1000 or so) are used. Above that number its shape will remain completely constant, though its height and width will continue to expand.

Now let's move on one more step. Suppose that IQ *is* in fact determined not by ten factors, but by a larger number, say 25. Let's assume further that there are several billion owners of IQs, not just 1024. Try to imagine the quincunx that would arise from this situation. Its rows would have to be very long, and the bottom row, where the cases finally pile up, would have a huge number of locations where the piles accumulate. The heights of the piles could vary by amounts that were so small, on the scale of the whole quincunx, that they could be considered

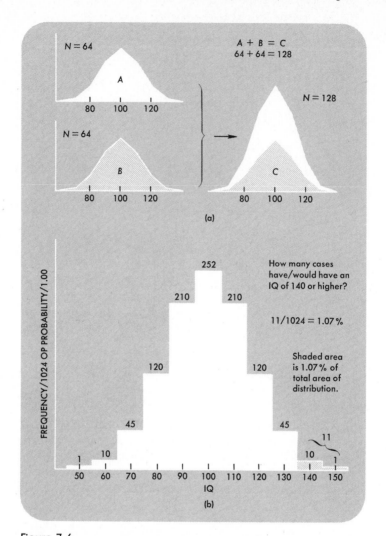

Figure 7.6

(a) The shape of a probability distribution does not depend upon number of cases. *A* and *B* are similar binomial distributions (from a seven-level Pascal triangle) each with the minimum of 64 cases. When the two are added on the same abscissa, the area under the combined curve is doubled but the shape of the curve is unchanged. *C* has the same shape as *A* or *B*. The total area under any probability distribution is 1.00.

(b) In our IQ example, what is the probability of finding an IQ 140 or higher? There are 11 ways out of 1024, which is 1.07% or 0.00107. This percentage or proportion is the actual area in question under the curve.

How can the scale of the ordinate and the area under the curve both indicate probability or relative frequency? To understand how, look at Figure 7.6(b). The height of each bar in the histogram is equal to frequency, or probability, as scaled on the ordinate to the left. Height therefore represents probability. Since the histogram bars are of a constant width, their *areas* are exactly proportional to their heights, so area must represent probability, just as height does. The same argument applies to polygons and curves, but not quite as obviously.

infinitesimal, because there are several billion cases involved (specifically, it would take 33,554,432 cases to finish the job for 25 levels). To carry this picture to its extreme, imagine that we have an infinite number of stalls and levels in the quincunx, all squeezed down to an infinitesimally small size so the whole thing won't be too large. The shape of the probability distribution would then be closely similar to that of the distribution in Figure 7.5, except that the scales on the ordinate and abscissa would be continuous rather than discrete and the straight lines forming the polygon would have melded into a curve. This is the curve with which we established an uneasy friendship in the previous chapter, the **Gaussian distribution**, sometimes (rather confusingly) called the standard normal frequency distribution or simply the normal distribution (there is nothing abnormal about other kinds of distribution). The Gaussian (pronounced *gow*-see-an) distribution is the most important theoretical distribution in all of statistics.[4] It would be better termed "the distribution of infinitely many scores, each de-

---

[4] There are really three phenomena known as Gaussian distributions:

1 The original, "real," one, which is defined mathematically. We won't bore you with this one.
2 The end result of infinitely extending the binomial distribution, as we just did. The result is essentially identical with the mathematical one but must be considered only a very close approximation, since it is defined differently.
3 The very close approximation to both of the first two that is frequently found in nature. Many measures, such as height, IQ, barometric pressure over a given spot, and flea population on wild foxes, are *approximately* Gaussian in their distribution. Or Gaussian distributions can be *derived* from such measures. Many other natural distributions are emphatically *not* Gaussian, so we must be careful.

The second kind is the one to pin your concepts on.

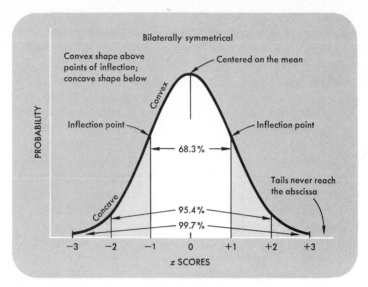

Figure 7.7

The shape and important characteristics of the normal, or Gaussian, distribution. It is bell shaped, but this alone does not make a distribution Gaussian.

termined by infinitely many chance happenings," but this is rather a mouthful. A Gaussian curve is shown in Figure 7.7. In this figure, a small problem is met and overcome: How do you scale the X and Y axes when the scores on each may go on indefinitely? For the Y axis, we work in proportions of the total area under the curve which is (as we have seen) 1.00. For the X axis we use z scores (having a mean, or central value, of 0 and a standard deviation (s) of 1 unit, which keep the distribution of scores from getting too wide.

Gaussian distributions have some other notable characteristics. They are bilaterally symmetrical and bell shaped. Depending upon the scale intervals on the ordinate and abscissa, they can appear tall or short. The tails of the curve never quite reach the abscissa, but keep going outwards indefinitely (but get infinitely close to the abscissa in the process). Most important, the Gaussian distribution has been so thoroughly studied that the proportion of the area under the curve that lies between the central (mean) value and *any z* score is known; tables (usually entitled "Areas under the Normal Curve") have been drawn up giving these proportions, which have tremendously important applications. There are three such proportions of area that you should know well (see Figure 7.7).

Between -1 and $+1$ standard deviations: 68.26% of the area
Between -2 and $+2$ standard deviations: 95.44% of the area
Between -3 and $+3$ standard deviations: 99.74% of the area

(These percentages are approximate.) By implication, only 0.26% of the area would lie beyond 3 standard deviations in both directions, and (since the distribution is symmetrical) only 0.13% of the area would lie higher than $+3$ or lower than -3. The remarkable implications of such knowledge will soon become evident.

Some hazards of using the term "normal distribution" should be noted first, however. Most people don't understand the concept well; as we have seen, it is a fairly difficult one. All too many types of measure are assumed to be normally distributed when in fact they are not, perhaps because "normal" is incorrectly taken to mean "not abnormal," or "as things ought to be." Many people believe mistakenly that any curve is normal if it is bell shaped. However, a bell-shaped curve that is too low and broad in the middle is **platykurtic**, not normal; one that is too high and narrow in the middle is **leptokurtic**, not normal. Finally, knowing what to *expect* based on probability does not let us assume that our actual data will *necessarily* fall neatly into place. Laws of probability are descriptions of what generally happens, not enforcers of what *must* happen. Many amateur gamblers have lost their shirts because they believed that probability theory guaranteed them something.

But now, let us return to the problem with which we began this chapter. We were wondering how probable it is that a 20:100,000 incidence of broken chromosomes reflects a harmful effect of LSD. (Note again that our data are hypothetical, for the purpose of discussion only.) We know that the mean incidence of broken chromosomes in the population of nonusers is 10:100,000. Let us now *assume* (from hypothetical evidence) that there is a Gaussian distribution of the reciprocal of chromosomal breakage frequency across the population of nonusers of LSD.[5] The mean of the distribution is 10,000, the reciprocal of the original mean value of 10 breakages in 100,000 chromosomes. The standard deviation of the Gaussian distribution might be

[5] As with many natural distributions, the incidence of chromosome breakage is skewed (since incidence cannot be less than 0 and the mean value is low, but scores sometimes run quite high), looking something like the lower panel in Figure 3.9. Transforming the raw incidences into their reciprocals will often turn a skewed distribution into an approximately Gaussian one. The following argument assumes that it *is* Gaussian, and this assumption must be kept in mind. It is probably quite reasonable. The reciprocal of 10:100,000 is 100,000/10 = 10,000, the reciprocal of 20:100,000 is 5,000 and so on.

known to be 2500. We now have the situation shown in Figure 7.8. Remembering that *area* under the curve represents *probability,* we see that 95.44% of nonusers will have breakage incidences between ±2 standard deviations of the mean, hence that 4.56% will have incidences outside these limits. In other words, *if* you are not an acid user, you stand a 4.56% chance of having an incidence whose reciprocal is less than 10,000 − (2500 × 2) = 5000 or more than 10,000 + (2500 × 2) = 15,000. Those are low odds. But here is an acid user with a reciprocal score of 5000!

As we said at the beginning of this chapter, unconscionably long ago, there are two possibilities for explaining such a peculiar event, and you can take your choice freely between them. Either you consider it just one of those coincidences that can happen anytime, concluding that acid usage has no effect upon the fracturing of chromosomes, or you can decide that such an event is too improbable to be mere coincidence; hence that LSD did increase the chromosome breakage. Either explanation *might* be correct. But, as we said earlier, it helps if we can state the odds more or less exactly. *If* we assume a Gaussian distribution of the characteristic in question (chromosome breakage), *if* we are sure of our measuring procedures, and *if* we have reason to believe that no unusual factor besides LSD is affecting our acid user, *then* we can state that if LSD has no effect, the probability of observing such an extreme incidence of chromosome breakage among nonusers is exactly 4.56%. The probability is 95.44% that nonusers will have a lesser incidence of breakage. If LSD does have the effect of breaking chromosomes, then a breakage incidence of 1 in 5000 might not be unusual at all among users. These actual probabilities are subject to some adjustment if the assumptions mentioned above are not perfectly valid, but at least we do have the figures on which to base some sort of rational conclusion. Probably we would feel that the most reasonable conclusion (though *not* an "absolutely proven scientific fact") is that LSD damages chromosomes. But research should continue, and more data should be obtained; eventually a clearer picture should emerge. Note, however, that complete certainty will never be possible. Even if a person was found with a reciprocal score 100 standard deviations lower than the mean, he or she is not necessarily excluded from our Gaussian distribution, because its tails never *quite* reach the abscissa (which would denote a 0 probability). But such a person would be *extremely* rare in that distribution. *If* acid really breaks chromosomes, a user might well be 100 s away from the mean.

This kind of logic underlies all scientific research. The most important point is that we deal with probabilities, not certainties. This may

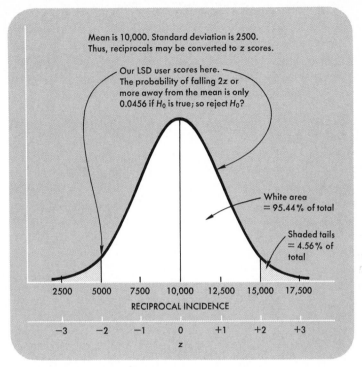

Figure 7.8
If the reciprocal of chromosome breakage incidence in the general population has a Gaussian distribution with mean 10,000 and standard deviation 2500, the distribution would look like this. The probability of a score falling more than $2z$ from the mean is only 0.0456. The shaded areas are 0.0456 of the total area. The score from a hypothetical LSD user falls $2z$ from the mean. The null hypothesis says that LSD makes no difference to chromosome breakage incidence. Should we decide that the null hypothesis is false?

be unpleasant to those who like to be definite, but we need a basis for translating probabilistic knowledge into positive action. We'll find this in the next chapter, but we've already made a good start here.[6]

Since in this book the more usual term "normal distribution" has been replaced by "Gaussian distribution" (to avoid the unwanted connotations of the word "normal"), it behooves the author to make you feel good about Gauss (remember, it rhymes with *house*). Carl Fried-

[6] If you want more information about actual research on the effects of LSD, see LeDain *et al.*, *Final Report of the Commission of Inquiry into the Non-medical Use of Drugs,* Ottawa: Information Canada, 1974, p. 382.

rich Gauss, one of the three or four greatest mathematicians of all time,[7] was born in 1777 and died in 1855 in Germany. He made almost all his basic mathematical discoveries while still a teenager and was noted for work in calculus, geometry, astronomy, magnetism, electricity, optics, crystallography, and mechanics—as well as statistics.

a direct use of Pascal's triangle

At this point in your patient pursuit of a subject matter so consistently theoretical, you should be rewarded with a useful application of it. Here is a statistical test that should be useful from now on for certain types of problems.

the binomial test

Imagine yourself a psychotherapist interested in a special technique of therapy for neurotics. We assume that half of the world's neurotics are male and the other half female and that there are equal numbers of both sexes in the world's population (not *quite* true, but close enough to it for our purposes). In the past year, you have treated many male and female patients, of whom only 10 showed complete recovery. Nine of them were men, one a woman. Question: does this constitute evidence that men are selectively more benefited than women by your therapy? As scientists like to phrase the question: assuming that men and women *are* the same in this respect, how likely would the occurrence of a 9:1 ratio by mere chance be? If this likelihood is too low, say less than 5%, we might well conclude that men *are* selectively benefited and direct our future efforts accordingly. The binomial test will tell us.

Remember that the hypothesis that there is actually no such difference between males and females is generally called the null hypothesis, or H_0 (H-zero, "hypothesis null"). We are seeking the exact probability, p, that our data could have happened by random accident, *given* that H_0 is really true. If this probability is unacceptably small, H_0 may be rejected in favor of the only alternative possible assumption, that males are preferentially more curable by this therapy.

In this book, and in all others dealing with fundamental statistics, you always begin with the *assumption* that H_0 is true. In our present

---

[7] The others were Archimedes, Newton, and perhaps John von Neumann.

example we therefore assume that the percentage of cured males should equal the percentage of cured females, except for random influences on a given bunch of data. The ratio *should* be 5:5; it *could* be 6:4 or 4:6; it *might* be 7:3 or 3:7; but the probability drops as the ratio gets increasingly unbalanced. At some point, you will exclaim in amazement, "I don't believe it! The probability of these data, given the null hypothesis, is just too small! This null hypothesis had better be rejected in favor of the hypothesis that men *are* selectively benefited by my therapy."

This is just like flipping 20 coins, believed to be ordinary legal tender. You would *expect* 10 heads and 10 tails; 9 of one and 11 of the other, or 8 of one and 12 of the other, would hardly surprise you. What about 20 of one, none of the other? Yes, it *could* happen, and once in a long while it *will* happen. (The binomial distribution for 20 cases tells you about how often.) But suppose it *does* actually happen: You might well prefer to believe that they are *not* customary coins after all, but rather two-headed offspring of the local Joke Shop. With coins, of course, you need only look at both sides to discover the truth. With the special technique of therapy for neurotics and many other questions, we must remain content with decisions based upon the probability of our data, given H_0. So let us find the probability of a 9:1 outcome of our neurosis therapy.

Now we see that 10 cases are involved altogether, so we symbolize this number as N. We also see that there are $N + 1$ *possible* outcomes, as follows:

> 0 males, 10 females cured
> 1 male, 9 females cured
> 2 males, 8 females cured
> 3 males, 7 females cured
> 4 males, 6 females cured
> 5 males, 5 females cured
> 6 males, 4 females cured
> 7 males, 3 females cured
> 8 males, 2 females cured
> 9 males, 1 female cured
> 10 males, 0 females cured

If we assume equal probability of cure between the two sexes, we have a picture similar to the Pascal triangle shown in Figure 7.3, on the lowest row, corresponding with $N + 1 = 11$ possibilities. There is just *one* way in which we could find 0 males cured: if all those cured were females. There are *ten* ways in which exactly one male might be found cured: any of the ten males could be the one. There are 45 ways in which exactly two males might be found cured, and so forth. Let's show the picture systematically.

Number of cured males	0	1	2	3	4	5	6	7	8	9	10
Number of ways in which this could happen	1	10	45	120	210	252	210	120	45	10	1

The same picture is shown in Figure 7.9, which (if necessary) you should try to relate to our earlier discussion of the quincunx. Adding all the ways in which all events could possibly occur, we have a total of 1024 ways, whereas only 10 + 1 of those ways would produce, by chance, *as many as* 9 cured males.[8] So there are only 11 ways out of 1024 of our data occurring by chance; this corresponds (recalling our definition of probability) to a probability of $\frac{11}{1024} = 0.0107$. We had decided to reject H_0 if our data were less than 0.05 probable, so we do reject it, replacing it with the conclusion that men *are* selectively helped.

That is basically the binomial test. It can also be used for more complicated problems, but these lie beyond our present level.[9]

scientific (exponential) notation

Occasionally it happens that very large or (as in this chapter) very small values must be expressed in decimal form. For example, the probability of getting ten tails in succession is

$$\frac{1}{2^{10}} \quad \text{or} \quad 0.0009765$$

When zeros between the decimal point and the first nonzero digit are very numerous, an alternative method of notation, called *scientific notation*, is often found. This is the use of exponents to denote how many times the "interesting" part of your answer (above, the digits

---

[8] If you have *ten* cured males, obviously you have *nine* cured males. We want the total number of ways of getting *at least* nine, so we must also count the ways of getting *more than* nine. That's why we add 10 and 1, the last two figures in the second line above.

[9] For further information, see Siegel, S., *Nonparametric Statistics*, New York: McGraw-Hill, 1956.

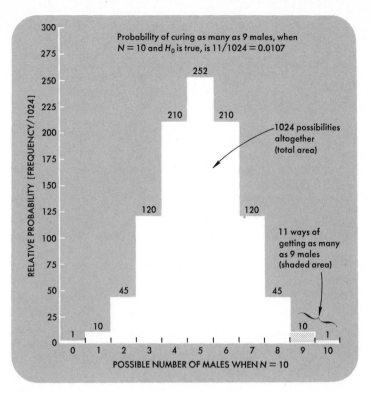

Figure 7.9
The binomial distribution of the probabilities that varying numbers of men will be cured by the special technique of therapy for neurotics, when $N = 10$ and we assume that men and women are equally curable. We see that the probability of nine or more men being found in a group of ten cured people is only 1.07%, or 0.0107. This probability is the same as the percentage that the shaded area is of the total. Should we decide that the null hypothesis (of no difference in curability of the sexes) is false? If we do so decide, the alternative hypothesis is that men are selectively more treatable than women.

9765) must be multiplied by 10 or by 0.10 (depending on whether the decimal point must be moved to the left or the right) to give it the correct value. To turn 0.0009765 into scientific notation, we could think of it as 0.9765 multiplied by 0.1 three times,

$$0.9765 \times 0.1 \times 0.1 \times 0.1 = 0.0009765$$

or as 9.765 multiplied by 0.1 four times,

$$9.765 \times 0.1 \times 0.1 \times 0.1 \times 0.1 = 0.0009765$$

or as 97.65 multiplied by 0.1 five times,

$$97.65 \times 0.1 \times 0.1 \times 0.1 \times 0.1 \times 0.1 = 0.0009765$$

or in many other different ways. The usual practice is to have one digit to the left of the decimal point, so here

$$p_{\text{ten tails}} = 9.765 \times 0.1 \times 0.1 \times 0.1 \times 0.1$$

The four terms on the right, all 0.1, equal

$$\frac{1}{10} \times \frac{1}{10} \times \frac{1}{10} \times \frac{1}{10} = \frac{1}{10,000} = \frac{1}{10^4}$$

Conventionally,

$$\frac{1}{10^4} \quad \text{is written as} \quad 10^{-4}$$

the negative exponent indicating the reciprocal with 1. So our value of 0.0009765 may be written as

$$9.765 \times 10^{-4}$$

The exponent -4 serves to shift the decimal point four places to the left. Were the exponent merely 4, it would serve to shift the decimal point four places toward the right. To express a very large number in scientific notation, a positive exponent would be used:

$$97,650 = 9.765 \times 10^4$$

It is possible that you will be encouraged by your instructor to use a slightly different notation that is now achieving popularity in the social sciences. Computer printouts use this style, because the printing machine has no superscripts and a long string of zeros is confusing. So you may find 9.765×10^{-4} printed as

$$9.765 \, D \, -04$$

for *D*ecimal 4 places to the $(-)$ left, or

$$9.765 \, -04$$

In both cases, the minus sign produces a negative (leftward) movement of the decimal point.

Similarly, 5432000000 might be printed as

$$5.432\,D\;09 \quad \text{or} \quad 5.432\;09.$$

Occasionally an E (for *exponent*) is found instead of a D (for *decimal.*) They mean the same thing. It is doubtful that you will be using computers right away, and when you do may be time enough for such an exotic way of writing numbers. However, it may help you gain new friendships and respect from those elitist kids in mathematics, physics, and chemistry. It should also be of some interest to know that some fantastically large numbers can be expressed economically by means of scientific notation. For example,

$$9^{\left(9^9\right)}$$

which is the number you get by raising 9 to the ninth power and then using the result (9^9) as the exponent for the lowest 9, is an almost inconceivably large number. Don't try to work it out; don't even challenge your calculator (or persecute it) with the computation: the answer has 369,693,100 *digits*. If you were to write the answer down, at one digit per second, it would take you 11 years and 8 months—without any break—to finish.[10] And it is probable that fewer than 10^{100} atoms exist, in total, in the universe.

Let's face it: The following are terms you can't get along without.

TWO-VALUED LOGIC

NULL HYPOTHESIS

DISTRIBUTION OF PROBABILITIES

THEORETICAL PROBABILITY

EMPIRICAL PROBABILITY

SUBJECTIVE PROBABILITY

THE ADDITION RULE FOR "OR"

THE MULTIPLICATION RULE FOR "AND"

MUTUALLY EXCLUSIVE . . .

    AND EXHAUSTIVE EVENTS

MUTUAL UNRELATEDNESS

---

[10] This and many other examples of fun with numbers may be found in a delightful book by A. Bakst: *Mathematics: Its Magic and Mastery* (3rd edition), New York: Van Nostrand Reinhold, 1967.

Vocabulary (*continued*)
QUINCUNX (yes, really!)
PASCAL TRIANGLE
BINOMIAL DISTRIBUTION
BINOMIAL EXPANSION
GAUSSIAN DISTRIBUTION
PLATYKURTIC DISTRIBUTION
LEPTOKURTIC DISTRIBUTION
BINOMIAL TEST

exercises for a rainy chemistry lab

1 It's time you were shown a little mercy. Get some heavy shirt cardboard and construct several cutouts based on Figure 7.7. Draw vertical lines at scores $\pm 1, \pm 2$, and ± 3, and then cut along them. Mystify your friends in chemistry lab by weighing the various pieces (nod wisely as you do so) and one entire graph. Assuming constant weight and thickness of cardboard, the weight ratios should correspond with area ratios. By this means, you can verify experimentally the figures shown in Fig. 7.7.

2 This one you can almost do in your head. Cut out of paper a fairly large normal distribution graph and pin it to the wall of the garage. Mark the mean and the various standard deviation points. Now stand back 12 feet, close your eyes, and throw darts at it. Get your friends to do the same. Eventually it will be full of dart holes. (If it isn't, try again with a much bigger graph.) Pay your kid brother 50¢ to count the total number of holes in the entire cutout and the total number between each adjacent pair of perpendiculars rising from the 2 scores. The ratios of the latter against the total *should* approximate the figures shown in Figure 7.7.

3 Do the same thing again, only with many more vertical lines evenly spaced across. (If in a hurry, simply draw more lines on the target already used for Exercise 2.) Count the holes in each vertical column and make a frequency polygon of the totals, using the same abscissa as on the target cutout. The polygon should look *something* like the Gaussian curve you were bombarding. (Why?)

   This exercise will be found cathartic as well as instructive and is good general exercise too.

self-test on chapter 7

Think of it this way: every time you reach one of these unspeakable Self-Tests, you have finished one more chapter in this (far from infinitely long) book.

Any number of answers may be correct.

1 If 10 coins were to be flipped and then examined for proportion of heads to tails,

   a 5 of them will show heads.

   b there is only about one chance in 1000 that all would show tails.

   c the probability of getting the exact sequence H T H H T T H T H T is 0.0009765.

   d there would be about a 25% chance of getting 5 heads and 5 tails.

   e any single outcome is mutually exclusive and exhaustive.

   f the multiplication rule is appropriate for calculating the probability of any one sequence.

2 In ascertaining natural truths in the sciences,

   a our practical idea is to state them as absolute certainties.

   b we must recognize that we can never be certain.

   c our strategy is to state them in probabilistic terms.

   d we accept predetermined probabilities of being mistaken in rejecting the null hypothesis, if we do so.

   e we usually decide upon a 50% chance of error.

   f we tend to lean over backward to avoid drawing a conclusion prematurely.

   g we leave a qualitative gap between fact and conclusions.

SCORING AND ANALYSIS of the multiple-choice questions. Award yourself 3 points and one smirk for each correctly circled choice. Subtract 2 points for each wrong one, and subtract 1 point for each correct choice you missed.

You can't win, can you? If you choose to ignore this ghastly business, you will have a negative score!

Answers to Question 1

a F   Five *might* be heads, but then again, who knows??

b T   That's true. Once chance in 1024, or $(\frac{1}{2})^{10}$.

c T   This is an interesting one. The sequence given looks marvelously random, hence very probable in a game of this sort. But the probability of *any given particular sequence* is still 1 in 1024 or 0.0009765.

d T   True. Here the particular sequence isn't specified. Turn to Figure 7.3; note that the result of 5 heads and 5 tails should be expected 252 times out of 1024.

e T   True. For the individual coin tosses, you get either a head or a tail. For the sequence, that's the one you got, other possible sequences being ruled out by the event.

f T   Because you want the probability of (say) H *and* T *and* T *and* . . . *and* . . ., which are independent events. (The multiplication rule, remember?)

Answers to Question 2

a F   It's an ideal, but hardly a practical one. Absolute terms, absolute cer-
b T   tainty, no. We can never be absolutely certain about any facts other
c T   then formal or mathematical ones. However, it helps if we know how to discover and state our degree of confidence in any conclusion we might make that the null hypothesis is incorrect.

d T   The predetermined chance that we accept of "discovering" something that really isn't there is usually either 5% or 1%. Occasionally other values are selected.

e F   Probabilities of 5% and 1% are a long way from 50%. To draw a general
f T   conclusion when we have a 50% chance of being wrong is really asking for confusion. So we become very conservative and hard to convince; the evidence really has to shout for us to accept it.

g T   It's an interesting thought, and a fairly deep philosophical problem as well. God *knows,* but human beings can only *decide* (with the help of statistics) and act *as though* they know the real facts.

3 In a class of 300 students, an experiment on ESP (extra-sensory perception) is performed. The instructor opens a textbook at random and looks at the page number. He concentrates first on the first digit and asks the class to write it down. He then concentrates on the second digit, and they write that one down too. The same with the third digit.

    a The probability of getting the entire page number correct is so small that if any student succeeded in doing so, it was probably ESP.

    b About 30 students should be expected to be correct on each single digit.

    c If it could be argued (as it often is) that ESP is an ability that comes and goes unpredictably, this experiment would *appear* to bring out considerable evidence for ESP.

    d The null hypothesis in this experiment should be that nobody will guess any number correctly.

4 In the spaces to the left, *lightly* pencil in the probabilities associated with the situations given to the right. When you have scored yourself (answers follow), erase your answers. Score 2 points for each correct answer.

    a _____ The probability of drawing *first* a heart and *then* a spade from a 52-card deck.

    b _____ The probability p of drawing one heart and one spade (in any order) in two draws from a 52-card deck.

    c _____ The probability p of drawing one heart *or* a face card from a 52-card deck.

    d _____ The probability p, in a game of 5-card poker with four players, of *somebody* holding at least one joker. (For the innocents among you, in poker there are sometimes two extra cards, called jokers, making a 54-card deck.) Each player holds 5 cards, face down. Include yourself among the probabilities if you are vicariously indulging in this hypothetical game.

Answers to Question 3

a F   The p is small—no doubt about that. There are 10 possible digits, giv-
b T   ing a $\frac{1}{10}$ chance of guessing any particular one correctly. With 300 students, each digit should be guessed correctly by one tenth, or 30, of
them. But not the same 30 people each time; the probability of *one* person guessing all three digits is $\frac{1}{10}^3 = 0.001$, or one chance in a thousand.
With 300 students, though, there is a $\frac{300}{1000} = 0.3$ chance that at least one
person will guess all three correctly. That is not a terribly low probability, since the $\frac{300}{1000}$ ratio is merely the center point of a binomial distribution of possible outcomes.

c T   Believing that ESP comes and goes unpredictably, one could (and probably would) conclude that correct guesses are due to its presence and
that incorrect guesses are due to its temporary absence.
It would be impossible for science to make a general conclusion about
ESP (or anything else) that came and went unpredictably; the predictable *is* the realm of science.

d F   No, the null hypothesis should be that chance *alone* is enough to explain any digit being correctly guessed.

Answers to Question 4

a 0.064 $p_{\text{heart}} = \frac{13}{52} = 0.25$, and $p_{\text{spade after heart}} = \frac{13}{51} = 0.255$. The probability of a
heart *and* a spade is then obtained by multiplying these: $0.25 \times 0.255 = 0.064$.

b 0.127 $p_{\text{heart or spade}} = \frac{26}{52} = 0.50$, and p of the remaining card is $\frac{13}{51} = 0.255$. Multiplying (because the two events are unrelated), we have $0.5 \times 0.255 = 0.127$.

c 0.043 $p_{\text{heart}} = 0.25, p_{\text{face card}} = \frac{12}{52}$, and $p_{\text{heart and face card}} = \frac{33}{52} = 0.0577$. To get the
probability of a heart *or* a face card, we compute joint $p = 0.25 + 0.2307 - 0.0577 = 0.423$.

d 0.608   This probability is best calculated indirectly. Twenty of the 54 cards
are dealt, so there is a $\frac{34}{54}$ chance that nobody has the first joker and a $\frac{33}{53}$
chance that nobody has the second one. The probability that *nobody*
has either of them is therefore
$$\frac{34 \times 33}{54 \times 53} = 0.392$$
Therefore, the probability that *somebody* has either of them is
$$1.000 - 0.392 = 0.608$$

e _____ The probability p, in a ten-level quincunx, of a ball bouncing left ten times in a row, to wind up in the far left stall.

f _____ The probability p, in the ten-level quincunx, of three balls in succession bouncing only left, to wind up in the far left stall.

g _____ The probability p of finding any score beyond $+3z$ from the mean of a Gaussian distribution.

The maximum possible score is 50. Since most of the choices were correct this time, it was hard to get too much of a negative score—unless, of course, you ignored the entire scene.

> 45 to 50   Great!
> 35 to 44   Not quite so great
> 25 to 34   Award yourself a C
> under 25   Do any combination of
> - this self-test in the first place
> - the chapter over again, if not in the first place.

e $\frac{1}{2^{10}}$ The probability can also be expressed as

$$\frac{1}{2^{10}} \text{ or } \frac{1}{1024} \text{ or}$$

$$0.0009765 \text{ or } 9.765 \times 10^{-4}$$

f This p is fantastically low. The multiplicative rule applies, so
9.3×10^{-10} $p = 0.0009765 \times 0.0009765 \times 0.0009765 = 0.00000000093$
better given as 9.3×10^{-10}.

g 0.0015 This question is answerable from what is known of the shape of any Gaussian distribution. Since 99.74% of cases lie within $\pm 3z$, only 0.26%, or 0.0026 of them can fall beyond three standard deviations from the mean in either tail. Since the Gaussian distribution is symmetrical, this 0.0026 is split equally between the two tails, giving 0.0013 *less than* $-3z$, and 0.0013 *more than* $+3z$.

*s*ome ordinary problem*s*

1 There was once a clever student who kept offering for sale a penny that (he said) had the remarkable ability to come up heads constantly. He offered, in fact demanded, to prove his claim by demonstrating the coin to the prospective buyer, stipulating only that the prospect had to buy the coin (for $20.00) once it was proven to be consistent, that is, to come up heads constantly. Most prospects decided that ten times in a row would be sufficient evidence. It was actually a fair coin. How did this clever swindle work?

2 List six everyday instances of two-valued logic improperly used.

3 Imagine a multilevel quincunx made up of doweling set in a sloping panel. Then imagine that some sneaky friend has shifted an entire row of pegs to the left just a small and invisible distance. What effect will this have upon the distribution of balls at the bottom?

4 The two left-most scores in a perfect binomial distribution add to 25. How many levels are there in the Pascal triangle of which this distribution is the bottom level?

5 Think of six everyday instances in which empirical probabilities can be known exactly, but in which subjective probabilities are very different. Do these instances have any common characteristics?

6 Find the exact probabilities of the following.

  a Getting the sequence H T T T H H T H H T with 10 coins.

  b Getting 5 heads and 5 tails when flipping 10 coins.

  c Getting exactly 3 heads among the first 5 coins flipped and exactly 3 tails among the last 5 flipped.

7 A class of surveying students goes on a field trip. They are directed each to measure independently the distance between two survey markers previously placed in position. Assume that there are 100 such students per year, that each makes ten measures, and that the results from ten years of the same procedure are all pooled.

  a Would all scores be the same? If not, why not?

  b What kind of score distribution would occur? Sketch it.

  c What are the characteristics of the distribution?

  d Mention three other general examples of ways in which such a distribution could be obtained. What do these examples have in common?

8 From a class of 20 students, an experimenter wishes to draw a sample of 5 at random.

  a What is any one student's probability of being drawn?

  b If the class contains only 5 women, what is the probability that the sample would contain all of them?

  c Suppose that any subject, once selected and experimented on, is then returned to the class again, eligible to be sampled again any number of times. What is the probability that the sample consists only of women?

  d Explain the difference between the probability found in (b) and that found in (c).

9 As everybody knows, leprechauns never part with the gold they bring in; they simply put it in a safe place called Faurt Gnaucks. Every spring each leprechaun has to confess the amount found the previous year (this custom is found elsewhere too). The mean amount reported last year was 500 beqa,[11] and the frequency distribution was Gaussian, with standard deviation 50 beqa.

  a If you were a leprechaun, how likely would you be to report more than 750 beqa?

  b What proportion of leprechauns would have reported finding between 500 and 550 beqa?

---

[11] The *beqa* is the world's most ancient unit for measuring the weight of gold.

    c What proportion of leprechauns would have found between 400 and 450 beqa?

    d What proportion of leprechauns would have found either less than 400 or more than 600 beqa?

    e Casey O'Grady, a leprechaun from the county of Tipperary, reported 850 beqa. Muldoon O'Sullivan, from Ballyporeen in Tipperary, reported 875 beqa. So did Francis McDuffy, who hails from Kilsheelah in Tipperary. There are two possible interpretations of their large hauls. What are they?

    f What would you do to find out which alternative explanation was the more viable one?

    g What name do we give to the hypothesis that Tipperary leprechauns do not find gold any more efficiently than other leprechauns?

10 A very improbable event, say a 1:1,000,000 long shot, will not necessarily occur only after 999,999 nonoccurrences. It is just as likely to be the first occurrence as any other. What implications does this have?

11 You are standing on the street corner with a friend, watching the world go by. The friend turns to you and says, "You know, a most unusual thing just happened. The car that just went by had the license number XYZ 123. Now the probability of that happening was very low, yet it happened. So that means that things with very low probabilities *are happening all the time,* which means that they couldn't really all be of low probability!"
Interesting?? Comment to your philosophical friend.

12 In our example on neurosis therapy, what would be the probability of finding 3 males and 7 females cured, if the null hypothesis is true?

chapter 8
samples and populations

In the first six chapters of this book, we concentrated only on *describing* the characteristics of data. So we know how to calculate the mean, standard deviation, and other describers. But you may ask, why should we want to? There are several reasons.

For example, imagine that you have become interested in the LSD question of Chapter 7 and, putting first things first, want to find out what the normal incidence of chromosome breakage is. How do you proceed? The obvious way is to assemble as large a group of "normal" people as possible, find the incidence of chromosome breakage for each person, and then find the mean and standard deviation of the reciprocal scores; and there you are.

But there is a serious problem with this approach. Exploring even *one* person's chromosomes is extremely laborious, so your "normal" group will have to be quite small. Whether it contains 3 people, 30 people, or 300 people, it's still quite small in comparison with the entire human population. And we want our "normal incidence" figures to refer to "normal people" in general, not just to the few people we actually manage to test.

In short, the difficulty is that while we can easily collect *samples* and find their means and standard deviations, *populations,* being much larger than samples, may be impossible to test in their entirety. And yet it is the *population* mean and standard deviation that we want to know. In the LSD example of Chapter 7 we assumed that these values were 10,000 and 2500, but how can one actually *know* these values? We can find out anything we like about a sample, but it is a whole population that usually interests us. Can we make inferences from samples to a population? If so, how safely, and within what boundaries of error? In this chapter we answer these questions. But first, a few definitions.

population

There are basically two kinds of population, with an essential idea in common. There is the human, or **demographic**, population, of the U.S., of China, of the world, and so on. Its most important feature is that it's quite large—limitless in size, actually, if we include people who are as yet unborn. (This would make sense if, for example, we wanted to talk about the effects of nuclear radiation upon "the human population." Obviously, if we contrive to make our world radioactive, it will affect people who aren't here yet.) But demographic populations can also be finite in number: for example, the present population of your home town. A demographic population is any group of people (or animals, insects, etc.) with some common characteristic such as living in the same town, or in China, or on this planet. Its size and characteristics depend on how you define it, which in turn depends on your research goals.

A **statistical population**, on the other hand, is a population of scores, not people. The scores may come from people, and in the behavioral sciences they generally do. Statistical populations are very frequently infinite in size, because there may be no limit to the number of collectible scores, but depending on your purposes, they may be finite in size too (when, for instance, you wish to consider the scores from a finite, here-and-now group of people, and not the world population in general). The essential feature of the statistical population is that it includes *all* scores that can possibly come from applying a defined measure in a defined way. Obviously, the scores must be numerical if we are to compute their mean, standard deviation, and so on. In our LSD example, the human population becomes a statistical one when each person is matched with his or her incidence of chromosome breakage.

sample

A sample is simply a selection of people from a human population or a selection of scores from a statistical one. Samples are usually much smaller than their populations, certainly finite in size, so we can easily calculate their means, standard deviations, etc. The procedures learned in the first six chapters are applicable only to samples rather than infinitely large populations. It is possible to conceive of the mean score (height, perhaps) of a population of several billion people, but measuring all those heights and computing their mean is an almost inconceivably difficult task. And for a truly infinite population, the job is impossible.

We can think of four types of sample: casual, biased, representative, and random.

casual samples

A **casual sample** is what it sounds like: one collected casually, with no attention given to randomness or representativeness. For example, a newspaper reporter who wants to find out what "the people" think of the latest scandal in City Hall may simply go out on the street and question the first six or eight people to come along. There is very little chance of such a casual sample accurately "speaking for the majority," because it is not selected to be an accurate cross-section of the population. Indeed, casual samples are often selected in line with the preconceptions of the sampler; not surprisingly, the resulting "evidence" is often in line with these preconceptions too. Except to be on guard against it, let's put casual sampling behind us.

biased samples

A **biased sample** is one collected by fair means or foul, so that its statistical characteristics will differ from those of the population. Obviously, casual samples are quite often biased in some way, as when a newspaper reporter collects a street sample when most people are at work, thus biasing it in favor of nonworking shoppers, outside workers, and the unemployed. But representative and random samples can, as we shall see, also be biased. Bias is not necessarily a bad thing, and sometimes it is inevitable; but we must be careful in generalizing from a biased sample to a population and take steps to remove the bias when necessary.

representative samples

A **representative sample** is one whose scores have the same mean, standard deviation, and distribution shape as those of the population from which it was taken. There are several methods used to make samples representative, but they need not concern us here. For the purposes of statistical methodology, the notion of a representative sample is not essential.

random samples

A **random sample** is one selected so that all members of the population have the same probability of being selected. Such a sample may or may not be representative. The definition applies to the procedure used, not to any characteristic of the sample itself, once it is selected.

Just how *do* you arrange things so that all members of a huge population have the same probability of being included in your sample?

Think about it. Too often the term "random sample" is used glibly but inaccurately, just because it sounds good. Whether you are sampling people or scores, it turns out to be incredibly difficult; you should be commensurately cautious about declaring a sample "random." With people, you are usually restricted to a small geographic location, and always to a short time span. Your sample will, of necessity omit the unfriendly, the uncooperative, and those who don't show up for their appointments—maybe even those who don't look ordinary enough to you, the sampler, for your "random" sample.

Sampling from statistical populations has similar problems; the scores obviously must come from somewhere, and all potential sources may not be equally accessible to you. This problem is serious in the social sciences, less so in the biological sciences, and nonexistent in the rarefied atmosphere of pure mathematical statistics, which deals with idealized (not real-world) samples, populations, and scores. Fortunately, it is possible, in the social sciences, to overcome the problem well enough to permit use of the mathematical abstractions. As we working people will see, such abstractions have a real payoff in statistical inference; they provide a way to draw conclusions regarding populations from facts about samples.

statistics and parameters

For every measure that can be applied to a sample, there is an equivalent one pertaining to the population. The sample has a mean value; so, obviously, does the population. The two values need not be the same (though they might be fairly close). But since populations are usually impossible to measure because they are so large (and we can never know the mean or standard deviation of an infinitely large population), we usually know these values only for samples.

A statistic (note that *this* is the only correct use of the singular form) is a measure, such as the mean or standard deviation, made on a sample. A parameter is an equivalent measure that exists in a population.

terminology

We may easily get into a tangle, now that we have two kinds of mean (statistic and parameter), two kinds of standard deviation, and so on. The easiest way to keep them straight is to adopt symbols to stand for and distinguish between them and to use those symbols consistently There are several different symbolic systems in use, (which creates some problems for those who must translate from one system to another), but we shall adopt the one most commonly found in psychological statistics.

The key to our system is that statistics, sample values, are shown in letters of the Roman alphabet (such as $\bar{X}$ for the mean and s for the standard deviation), while parameters, population values, are represented by Greek letters. The most important ones are μ, *mu* (pronounced "moo" or "mew," the Greek lower case m, for *m*ean), and σ, *sigma* (the Greek lower case s, for *s*tandard deviation.) We use the name "sigma" to refer to σ, not to our old friend Σ, the summation sign.

estimating parameters from statistics

We would like, of course, to *know* the values of parameters for various purposes. But unless the population is finite and totally accessible, it is impossible to know them directly. Even when it is possible, it would be tremendously difficult to look at every potential score in a population. However, it is quite easy to *estimate* a parameter from what is known about a statistic.

estimating μ from $\bar{X}$

If $\bar{X}$ has been obtained from a *random* sample, the value of $\bar{X}$ is a *best estimate* of μ. The term **best estimate** does not necessarily mean an *accurate* estimate; it is simply the value with the greatest probability (given your data) of being μ, and having (again given your data) equal probabilities of being less than the real value of μ and of being greater than μ. If your sample is really random, the value of $\bar{X}$ will differ from μ only by chance. Another random sample would probably yield a different $\bar{X}$, another "best estimate." Indeed, by taking many random samples and determining a mean $\bar{X}$ for each, we could create a *distribution of means,* which would also be a distribution of best estimates. We will discuss this distribution soon, but note for now that the mean of a randomly obtained sample is an unbiased best estimate of the population mean.

We need more terminology already. Whereas $\bar{X}$ indicates a *sample*

mean and μ indicates the actual (but unknowable) population parameter, we lack a symbol for a "best estimate" as such. We use a **circumflex** over the parameter symbol, where necessary, to show that its nu-
mercial value is neither a statistic nor the actual parameter, but an
unbiased estimate of the parameter. If $\bar{X}$, a sample mean, is our best
estimate, we can write

$$\bar{X} = \hat{\mu} \cong \mu$$

$\bar{X}$ is what we actually found for a random sample, $\hat{\mu}$ is what we estimate for the population, and μ is what is really there in the population.

estimating σ^2 from s^2

If the value of s^2 has been taken from a random sample, it is a valid
starting point for estimating σ^2, but it is a **biased** estimate. The value
of s^2 will tend to be smaller than the value of σ^2. Figure 8.1 illustrates
why. The value of s^2 is calculated from $\bar{X}$, which was obtained from the
sample and which is probably not identical with μ. Thus, μ will not be
quite in the center of the sample of scores, therefore, the sum of squared
deviations of scores from μ will be greater than the sum of squared
deviations from $\bar{X}$. A moment's thought, or very simple do-it-yourself
calculation, will disclose that the sum of squared deviations is smaller
from $\bar{X}$ than from any other value. If the sample size is increased, the
accuracy of $\bar{X}$ as an estimate of μ improves, so the bias is reduced; the
amount of bias is tied to the size of n.

Unfortunately, we never know the value of μ in real-world situations, so we must correct the variance size in some way not involving
knowledge of μ itself. Fortunately, though, this is very easy to do.

The bias is removed by increasing the value of $\hat{\sigma}^2$ (the best estimate
of σ^2) so that it becomes somewhat greater than s^2. Clearly the amount
of increase must be based upon the size of the sample, since a greater
increase is needed for smaller samples. The correction is quite simple:

$$\sigma^2 \cong \hat{\sigma}^2 = s^2 \frac{n}{n-1} = \frac{\sum (X - \bar{X})^2}{n-1}$$

The correction can also be built into the formula for the standard deviation merely by replacing the denominator n by $n - 1$, which gives us

$$\hat{\sigma} = \sqrt{\frac{\sum (X - \bar{X})^2}{n-1}}$$

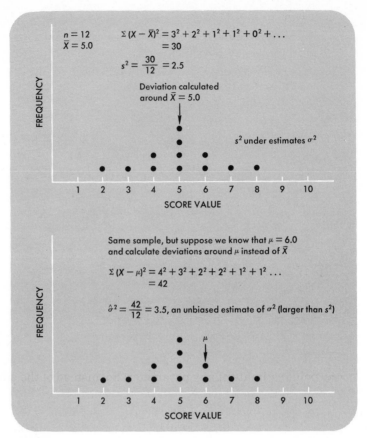

Figure 8.1

A sample variance (s^2) is an underestimate of population variance (σ^2). This illustration shows why. The value of $\overline{X}$ is at a minimum total distance from its scores, whereas μ, being different from $\overline{X}$, is not. Note here that the values of s^2 and $\hat{\sigma}^2$ work out a little differently. Remember that $\hat{\sigma}^2$ is merely an *unbiased* estimate of σ^2, not *necessarily* an accurate estimate!

This is the "other kind" of standard deviation mentioned in Chapter 5. It is the best estimate of the population standard deviation.

We can now see how the LSD-chromosome research in the previous chapter managed to know the healthy population incidence of chromosomal breakage, whose mean reciprocal value was 10,000 and whose standard deviation σ was 2500. Perhaps a random sample was taken of 10 people, and chromosome breakage scores found for each person. With the scores transformed to their reciprocals, the mean value turned out to be 10,000 and the value of s turned out to be 2372. Work this one out yourself. When $n = 10$ and the sample is random, an s of 2372 becomes a $\hat{\sigma}$ of 2500. Repeated sampling and the same procedure might then yield a mean (of all the sample means) of 10,000; and a mean standard deviation of 2500. This way we would be sure. But in practice, usually only *one* sample is taken.

a small question of accuracy

Perhaps by now you are screaming silently (or even loudly) that a best estimate is *only* an estimate and probably does not exactly match the parameter, so how can we justify behaving as though it does?

It's good that you asked. To begin with, a **point estimate** of a parameter, as we call an estimate coming from just one sample statistic, is merely one point on a continuum of possible estimates; this is because if different random samples were taken, the different values of $\bar{X}$ (or s) obtained would occupy different points on the continuum. We don't know that our *one* obtained statistic is correct; indeed, we don't *expect* it to be exactly correct. The same comment could be made regarding *any* point estimate of any parameter. So, instead of the single estimate, we need something like a *zone* along the continuum, within which we are rather confident (willing to state odds) that the real value of μ lies (or the true value of σ, except that such a zone for σ is not very important for our purposes).

We will consider only such a "zone of possibility" for the mean. In order to do so, we must return to our friend the Gaussian distribution.

Let's imagine that instead of taking just one random sample from a population and calculating its mean, $\bar{X}$, we took a very large number of samples (ideally an infinite number of them), all of the same size, n. If we take (or imagine taking) *all possible* samples of size n from our population, and calculate the mean of each sample, we can make a frequency distribution of all those means. This distribution is called the **sampling distribution of the mean**, and we would find that it was extremely close to being Gaussian in shape. This happens because the values of $\bar{X}$, being randomly sampled, but from the same population, differ from each other purely on the basis of chance, and the Gaussian distribution is the result of chance operating in an infinitely long and wide quincunx. Thus, it is not surprising that the sampling distribution of $\bar{X}$ should fall into an almost Gaussian distribution. The closeness of approximation is really good if there are many scores in each of the samples, not so good if the samples are too small. As a rough rule of thumb, samples of 30 or more will have means distributed very close to the Gaussian curve. This fact is referred to as the **Central Limit Theorem**.

Make a mental note to distinguish between the population **distribution of scores** (which can be any shape, not necessarily Gaussian) and the **sampling distribution** of all the values of $\overline{X}$ that can possibly come from the population. The latter will be spread out far less than the former, because mean values tend to stay clustered in the middle of the range of raw scores.

In case you're turning a bit brown at the edges, remember that we are doing all this in imagination only. Don't let the thought of actually *having* to calculate an infinite number of $\overline{X}$ values from infinitely many samples inhibit your imagination. Nobody has to do it for real, as we will now discover.

The mean value of the Gaussian distribution of means will be identical with μ (unless few sample mean values went into it—but we assumed a very large or infinite number of samples). In fact, no matter what the size of each sample is, the mean value of all possible means (which we can call $\mu_{\overline{X}}$) is identical to the mean of the population from which those mean values came. In practice, $\mu_{\overline{X}} \cong \mu$ even with very few values of $\overline{X}$, assuming their samples were random.

The standard deviation of the Gaussian distribution of sample means, $\sigma_{\overline{X}}$, is called the **standard error of the mean**. Its value is directly related to the value of σ, the standard deviation of the original population. The relationship is described by the equation

$$\sigma_{\overline{X}} = \frac{\sigma}{\sqrt{n}}$$

From this we see that, as we should expect, the standard deviation of the means is much smaller than the standard deviation of the raw scores, because means are not as spread out as the scores they represent. The equation above makes it clear that the larger the value of n (the number of scores in each of the samples whose means form the distribution), the smaller $\sigma_{\overline{X}}$ will be.

The parametric value $\sigma/\sqrt{n}$ is closely approximated by $s/\sqrt{n-1}$, which we obtain easily from just *one* random sample. We now have all we need to establish (with a known probability of being correct) the *range* of values within which μ is located. From just one random sample! Press on.

confidence intervals

Let us decide that we want a 99.74% probability that μ will fall within the particular range of values we will eventually find. To begin with, we know that 99.74% of the scores in a Gaussian distribution will be within 3 standard deviations of the mean, that is, less than $3z$ away from the mean in either direction. Now, we know that the sampling distribution of means is Gaussian, or nearly enough so; we also know what the standard deviation of that Gaussian distribution is: the standard error of the mean! And we can easily estimate the standard error of the mean as $s/\sqrt{n-1}$. We know then that *if* our one obtained sample mean is one of the 99.74% of those that lie within $\pm 3z$ of μ, then μ cannot be more than $3z$ away from our obtained $\bar{X}$ in either direction. So we multiply our value of z by 3 and mark on our raw-score abscissa points corresponding to $\pm 3z$ away from the sample mean, $\bar{X}$. We are now 99.74% confident that μ falls within these limits and (correspondingly) only 0.66% fearful that μ falls outside them.

an example worked out

Our researcher on LSD and chromosomes works very hard, and winds up with a random sample of 50 normal people. Incidence of chromosome breakage is determined for each, and each score is converted to its reciprocal. The mean reciprocal is 9570, and the value of s is 2475. The standard error of the mean is estimated as

$$\hat{\sigma}_{\bar{X}} = \frac{s}{\sqrt{n-1}} = \frac{2475}{7} = 353.6$$

$3(353.6) = 1060.8$, so the researcher is 99.74% sure that μ falls somewhere between 8509.2 and 10,630.8. Now this may not look very impressive at first. But consider that the real value of μ is inherently unknown and unknowable. To locate it within a given range, with a known degree of accuracy, is something of a triumph. Far better than guessing, and infinitely better than confessing ignorance and giving up. Figure 8.2 may help you visualize the concept of confidence intervals.

The "level of confidence" associated with the interval to be found can be set at will. If you were content with a 68.26% chance of μ lying within your interval, the interval would be much narrower: between $\bar{X} + z$ and $\bar{X} - z$.

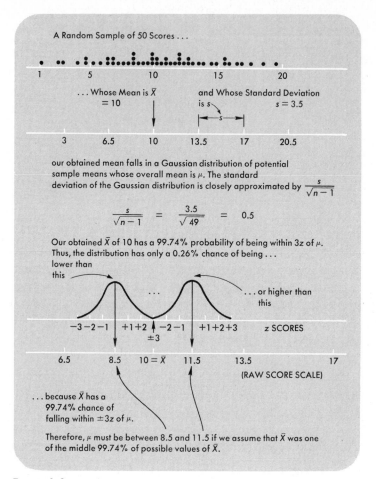

Figure 8.2
A random sample of 50 scores from a population. We know the value of $\bar{X}$ and s. Where is μ?

You can work in either of two ways:

1 decide initially on the *probability* that you want of having μ somewhere in the interval you will find, and then find the interval, or

2 decide on the *interval* that you want, and then find the probability that the interval includes μ.

In order to take either of the choices, you'll require the information given in the Appendix Table of areas under the Gaussian curve.

a couple of cautions and a bonus

Bear this in mind: the entire argument of this chapter assumes random sampling, perfectly random sampling from a population that may not be perfectly accessible. Perfectly random sampling is generally an ideal rather than a practical objective. But you can still try for it; you must, in fact. Remember too that we are dealing in *probabilities*; as we saw in the preceding chapter, a probability can let you down unexpectedly. Moreover, we have been assuming that our Gaussian model is a perfect one, but sometimes we are actually stuck with only a close approximation. In the light of these observations, we find that the confidence intervals that we obtain have slightly fuzzy borders. This does not matter in practice, because we usually insist upon a very high probability (95% or 99%) for our confidence intervals; if, invisibly, they are really 93% or 98% confidence limits, it still does not matter in a practical sense. We can still make a strong and rational statement about where μ is likely to be located.

These problems are the reasons why scientists annoy lay people by refusing to make absolute statements of fact. They deal in probabilities that are closely (but not perfectly) known, and they are especially unwilling to make sweeping generalizations from just a few samples or experiments. Think of all the "scientific breakthroughs" touted briefly by newspapers but never heard of again; a reporter who lacks the scientific spirit of caution is usually responsible for such quick splashes.

The bonus pertains to the Central Limit Theorem (which states that an infinite number of randomly sampled means will fall into a Gaussian distribution). The shape of the population of scores does not matter; the sampling distribution of means will be Gaussian as long as a reasonable number of scores, n, is included in each of the hypothetical (or actual) samples whose means make up the distribution. And if the population of scores happens to be approximately Gaussian itself, the sampling distribution of $\bar{X}$ will be close to being Gaussian even if n is quite small. There is an important payoff from all this: since a great many of the populations in which social scientists are interested have bell-shaped, roughly Gaussian distributions, we can tell a lot about these populations on the basis of small samples. Even if the populations distributions are skewed, as is the incidence of chromosome breakage, the scores can often be transformed to form a more nearly Gaussian distribution. Thus, useful and valid research can occur with quite small samples.

With the knowledge now at hand, we are able to answer certain types of research questions with no need of any nail chewing whatsoever. Let's work an example to see how.

the z test

In survey-type research, it frequently happens that the parameters for a population are known with reasonable precision, and the question arises whether a certain sample belongs to that population. For example, we know that the white population has a mean IQ $\mu = 100$, and standard deviation $\sigma = 15$. Suppose that in a random (urban and rural) sample of 10,000 black children we find a mean performance IQ of $\bar{X} = 101.5$, and a mean verbal IQ of $\bar{X} = 98.0$ Does this mean that blacks are, on the average, less intelligent than whites? Or more intelligent? Not at all. It may mean, however, that blacks form a *different* population from whites, more capable in some respects (such as "performance" skills) than whites and less capable in other respects, (such as verbal skills).[1] But on what grounds and by what method can we decide whether the blacks do or do not share the white parameters on the (white-oriented) intelligence measures?

The z test is appropriate for answering such questions, in which the parameters and a sample mean are both known. The sample should be as large as possible, in order to maximize the sensitivity of our test, but any size of *random* sample is legitimate.

The z test rationale is simple. Recall that the sampling distribution of $\bar{X}$ around μ is Gaussian and that the standard error of the mean, $\sigma_{\bar{x}}$, is equal to $\hat{\sigma}/\sqrt{n}$ (where n is the number of scores in each random sample and $\hat{\sigma}$ is known either directly or by inference from s). Recall too that only 0.26% of those $\bar{X}$'s will fall further than $3\sigma_{\bar{x}}$ away from μ. All we need to do is determine the standard error of the mean, $\sigma_{\bar{x}}$, for means of samples of size $n = 10,000$ and compare it with our obtained sample mean. If the latter is too far away from μ to be plausible, we have grounds for rejecting the null hypothesis and retaining instead the alternative hypothesis. In our example, H_0 is that blacks and whites have the same parametric values for verbal and performance IQ, the sample statistics being discrepant only by chance. The alternative hypothesis is that the black and white populations have different parameters. To accept the latter hypothesis would be equivalent to

---

[1] Performance skills are such things as picture recognition, perceptual-motor integration, and dexterity. Verbal skills involve facility with language.

making a general statement about a population on the basis of a sample, the objective of statistical inference.

Let us try to reach a decision about verbal IQ.

For samples of 10,000, and for σ of 15, we can readily see that the standard error of the mean is

$$\sigma_{\bar{X}} = \frac{\sigma}{\sqrt{n}} \cong \frac{\acute{\sigma}}{\sqrt{n}}$$

$$= \frac{15}{\sqrt{10,000}} = 0.15$$

The difference between μ and $\bar{X}$, between 100 and 98.5, is 1.5: ten standard errors! Since we know that only 0.3% of randomly sampled means would fall more than three standard errors from μ, we would feel very safe in concluding that the parameter for the black population is different from that of the white population.

Expressing the preceding as a formula, we have

$$z_{\bar{X}} = \frac{\mu - \bar{X}}{\sigma_{\bar{X}}} \cong \frac{\acute{\mu} - \bar{X}}{\acute{\sigma}_{\bar{X}}}$$

Note that in this usage, $z_{\bar{X}}$ is somewhat different in detail from the z scores discussed in previous chapters; it involves parameter estimates rather than statistics and is based on the standard error of the mean rather than the standard deviation of the population.

Table 8.A shows the z values that must be exceeded if we are to reject the null hypothesis at varying degrees of risk.

Why don't *you* perform the z test for performance IQ? Remember, in a probability curve, amount of *area* between any two vertical lines represents *probability* of a score falling between the two abscissa values from which the lines rise. When an obtained $z_{\bar{X}}$ falls beyond the **critical value** (such as 1.645 in Table 8.A) which has been decided upon by the researcher, it is said to lie in the **rejection region** of the Gaussian curve, because the consequence is rejection of the null hypothesis. If the obtained $z_{\bar{X}}$ falls short of the critical value, it is in the **retention region** of the curve, because the null hypothesis would be retained. Note the terminology: *rejected,* not disproven; and *retained*, not proven. We are merely making a decision, one way or the other, on the basis of probability. We are certainly *not* making flat statements of fact.

There is a major precaution to be kept in mind in connection with the z test. Very rarely are the data from a properly controlled experiment

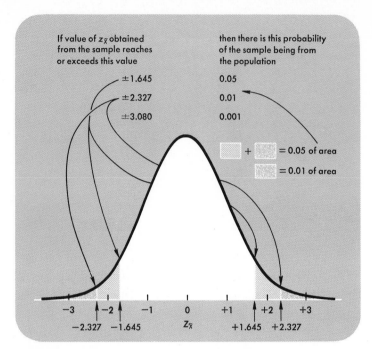

If value of $z_{\bar{x}}$ obtained from the sample reaches or exceeds this value

then there is this probability of the sample being from the population

±1.645	0.05
±2.327	0.01
±3.080	0.001

+ = 0.05 of area

= 0.01 of area

−3 −2 −1 0 +1 +2 +3

−2.327 −1.645 $Z_{\bar{x}}$ +1.645 +2.327

Table 8.A

involved in the z test calculations. Only one sample is involved, being compared with the parameters of a population made up of individuals who may differ in some *other* important respects from the sample we have collected. The black children in our hypothetical study may have differed from the white parameters not because they were black, but because their native dialect of English was not the same as the dialect in which the verbal IQ test was couched. Be very careful about making conclusions about the *causation* of any parametric differences between populations—especially when, as in this example, emotions may cloud thinking.

In summary, we have made a great leap forward. In the previous chapters there was little concern about going beyond actual measures. Anything that we found out pertained to the data actually present. But now we are able to go beyond our data and draw conclusions about matters about which we may never have direct knowledge. It is a form of democracy. If all the potential scores can be induced to send us typical representatives, then the characteristics of the majority may be known with great accuracy. All that is required for this mathematical electoral process to work is

1 The population characteristic in question must be quantified at least by interval scaling. It must be measured or measurable.

2 The electoral process must be random sampling. Truly random, not just a ritual mouthing of the word.

3 The number of sampled values, *n,* should be as large as possible while still being practical, in order to maximize precision.

4 A small but known probability of error must be accepted in any conclusion we make about population parameters on the bases of simple statistics.

This population of terms is not very large, so instead of sampling it, be sure you know every member of it *very* well!

POPULATION and SAMPLE

DEMOGRAPHIC POPULATION

STATISTICAL POPULATION

CASUAL SAMPLE

BIASED SAMPLE

REPRESENTATIVE SAMPLE

RANDOM SAMPLE

PARAMETER and STATISTIC

POPULATION MEAN μ

SAMPLE MEAN $\bar{X}$

POPULATION VARIANCE σ^2

SAMPLE VARIANCE s^2

STANDARD ERROR OF THE MEAN, $\sigma_{\bar{X}}$

ESTIMATED STANDARD ERROR OF THE MEAN, $\hat{\sigma}_{\bar{X}}$

SAMPLING DISTRIBUTION OF $\bar{X}$

BEST ESTIMATE

BIASED ESTIMATE

POINT ESTIMATE

CENTRAL LIMIT THEOREM

CONFIDENCE INTERVAL

z TEST

z and $z_{\bar{X}}$

CRITICAL VALUE

REJECTION REGION

RETENTION REGION

exercise for a rainy st. swithin's day

Find your set of darts and the target you fashioned after Figure 5.4. (It is believed by some that

> St. Swithin's day, if thou dost rain,
> For forty days it will remain.
> Anonymous

so maybe you'd better have them waterproofed first.) Stand back a few feet and hurl the darts, one at a time, to hit the line labeled 0. You will miss, of course, by varying distances. You'd better. That's the whole point.

1 Now calculate the *mean* score, by averaging the values of the vertical lines where your darts hit. This mean value, calculated from the scores themselves, corresponds to $\bar{X}$, your sample mean. The target line, 15, would be μ. Unless you are suffering from the Cross-Eyed Leprechaun Phenomenon, the $\bar{X}$ should be an unbiased estimate of μ, but the two won't be identical, will they?

2 Retrieve your darts, and keep on throwing them in batches of 4, calculating $\bar{X}$ for each such batch. Keep a record of the values of $\bar{X}$; after an hour or two, or when you get tired, draw a frequency distribution of the means. You will then have created a sampling distribution of $\bar{X}$. Find the mean of the latter. It should be very close to 0, which is μ. The two should be identical if you gathered an infinite number of sample means, but to do that would take longer than one rainy St. Swithin's Day. So *pretend* that the mean of your sampling distribution is 0, if it isn't quite.

3 Find the standard deviation of your sampling distribution. That's equivalent to the standard error of the mean, $\sigma_{\bar{X}}$. Pick one of your *samples* of four darts at random, calculate *its* standard deviation s, and divide s by $\sqrt{n-1}$, which is $\sqrt{3} = 1.732$. This gives an unbiased estimated of $\sigma_{\bar{X}}$, right? In fact, how well do they match? Take another five or six of your samples of size $n = 4$ and find their standard deviations, repeating your estimation of $\sigma_{\bar{X}}$ with each.

4 Take any of the samples for which you have already calculated the standard deviation and calculate it again, only this time in relation to the value of $\mu = 0$. Work as in Figure 8.1. What do you notice about the size of your obtained value, relative to the previously obtained one? Yes, s is smaller than σ.

5 Compare your obtained value of σ with your standard error of the mean, obtained in (3). Which of them is smaller?

If your accuracy at dart throwing is so excellent that you hit line 0 every time, or nearly so, merely stand further away. Or take a few short pauses, partaking of Irish Dew. *That* should increase your variable error!

It's best to curtail your sampling procedure if anybody bigger, stronger, or more important than you tries to walk between you and your data during the sampling process.

self-test on chapter 8

The usual rules prevail. Any number of the answers may be correct. Circle lightly in pencil, and score yourself in the usual way.

1 A *statistic* is

    a any characteristic of a population which is translatable into scores and which we study using statistics.

    b a numerical value pertaining to a sample's characteristics.

    c invariably an unbiased estimate of a parameter.

    d exemplified by μ and σ.

    e the basis for estimation of a parameter.

    f exemplified by $\bar{X}$ and s.

    g the basis for exact specification of a parameter value.

2 "$\bar{X}$ is an unbiased estimate of μ." This statement

    a is true, but depends upon random sampling.

    b implies that $\bar{X}$ will always have a value equal to μ.

    c is not true.

    d implies that $\bar{X}$ will be smaller than μ about half of the time.

    e is sometimes true and sometimes false, when the sampling method is random.

    f depends upon every member of the population, no matter how large the population, having the same chance of being sampled.

3 "The value of s^2 underestimates σ^2 as a function of decreasing n." This statement

    a is generally true.

    b is necessarily always true.

    c while true, is not of practical importance when n exceeds 100 or so.

    d would not be true if the deviations were calculated from μ rather than $\bar{X}$.

    e implies that s^2 is not a very useful statistic.

SCORING AND ANALYSIS of self-test questions

For Questions 1 through 5, score $+3$ for each correct choice, -2 for each incorrect, and -1 for each correct choice not so identified.

Answers to Question 1

a F   If you marked this one, you were thinking of *parameters*.

b T   Correct; a *statistic* refers to a sample, $\bar{X}$ and s being the major examples.

c F   A statistic is not *invariably* unbiased; it is unbiased for estimating μ but not σ.

d F μ and σ are parameters, not statistics.

e T   Certainly, whether biased or not, a statistic is the starting-point for

f T   estimation of a parameter.

g F   But the parameter can never be specified exactly, only estimated with varying degrees of precision, depending upon n.

Answers to Question 2

a T   (a) is true in every respect.

b F   Lack of bias does not ensure that $\bar{X}$ will always equal μ, merely that the errors will fall equally on each side of μ.

c F   It is simply not true to say that the statement is not true.

d T   Think of the consequences of having a distribution centered on μ; the scores of that distribution fall equally on both sides.

e F   If sampling is truly random, $\bar{X}$ is *always* an unbiased estimate, but not necessarily an accurate one.

f T   Yes: this is the functional definition of randomness in sampling. Randomness refers to the *method* of sampling rather than the result.

Answers to Question 3

a T   Yes, it is generally true. Specific exceptions will occur, however, on

b F   account of the unreliability of s^2 when n is very small, so it is not always a true statement.

c T   When n is 100, the average degree of underestimation is only about 1%; s^2 is, however, quite reliable with an n of this size.

d T   True. Using μ rather than $\bar{X}$ gives an unbiased estimate of σ^2: still just an *estimate,* unless n approaches the population size.

e F   Size of n affects degree of underestimation of σ^2 as well as the reliability with which σ^2 is estimated. s^2 is a very useful *statistic*. Without it, this entire discussion would be nowhere.

4 If, from an extremely large population, a very large number of samples were drawn randomly and their mean values calculated, these many means

a would each be equal to the population mean μ.

b would vary from μ only by chance, and hence would form a Gaussian distribution.

c would have a grand mean equal to μ.

d would form a distribution whose standard deviation is equal to s.

e would form a distribution whose variance would, if multiplied by n, be equal to σ^2.

f would be almost identical if the sample n's were very large.

5 An adequate random sample of freshman college students in the state of Wisconslahoma could be acquired by

a serial numbering each eligible student in one or two typical colleges, and picking numbers out of a hat until the desired n is achieved.

b selecting one college at random, and using its entire freshman body as the sample.

c within every college, dividing the freshmen according to surname initial (A, B, . . ., thru Z) and randomly selecting a subject from each initial in turn, until the desired n is reached.

d requiring every student in Psychology 1 to volunteer in each college and taking these "volunteers" until the desired n is reached.

e getting a list of Social Security numbers of all freshmen in the state and pulling these numbers out of a hat until the desired n has been reached.

f getting a complete list of freshman names from each college, putting each name on a slip, and pulling the required number of names blindly out of a hat.

Answers to Question 4

a F No, these many means would differ from μ by chance alone, so they
b T would fall into a Gaussian distribution.

c T Yes, in principle their grand mean (the mean of the means) would be
equal to μ. It might differ from μ by chance, but to an infinitesimal extent, and not if the total number of samples were large enough.

d F The standard deviation of the distribution of means would be much
smaller than the s of any of the raw score distributions. Remember:

$$\sigma_{\bar{X}} \cong \acute{\sigma}_{\bar{X}} = \frac{s}{\sqrt{n-1}}$$

e T Yes: if $\acute{\sigma}_{\bar{X}}^2 = \dfrac{\acute{\sigma}^2}{n}$, then $n\acute{\sigma}_{\bar{X}}^2 = \acute{\sigma}^2 \cong \sigma^2$

f T If n were so large as to include the whole population, then all the
values of $\bar{X}$ would be equal to each other as well as μ. If n is less than
infinite but still very large, $\bar{X}$ will be very close to μ and to any other
possible value of $\bar{X}$.

Answers to Question 5

Remember that a random sample is one in which *every* case has an
equal chance of being drawn. That is to say,

$$p = \frac{1}{n}$$

where p is the chance of any case being sampled and n is the number
of cases in the population. So let's see how the following sampling
methods shape up to this ideal.

a F If only one or two colleges are drawn, there is 0 probability of students
in any of the other colleges being drawn.

b F Sampling of each case must be independent of any other case. If one
college is drawn, then all the sampled students within it are related in
their having been sampled.

c F This method would be all right, *if* an equal number of students had
surnames beginning with each letter of the alphabet. But there are few
people whose surnames begin with X, so such a person would have a
good chance of being sampled, a chance unequal to that of somebody
named Smith.

d F This method rules out the sampling of any student not in Psychology 1.
Answers to Question 5 are continued on p. 233.

Answers to Question 5 (*continued*)

e F This method would be all right *if* every student had a Social Security number; but some probably would not and would thus have a 0 probability of being sampled.

f T This method is the best. Also the simplest in principle.

6 A population is known to have the following parameters:
$$\mu = 24.400 \quad \sigma = 5.400$$

Below are given the values of $\bar{X}$, s, and n for a series of samples which may or may not have come from the population defined above. In the spaces provided, lightly pencil in the value of z and the decision which you would make. Give yourself three points for each correct score. Use the 0.01 probability level in Table 8.A.

a $z = $ _____

$\bar{X} = 24.0 \quad s = 5.38 \quad n = 121$

Decision: _____

b $z = $ _____

$\bar{X} = 22.8 \quad s = 5.20 \quad n = 4$

Decision: _____

c $z = $ _____

$\bar{X} = 24.3 \quad s = 5.400 \quad n = 1,000,000$

Decision: _____

d $z = $ _____

$\bar{X} = 25.2 \quad s = 5.370 \quad n = 2$

Decision: _____

e $z = $ _____

$\bar{X} = 24.40 \quad s = 5.399 \quad n = 10,000$

Decision: _____

Answers to Question 6

The general formula for calculation of z when μ and σ are known is

$$z_{\bar{X}} = \frac{\bar{X} - \mu}{\sigma_{\bar{X}}} = \frac{(\bar{X} - \mu)\sqrt{n}}{\sigma}$$

a -0.81
retain

a $\quad z = \dfrac{[24.0 - 24.4] \times \sqrt{121}}{5.4} = -0.81$

retain null hypothesis

b -0.5925
retain

b $\quad z = \dfrac{[22.8 - 24.4] \times \sqrt{4}}{5.4} = -0.5925$

retain null hypothesis

c -18.518
reject

c $\quad z = \dfrac{[24.3 - 24.4] \times \sqrt{1{,}000{,}000}}{5.4} = -18.518$

reject null hypothesis

d $\quad 0.209$
retain

d $\quad z = \dfrac{[25.2 - 24.4] \times \sqrt{2}}{5.4} = 0.209$

retain null hypothesis

e $\quad 0$
retain

e Just *look* at this one; if $\bar{X} = \mu$, z cannot be other than 0. So retain null hypothesis.

7 On strictly theoretical grounds, the population mean μ of the differences between spouses' IQs is considered to be 0. The value of σ is not known but can be estimated. For the values below, in which n is the number of pairs of spouses, decide whether the null hypothesis (that IQ differences should average out to 0) should be rejected. The direction of difference is always husband's − wife's. Use the 0.01 probability level in Table 8.A. Score three points for each correct answer.

a $z =$ _____

$$\bar{X}_{\text{diff}} = 1.0 \quad s = \quad 1.0 \quad n = 101$$

Decision: _____

b $z =$ _____

$$\bar{X}_{\text{diff}} = 1.0 \quad s = \quad 1.0 \quad n = 5$$

Decision: _____

c $z =$ _____

$$\bar{X}_{\text{diff}} = 1.0 \quad s = 10.0 \quad n = 26$$

Decision: _____

In Questions 6 and 7, note carefully the effects of varying s or σ, the amount of difference $\bar{X} - \mu$, and the size of n. In particular, note that the size of n is totally under the control of the experimenter, so even a small difference between $\bar{X}$ and μ can enable us to detect a difference between a sample and a population.

EVALUATION
For Questions 1 through 5, + 3 for each correct, −2 for each mistake, −1 for each omitted that would have been correct. For Questions 6 and 7, +3 for each correctly calculated z *and* correct conclusion.

50 to 66	
40 to 49	B
25 to 39	C
10 to 24	D
Under 10	P (poor)

Answers to Question 7

When σ is unknown, it must be estimated from s. In practice, this is usually necessary.

a 10.0
reject

$$\text{a} \quad z = \frac{[1.0 - 0] \times \sqrt{101 - 1}}{1.0} = 10.0$$

reject null hypothesis

b 2.0
retain

$$\text{b} \quad z = \frac{[1.0 - 0] \times \sqrt{5 - 1}}{1.0} = 2.0$$

retain null hypothesis

c 0.5
retain

$$\text{c} \quad z = \frac{[1.0 - 0] \times \sqrt{26 - 1}}{10.0} = 0.5$$

retain null hypothesis

ſome ordinary problemſ

1 Shown in Table 8.B is a population of $n = 1024$ scores, arranged randomly. Perform the following operations on these data:

   a Take five separate samples of 30 successive scores each, and calculate $\bar{X}$ and s for each.
   b Estimate the values of μ and σ.
   c Find the 95% and 99% confidence intervals for μ.
   d For the scores in the sample below, find the value of z relative to the population. What is the probability that the sample is merely a random subset from the population?

50	60	80	70	40	50	70	80	60	40	90	60	60	70	50
30	50	60	80	60	40	90	80	60	80	90	70	50	60	80

2 In Question 1(a), if an infinite number of random samples of $n = 30$ had been drawn and their means calculated, into what shape would the distribution of those means have fallen?

3 What would be a reasonable estimate of the standard deviation of those many means?

4 Sketch a Gaussian distribution and indicate its important features.

5 State the importance of n in the decision whether a sample does or does not consist of a subset from any given population.

6 Distinguish between *random* and *representative* in regard to sampling.

7 What is the importance of the Central Limit Theorem?

Table 8.B
APPROXIMATELY GAUSSIAN POPULATION

26	28	28	28	56	54	52	49	10	62	58	26	48	60	59	36	32	59	58	49
49	41	59	52	62	48	52	52	36	46	34	41	52	93	40	61	40	36	49	32
28	49	34	74	20	28	46	66	62	74	66	58	56	24	62	87	72	41	24	61
84	66	68	26	64	41	70	34	74	54	56	50	64	62	76	58	50	87	36	41
51	62	50	32	61	72	34	49	46	54	68	62	60	42	52	20	90	16	32	41
56	76	50	26	59	61	61	38	30	41	66	66	13	44	52	76	64	13	58	36
60	26	24	44	42	66	56	51	61	60	26	44	52	61	54	70	66	38	34	74
70	62	54	59	84	40	62	24	49	60	51	20	61	50	90	28	56	32	52	72
52	28	38	64	40	54	28	28	50	80	58	49	68	58	68	54	40	39	61	24
42	41	41	44	64	56	44	64	48	40	39	49	34	38	56	16	59	44	16	49
80	70	84	68	76	61	54	99	59	42	61	56	72	54	39	39	58	44	32	52
38	46	49	32	49	58	59	39	42	46	46	40	62	39	44	64	36	34	26	61
49	32	16	39	40	56	58	70	32	59	24	58	60	62	39	39	48	59	72	28
60	41	68	51	38	68	64	70	76	48	30	60	49	56	76	58	26	76	62	51
48	59	46	30	41	74	39	48	60	30	64	16	70	74	54	38	44	30	48	38
36	42	40	66	60	38	52	76	56	42	60	41	51	20	72	36	62	80	46	30
62	72	10	70	66	38	41	39	46	44	62	49	64	50	50	48	50	32	44	58
76	42	56	13	48	26	50	51	64	38	72	84	70	41	60	70	66	32	34	34
54	60	56	59	66	56	62	62	68	87	54	50	44	41	80	36	74	64	16	60
32	62	64	46	48	49	59	51	52	36	56	68	26	58	56	48	42	40	56	42
38	54	59	66	50	40	48	70	52	42	68	46	66	59	32	72	28	66	42	52
50	36	38	48	41	24	51	42	30	76	40	50	36	66	52	30	54	46	70	68
51	62	51	48	51	87	59	39	52	50	72	68	58	72	51	39	64	38	52	52
64	39	26	46	51	60	36	74	59	52	24	68	64	64	24	42	80	70	38	61
48	42	24	50	52	26	80	32	46	51	49	61	50	58	50	54	74	60	74	56
48	42	84	51	28	72	46	24	59	41	41	50	84	68	64	70	58	44	44	52
49	66	51	36	60	50	54	44	39	68	59	46	76	40	74	34	72	68	40	50
36	42	74	46	48	41	26	34	34	49	30	28	41	74	70	42	46	16	48	20
16	76	28	50	64	59	50	59	68	59	30	59	40	56	61	40	20	48	58	56
80	38	50	64	44	42	49	42	42	40	76	54	64	66	49	51	76	41	42	36
68	38	60	62	44	51	30	40	44	46	46	54	49	41	41	64	62	52	48	20
60	40	70	20	28	66	52	38	50	62	26	60	42	52	54	49	44	38	24	72
46	60	32	59	54	60	60	34	30	58	80	46	36	41	70	49	70	40	36	1
64	46	48	54	51	56	59	74	64	52	30	61	34	32	40	41	51	49	13	60
51	59	36	66	34	28	34	52	50	56	39	50	64	46	40	44	84	38	68	24
30	58	52	44	7	50	36	49	40	51	58	30	74	34	32	76	66	38	34	26
39	48	30	49	50	72	80	59	49	36	32	58	36	48	42	16	41	24	41	60
80	54	46	36	56	72	44	32	72	51	58	66	61	61	56	36	42	30	54	62
48	84	80	34	28	48	56	44	74	20	44	40	49	20	20	24	84	60	76	13
44	10	50	50	42	59	51	61	40	42	46	39	39	39	32	46	50	36	20	30
66	39	50	26	32	93	16	38	61	51	56	48	50	34	48	54	28	52	34	61
32	26	51	68	24	56	51	41	56	51	59	58	72	72	40	68	54	50	56	72
49	56	51	64	62	44	70	66	34	42	40	52	39	58	51	66	46	64	32	20
70	58	61	58	40	66	39	52	50	38	84	74	70	30	68	52	54	59	60	68
34	58	56	61	41	61	76	30	58	66	34	16	68	38	36	62	38	84	38	41
39	46	50	61	70	42	54	42	54	39	49	66	24	26	54	30	51	51	60	46
51	28	44	70	44	80	30	70	44	68	54	39	51	48	51	54	20	34	60	28
62	61	50	54	26	48	49	36	30	80	80	66	74	70	32	24	62	34	26	52
46	62	52	60	74	58	52	49	44	59	36	34	41	90	40	32	44	87	42	44
61	84	61	49	49	61	28	74	42	39	49	46	39	49	58	48	80	32	20	50
30	30	38	48	39	49	68	48	64	72	50	51	16	40	48	51	48	46	54	76
38					31					27					7				

To use this table, close your eyes, stand well back, wave your arm around, and bring your fingertip or pencil point down onto the page. Start sampling at the point touched. You may move vertically upward or downward (flip a coin) or horizontally in either direction. Within one "sample" you must be consistent in this choice. If the end of a row or line is reached, go to the next one consistent with the method chosen. The scores are already arranged in a random order, so do not hop around the table. Continue to sample until the required n is reached.

chapter 9
correlation: measuring relatedness

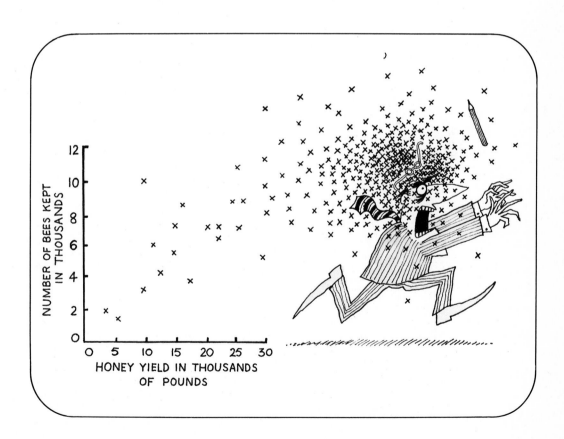

The term **correlation** is very widely used in statistics and in life generally. It is frequently misused as a more technical-sounding equivalent of **relationship**. Correlation actually involves two concepts, and relationship is one of them. The other is the concept of **quantification** of the *strength* or *degree* of relationship; in statistics, a correlation is actually measured and expressed mathematically. Let's now examine these two concepts more closely.

relationship

There are relationships between people when they have something to do with each other. There are relationships among everyday events when there is something connecting them. For example, there is a relationship between brother and sister: they have in common the same parents. There is a relationship between incidence of sunstroke

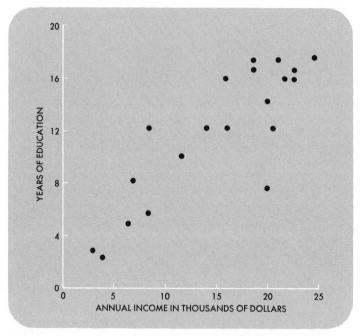

Figure 9.1

A positive correlation: the relationship between educational level and annual income in a group of 20 randomly selected men. In general, the longer the educational period the greater the income, but there are specific exceptions to this generalization.

and season of the year: sunstroke is more prevalent in summer and less prevalent in winter. Such things have "something to do with" each other. The quality of intellect that allowed humanity to swing down from the trees and develop knowledge and mastery over nature was the ability to perceive and use the relationships present in nature's daily operations. It is quite possible and usual to think of natural relationships without any recourse to mathematics.

There is often an implication of **direction** in any relationship, though this may be so obvious that it is not noticed. It is clear in the sunstroke–season relationship that sunstrokes do *not* occur primarily in the winter. In other instances the notion of a relationship's direction is important rather than trivial. We speak of *positive* relationships when the presence of two things is associated and when the absence of one goes with absence of the other. A *negative* relationship is one in which the presence of one thing is systematically associated with the absence of the other, and vice versa. Figure 9.1 shows a posi-

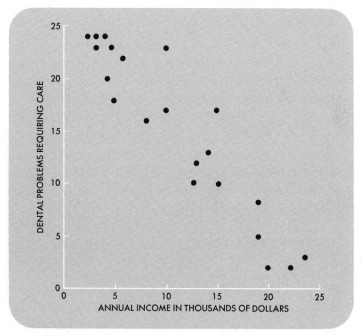

Figure 9.2
Relationship between average annual income (in groups of 100 families) and incidence of dental diseases within these groups of families. An example of *negative correlation.*

tive relationship, in which a lot of education is associated with a lot of dollars. Figure 9.2 shows a negative relationship, in which a lot of dental disease is associated with relative absence of dollar income.

quantification of relationship

Figures 9.1 and 9.2 would have been very difficult to draw had we lacked the measures shown on the X and Y axes. It seems clear that expressing the existence of a relationship is much easier if we use numbers to express our related observations. And if we use numbers to express the *strength* of the relationship as well, we have a *correlation*. The numerical end result of this undertaking, symbolized as r, is a **correlation coefficient**, or **coefficient of correlation**. It may be calculated in several ways, all somewhat different in their implications, but alike in two ways:

1 Sign (+ or −) denotes the *direction* of relationship, whether positive or negative.

2 All the various possible varieties of correlation coefficients use the size of their numerical values, irrespective of sign, to denote the strength of relationship. The limiting value is always 1.00, with −1.00 expressing perfect negative correlation and +1.00 expressing perfect positive correlation.

Obviously, the midpoint, 0, having neither magnitude nor sign, must denote the lack of *any* relationship (or more exactly, the lack of any relationship to which the coefficient and its method of calculation are sensitive).

What is meant by the terms **positive correlation** and **negative correlation**? A common problem in these days of runaway verbiage (in which a beach is an earth–ocean interface) is the use of big words for small ideas. A **positive correlation** is the name given to the situation in which relatively large scores in one distribution are associated with relatively large scores in the other and relatively small scores in the one are associated with relatively small scores in the other. As an example, there is a positive correlation between age and height in children. The older a child, the taller the child—in general. The younger the child, the shorter—in general. There are exceptions, of course, but they only indicate that this particular correlation is not a perfect one.

A **negative correlation** is one in which relatively large scores in one distribution are associated with relatively small scores in the other. Among adults there is a negative correlation between age and muscular strength—in general. As adults get older, they generally get weaker. Again, the exceptions merely show that this correlation is not perfect.

Be careful not to confuse **negative correlation** with **0 correlation**. A 0 correlation means simply *no* relationship between two sets of scores obtained from the same individuals; with a negative correlation there is a definite relationship, the strength of which is indicated by the size of the negative coefficient.

Clearly, to express a relationship as a correlation, one must first translate the things that are related into numerical scales. At least ordinal scaling is necessary, and cardinal scaling is usually preferable if possible. (There are, however, some exceptions to this latter ideal.) Moreover, each individual case must have two actual numerical scores, one for the X axis and one for the Y axis. The location of each single point in Figures 9.1 and 9.2 refers to the two scores of one individual. This is better seen by dropping perpendiculars from any point to each of the graph coordinates, as in Figure 9.3.

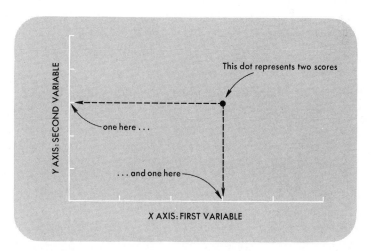

Figure 9.3
Construction of a *correlation scattergram*. The score plotted actually refers to *two* scores, one on the X axis and the other on the Y axis. The two scores are related, having come, for example, from one person.

showing correlation graphically

Figures 9.1 and 9.2 are examples of **correlation scattergrams**. They are constructed merely by repeating the operation shown in Figure 9.3 until all the data are displayed. Much can be told about correlations simply on the basis of their scattergrams. The **direction** of a correlation (positive or negative) can be seen instantly by whether the scattergram slopes up or down from left to right. Figure 9.4 shows several idealized scattergrams; you can see that the shape of the collection of points is tied to the value of r: The fatter the scatter, the smaller the r, the extremes being when r is ± 1.0 (then the scattered points fall exactly along a straight line) and when r is 0 (then the points fill out a circle). Correlation coefficients between 0 and 1 are associated with scattergrams that are more or less cigar-shaped or oval.

When any degree of correlation exists, it is possible to run a straight line through the scattergram to represent the general relationship shown in the scattergram. It is called a **line of best fit**, though clearly the goodness of fit depends upon the size of r.

With the general concept of correlation firmly in hand, we can now thoroughly examine two of the most frequently used coefficients of correlation, the **Pearson product-moment coefficient** and the **Spearman rank-order coefficient**.

the Pearson product-moment coefficient of correlation

The information given in Figure 9.1 is shown again in Figure 9.5, but with the coordinate changed to the z scores corresponding with the raw scores in the original graph. Now look closely at Figure 9.5. Toward the upper right side (to the right of 0 on the X axis and above 0 on the Y axis), the z scores are positive for both education and income, while to the lower left both sets of z scores are negative. What would happen if, for each point, we multiplied its two z scores together and then found the mean value of all those cross-multiplied values? Since positive numbers times positive numbers give positive products, and negative numbers times negative numbers also give positive products, the *total* of these products would be relatively large and positive, and so would their mean value. Try it and see. The

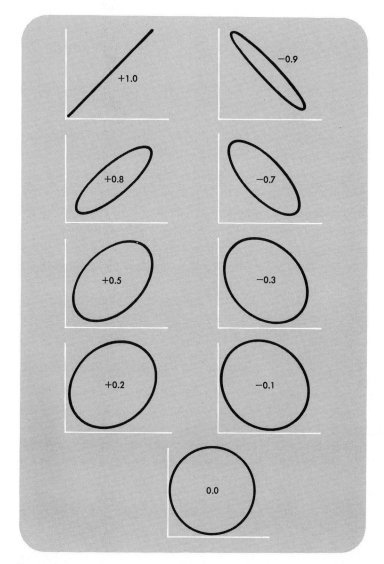

Figure 9.4
The shapes and directions of slope of the score clusterings in scatter-grams associated with rs from 0 to 1.0. These "envelopes" are idealized; actual distributions of plotted points within them would probably not look so smooth.

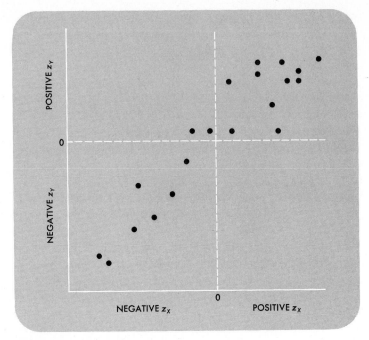

Figure 9.5

The data of Figure 9.1, with coordinates relabeled as z scores. In this relationship, which is positive, note that positive values of z_X are associated with positive values of z_Y and negative z_X with negative z_Y. Thus the product of each pair of scores z_X and z_Y will be positive (except for only two scores) and the total of all such products will be relatively large. The mean of those products will be close to the maximum value, $+1.00$, so we have a high positive correlation coefficient.

mean value is positive and only somewhat less than 1.00. You have just calculated the Pearson product-moment coefficient of correlation, whose abbreviation symbol is r.

$$r = \frac{\sum (z_X z_Y)}{n}$$

This is the simplest formula for the Pearson r.[1] It can yield only values between -1.00 and $+1.00$ inclusive, provided that the z scores are

---

[1] Note the use of subscripts to avoid ambiguity. z_X denotes the z scores for values of X, and z_Y denotes them for values of Y. Don't confuse z_X with $z_{\bar{X}}$ (z scores of values of $\bar{X}$ in the previous chapter).

correctly calculated for both the X (income) and Y (education) variables. The limit of 1.00 is reached when the z scores are identical within each cross-multiplied pair. In this extreme instance the formula above could be written as

$$r = \frac{\sum z_x z_y}{n} = \frac{\sum z_X^2}{n} = \frac{\sum z_Y^2}{n} = 1.00$$

What happens in the event of a negative correlation? In such a case, positive z values would be associated with negative z values between the two scales, giving predominantly negative cross-products. Thus, the total (and consequently the mean) of the cross-products would be a negative number, as befits a negative correlation. In the event of a 0 correlation, positive z values would be associated with negative z values only about half of the time; hence the total, $\Sigma z_X z_Y$, would balance out around 0 and so would $\Sigma (z_X z_Y)/n$.

It is fun to work out your own examples to test the truth of the foregoing, but observe one important point. You cannot make up your own arbitrary z scores; you must calculate them from raw scores (which *may* be invented), following the method given in Chapter 6. It is especially enjoyable to prove experimentally that no higher value of r than +1.00 is possible. You should actually try this exercise for the following data, working within the spaces provided in the table.

X values	z_X values	Y values	z_Y values	$z_X \times z_Y$
1		10		
3		11		
5		12		
7		13		
9		14		
$\bar{X} = $ ____ $s_X = $ ____				____
$\bar{Y} = $ ____ $s_Y = $ ____			$\Sigma (z_X z_Y) = $	

The summation above should be exactly 5.0, so r should be exactly +1.00. Slight inaccuracies may result from rounding errors.

If you have skipped this exercise, shame on you. Go back and do it. Statistics cannot be learned just by reading about it.

This entire treatment of the Pearson r may look extremely, suspiciously, simple. Especially if you have already been suffering from more impressive-looking treatments of it given elsewhere. Your unease is baseless; this is all there is to Pearson's *r*, except for ways of making it even *easier* to calculate.

The **Pearson r,** or **Pearson product-moment correlation coefficient**, is named to honor Karl Pearson (1867–1936), the founder of modern statistical theory. The term *product-moment* reflects the fact that each cross-product of *X* and *Y* scores exerts a leverage, or more technically a *moment,* in the same way that we've already seen scores affecting the mean. Both the mean and the Pearson *r* may be unduly affected by the leverage of extreme and unbalanced scores.

calculation of *r*

It is a grievous possibility, but you may sometime find it necessary to calculate *r* from actual data, and the *z*-score method above is rather awkward to use (though not to memorize). So here is a fairly simple working formula, derived directly from the original one and almost as easy to memorize. For the *X* and *Y* data, however, you will need the values of $\bar{X}$, $\bar{Y}$, s_X, and s_Y,[2] all of which may be available anyway. And, of course, the raw scores must be available so you can find ΣXY.

$$r = \frac{(\Sigma XY/n) - \bar{X}\bar{Y}}{s_X s_Y}$$

derivation of the working formula

The derivation is very easy, since it is merely an expansion and simplification of the theoretical formula. To begin with,

$$r = \frac{\Sigma z_X z_Y}{n} = \frac{\Sigma(X - \bar{X}/s_X)(Y - \bar{Y}/s_Y)}{n}$$

$$= \frac{\Sigma(X - \bar{X})(Y - \bar{Y})}{n s_X s_Y}$$

[2] s_X is the standard deviation of the X scores, and s_Y is the standard deviation of the Y scores.

It is now necessary to multiply out the numerator on the right. Remember that each term must multiply each of the terms in the other set of parentheses.

$$\sum (X - \bar{X})(Y - \bar{Y}) = \sum XY - \sum X\bar{Y} - \sum Y\bar{X} + \sum \bar{X}\bar{Y}$$

but

$$\sum X = n\bar{X}, \quad \sum Y = n\bar{Y}, \quad \text{and} \quad \sum \bar{X}\bar{Y} = n\bar{X}\bar{Y}$$

$$\sum (X - \bar{X})(Y - \bar{Y})$$

so we can rewrite $\sum (X - \bar{X})(Y - \bar{Y})$ as

$$\sum XY - n\bar{X}\bar{Y} - n\bar{Y}\bar{X} + n\bar{X}\bar{Y}$$

It is easily seen that the last two terms cancel out, so

$$r = \frac{\sum XY - n\bar{X}\bar{Y}}{ns_X s_Y} \quad \text{or} \quad \frac{(\sum XY/n) - \bar{X}\bar{Y}}{s_X s_Y}$$

Perhaps it's not essential for you to follow this derivation, but do try! You'll feel good when you realize that you do understand it!

sample calculation of r using raw data, given $\bar{X}$, $\bar{Y}$, s_X, and s_Y

X scores	Y scores	XY	
5	8	40	$s_X = \sqrt{\dfrac{5^2 + 9^2 + 1^2}{3} - 5^2}$
9	7	63	
1	9	9	$= \sqrt{10.667}$
$\sum X = 15$	$\sum Y = 24$	$\sum XY = 112$	$s_Y = \sqrt{\dfrac{8^2 + 7^2 + 9^2}{3} - 8^2}$
$\bar{X} = 5$	$\bar{Y} = 8$	$n = 3$	
$s_X = 3.265$	$s_Y = 0.816$		$= \sqrt{0.667}$

$$r = \frac{(\sum XY/n) - \bar{X}\bar{Y}}{s_X s_Y} = \frac{(112/3) - 5 \times 8}{3.265 \times 0.816} = \frac{-2.67}{2.67} = -1.00$$

This example of a perfect negative correlation emphasizes a warning mentioned previously: errors may occur from rounding down. If the values of s_X and s_Y had been taken to only two decimal places, the denominator would have been slightly smaller than actually found. This would have made the correlation coefficient slightly greater than -1.00 (in a negative direction: perhaps -1.10), thus giving an impossible answer. Rounding s_X and s_Y can very slightly inflate the values of r obtained by this formula unless care is taken to use sufficient significant figures.

You should study this example carefully and practice the procedure on your own data.

machine calculation of r

Occasionally it is necessary to calculate correlations on an electronic calculator. While it is possible to utilize the working formula above, it is sometimes easier to translate the denominator values of s_X and s_Y into raw-score form. Accordingly, you might wish to rewrite the formula as follows, despite its even greater apparent complexity.

$$r = \frac{(\sum XY/n) - \bar{X}\bar{Y}}{\sqrt{[(\sum X^2/n) - \bar{X}^2][(\sum Y^2/n) - \bar{Y}^2]}}$$

Any theoretical formula has very many valid working versions. In principle an infinite number exist. So here is a word of advice: do not preoccupy yourself with more than one or two working formulas (the two given above are as good as any), and do not try blindly to memorize them. Of course, you should have a sound appreciation of the theoretical formula, but embarking upon a lengthy calculation on the basis of a poorly memorized working formula is an invitation to high head temperatures.

One limitation of Peason's r must be remembered: it is restricted to straight-line or linear relationships, that is, relationships that take roughly the shape of a straight line or unbent cigar when plotted in a scattergram. A strongly linear relationship among the data will be reflected in a high (positive or negative) value of r, but if the data are related in some nonlinear fashion, the Pearson r will not reflect this adequately. There are several types of nonlinear relationships: curvilinear monotonic ("monotonic" means "pulling in only one direction"), curvilinear arched or U-shaped, and irregular. Examples of these are shown in Figure 9.6. In deciding whether or not a relationship is linear it is difficult to reach a decision except when the plotted points lie very close to some such line. Generally, when the value of r

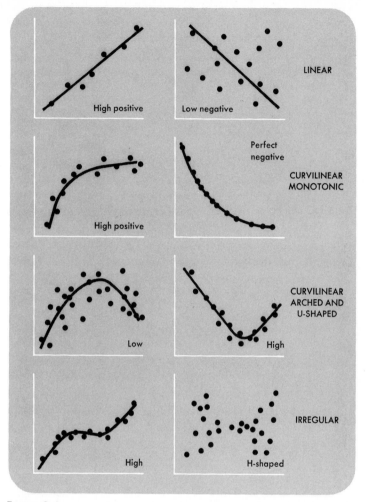

Figure 9.6
Scattergrams and lines or curves of best fit for various types of relationship. The top pair show the only type for which the Pearson r is appropriate. The Spearman r_S may be used for the linear and curvilinear monotonic types of relationship. The two lower pairs of relationships cannot be quantified except by advanced methods.

is large (positive or negative), this is an easy decision. When there is little or no obvious relationship in the data, the value of the correlation coefficient will be close to 0 anyway. However, referring to the right-hand panels of Figure 9.6 that show curvilinear relationships, you will see that a high degree of relationship does exist, but not on a straight line. A Pearson r would seriously understate the degree of relationship in these two instances, so we use other indices of correlation instead.

the Spearman rank-order coefficient of correlation

Three problems exist with the Pearson r. It must assume linearity, it tends to desire cardinal scaling, and it is relatively tedious to calculate. The Spearman coefficient, usually called the **Spearman** r, r_S, is a fast, easy, and acceptable choice for linear, monotonic curvilinear, and roughly monotonic irregular relationships. It requires ordinal scaling; cardinal values must be converted to their rank orders. The formula for r_S is simple. For each pair of scores X and Y, first we calculate the quantity D, which is the rank of X minus the rank of Y. Then

$$r_S = 1 - \frac{6 \sum D^2}{n(n^2 - 1)}$$

The derivation of this formula is intuitively difficult, but working out three simple examples (for perfect negative, perfect positive, and 0 relationships) should help you understand it better.

Example 1—Perfect negative (inverse) relationship

Cardinal X scores	Ranks	Cardinal Y scores	Ranks	Diff	Diff²
196	1	12	5	−4	16
140	2	110	4	−2	4
101	3	129.5	3	0	0
62	4	359.0	3	2	4
60	5	368.5	1	4	16

$$r_S = 1 - \frac{6 \times \boxed{40}}{5 \times 24} = 1 - 2.00 = -1.00 \qquad \sum D^2 = \boxed{40}$$

Note especially that the value 6 in the numerator is constant and does not depend upon any of the values in the data. Remember, in any correlation n refers to the number of *pairs* of observations; thus in this example, n is 5.

Example 2—Perfect positive relationship
With the preceding example before us, the result for a positive relationship can be found mentally. We know that the r_S value for a perfect positive relationship should be $+1.00$. Now clearly, in such a relationship, the X and Y ranks within each pair should be identical, so the difference between them will be 0. Thus, the D column will consist only of 0s, and so, obviously, will the D^2 column. The sum of D^2, ΣD^2, will thus have to be 0, so in the formula

$$r_S = 1 - \frac{6 \Sigma D^2}{n(n^2 - 1)}$$

the numerator of the right-hand term, $6 \Sigma D^2$, will be 0. A fraction has the value 0 when its numerator is 0, so the entire term, $6 \Sigma D^2/n(n^2 - 1)$, will be 0. The formula then reduces to

$$r_S = 1 - 0 = 1$$

so our answer is $+1.00$, as expected.
Example 3—0 relationship

Rank of X score	Rank of Y score	D	D²
4	3	1	1
5	2	3	9
3	4	−1	1
2	5	−3	9
1	1	0	0

$$r_S = 1 - \frac{6 \times 20}{5 \times 24} = 1 - \frac{120}{120} = 0 \qquad \Sigma D^2 = 20$$

Inspection of the ranks in Example 3 shows no systematic relationship between the two sets, and this lack of relationship is properly reflected in the value of r_S, 0.

Charles Edward Spearman was a British psychologist whose main contributions were in the field of intelligence measurement. He applied Pearson's techniques, and later his own, to the measurement of intelligence; one of his techniques, which we have just met, bears his name. Spearman died in 1945.

how (not?) to lie with correlation statistics

The English statesman Disraeli has been quoted as saying that ". . . there are lies, damned lies, and statistics." Unfortunately, we are lied to constantly with the help of statistics; here are some examples of such lies.

1 "Buy an Overdale Typewriter. Studies have proved that a relationship exists between typewriter use and grades. Get a typewriter now, and improve your grades."

2 "The most desirable and sought-after men and women use The Magic Comb electric hairbrush and dryer. It's the easy way to keep hair looking great. Buy a Magic Comb and become more popular."

3 "My child, your grades were failing at Christmastime. Now I don't think you've been applying yourself, have you? The most successful students study at *least* 9 hours a day. Now, if you apply yourself to your studies, I'm sure you will do much better by the end of the school year!"

Can you detect the flaws in these arguments? In each case, there may be an undisputed *fact* of the relationships implied. Typewritten essays *do* tend to get better grades. Electric hairbrushes may be used most by the most popular people. Successful students often do study long hours. The flaws are not in the facts, but in the inferences of *causation*. Mere correlation does not imply causation. Good students are the ones who buy most of the typewriters, but they would do well anyway. People who are very popular (for whatever reasons) may have more preoccupation with their appearance, and maybe more to lose; popularity is more likely to be an indirect cause of use of electric hairbrushes than an effect of it. The failing student may be stupid, or badly taught, or in the wrong kind of school, rather than lazy; a lot of study is done by bright students, but study is not the only requirement for good grades.

Illegitimate inferences of causation are often very subtle; they also become socially important when, for example, a high correlation is noted between poverty and delinquency of behavior. If the conclusion is drawn that poor people naturally tend toward delinquency (and it often has been drawn), the mere existence of the correlation *does not* support that conclusion. Many other possible explanations besides poverty could be found for delinquent behavior. The implications for commercial exploitation, social pathology, behavioral manipulation of the statistically inept, and downright logical goofiness are endless.

Another misuse of correlation involves nonrandom selection of data. Consider Figure 9.7. It appears that a *strong* negative relationship exists between annual income and unmarried pregnancy rate. Such a scatterplot might have been drawn by a naive researcher who had certain preconceptions about the topic and wanted to economize

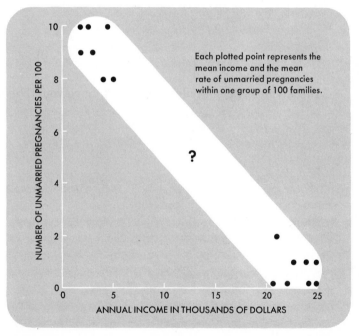

Figure 9.7
The relationship between mean income within groups of 100 families and rate of unmarried pregnancies within those groups of 100 families, when attention is restricted to families whose incomes are below $5,000 or above $20,000 per year. *Apparently* a strong negative correlation exists, but what about those missing middle-income families?

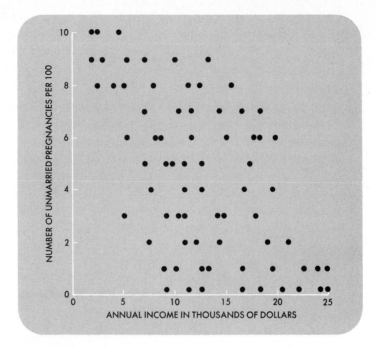

Figure 9.8
The data that *should* have been included in the research for Figure 9.7.
Perhaps there is still some degree of relationship between poverty and
unmarried pregnancies, but the correlation is not as strong as suggested
by the previous figure. Correlations should involve cases drawn from all
levels of one of the two variables (in this example, income level).

on effort by dealing only with the "interesting extremes" of the population. Since Figure 9.7 probably confirmed the researcher's preconceptions, he or she would have no incentive to question it and might
well conclude that a strong linear correlation exists. Let us examine
the question more carefully, however, with the aid of an industrious
researcher (yourself, perhaps), who includes the missing (and "un-
interesting") middle-income families. The results are as seen in Figure 9.8. There may possibly be a *slight* degree of relationship, but not
as high as suggested in Figure 9.7. Quite different conclusions would
be drawn from the two scatterplots.

A third frequent misuse of correlation occurs when the subject
sample is *attenuated*. Again, an example will illustrate. Suppose (as is
the case in some places) that educators feel that IQ measurements
are almost unrelated to academic ability and success. In order to

show this, they might take a sample of 25 tenth-grade students, measure their IQs, and plot these against grade point average. The scatterplot might appear as in Figure 9.9, thus apparently justifying their belief. Can you detect the flaw here? Tenth grade is a very narrow segment of academic life, which ranges from kindergarten to graduate studies. If the entire range were looked at, as in Figure 9.10 (in which the ordinate is necessarily relabeled), then a very different picture might emerge; IQ then seems to be highly correlated with academic attainment. Moral: take a close look at the breadth of your sample, apropos the conclusions you wish to draw.

An important point emerges from the foregoing. Anyone who knows how can go through the motions of determining correlation coefficients and getting arithmetically correct results. However, it need not follow that such results have any meaning, or the meaning imputed to them. Besides knowing the mathematical drill, the user (and consumer–critic) of statistics must know when (and when not) to use statistical procedures. Mathematical ability is not a substitute for wisdom.

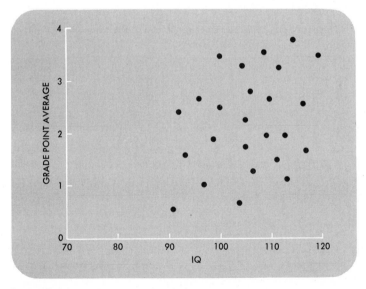

Figure 9.9
Does Grade Point Average have anything to do with IQ? This scattergram shows at best a weak correlation within a tenth-grade class of 25 students. But looking only at tenth-graders is rather limited. The sample is *attenuated* with regard to the entire range of schooling.

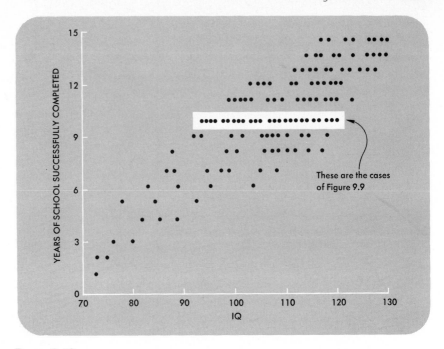

Figure 9.10
The low correlation of Figure 9.9 becomes a strong correlation when the entire range of schooling is included.

correlation and experimentation

Although correlation *techniques* do not justify conclusions about causal relationships, the possibility of causation is not ruled out just because the experimenter expressed a data relationship in terms of correlation. When we stress that *correlation does not necessarily show causation,* we are *not* saying that *correlation necessarily does not show causation.* The word order is not commutative! A correlation *can* show causation when the experimenter wants it to and designs the experimental part of the job properly. The formula itself is indifferent to whether a causal relationship is present.

Correlation is the concept that underlies the philosophy of *experimenting* to test hypotheses. Consider the interesting idea of letting the X axis represent scores *manipulated* by the experimenter instead of merely observed field data. If we also let the Y variable be some observable, measurable effect of the manipulation of the X variable, we are back in the domain of the dependent variable and the independent variable. Figure 9.11 shows a most interesting comparison. In fact, there are only two differences between correlation and experimentation: (1) In experimentation, one of the variables (shown on X axis) is manipulated and the other (on Y axis) is observed, and thus

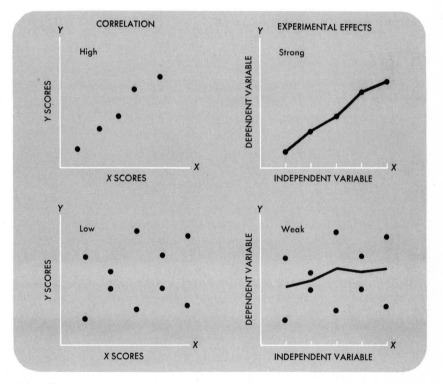

Figure 9.11
There is apparently a close similarity between the concept of correlation, shown in the left-hand panels, and the concept of experimentation (right-hand panels), in which the value of the independent variable is manipulated and the effect on the dependent variable is observed. A strong experimental effect and a high correlation are closely equivalent (top panels), while a weak experimental effect is equivalent to a low correlation (lower panels). The interpolation lines in the lower right-hand panel connect the means of each of the two scores at each level of the independent variable.

an implication of causality is possible. (2) In correlation, scores may lie anywhere along the X axis, while in experimentation, the X scores are more properly thought of as manipulations that have merely been scaled in a certain way. From the mathematical point of view, neither of these differences is very important, so if you thoroughly understand correlation, you are already in good shape for the remainder of this book.

generalizing about correlations

Let us imagine that you've followed the tortuous path of correlation without getting lost and are now the proud owner of a shiny new correlation coefficient of $r = +0.69$ between extroversion and social popularity—at least among the 50 students you managed to test at your college. Wonderful, but then what?

You have demonstrated a correlation that pertains to 50 people, 50 pairs of observations, only. Even assuming that the 50 chosen were typical people, can we feel secure in concluding that a correlation between extroversion and popularity exists in general? If only because you're well inured to rhetorical questions by this time, you'll allow that we can't feel at all secure.

After all, you did not manage to test *everyone* in the world; it's conceivable that if you had, the correlation might have been 0. Accidents do happen, and the 50 people observed *might* have displayed their correlation as a long-shot coincidence, not as an indication of any general rule. In other words, you have a choice: you must pick one of the following explanations for your obtained r.

a It happened because there really is a correlation out there, and it showed up in your sample.

b There really isn't any general overall relationship between extroversion and popularity, but accidents do happen, especially when one tries to formulate a general rule about some 4 billion people (with more to follow) on the basis of only 50.

Can we ever conclude that our *obtained* (sample) r does reflect a population correlation?

An affirmative answer, which we would certainly enjoy having, depends upon two factors: (a) the size of the correlation coefficient obtained in the sample, and (b) the size of the sample. We must also (as always) be able to assume, with some good reason, that the sample is a random one.

The need for these two factors should be clear. If a correlation coefficient is small, it could be merely an accident of sampling, and such an accident will be relatively likely if the sample is small. If

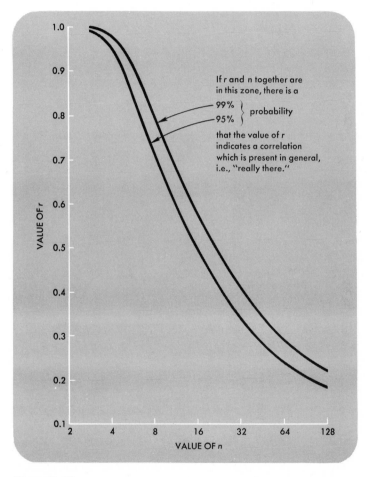

Figure 9.12

As *n* (number of pairs of scores) increases, the *critical value* of *r* decreases. Only two critical levels (1% and 5%) are shown. Note the logarithmic scaling of the *X* axis. Random sampling of cases must be assumed.

either the correlation coefficient or the sample is very large, the probability of such an accident is reduced.

Now in any population of X and Y scores, there is some degree of relationship between X and Y, even if the relationship is 0. And for every *population* correlation coefficient, there will be a range of *sample* correlation coefficients, which will fall into a sampling distribution. The shape of this distribution is approximately Gaussian only when the population correlation coefficient ρ is 0. (ρ is the Greek letter *rho*, or r, in keeping with our use of Greek letters to refer to

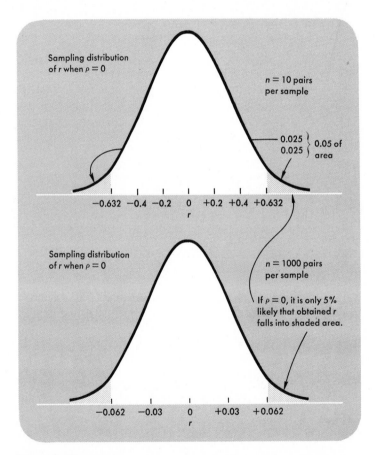

Figure 9.13

The sampling distribution of r is Gaussian when $\rho = 0$. However, the variability of the sampled cases of r depends upon n. Try to relate this figure to the information in Table 9.A.

parameters and Roman letters for statistics.) When the population co-efficient is other than 0, the sampling distribution is skewed. Usually though, it is the null hypothesis that $\rho = 0$ that we would like to reject. So, for any obtained sample value of r, we can ask, "Assuming that $\rho = 0$, what is the probability that this sample r came from the sampling distribution?" If the probability is too small, less than 5% or 1%, we reject the null hypothesis and adopt instead the hypothesis that the population itself contains a correlation between X and Y. Figure 9.12 shows that the critical value of r, whether for the 1% or 5%

Table 9.A

CRITICAL VALUES OF r FOR
5% AND 1% LEVELS OF SIGNIFICANCE

n	5%	1%
3	0.997	0.999
4	0.950	0.990
5	0.878	0.959
6	0.811	0.917
7	0.754	0.874
8	0.707	0.834
9	0.666	0.798
10	0.632	0.765
20	0.444	0.561
30	0.361	0.463
40	0.312	0.403
50	0.279	0.361
100	0.195	0.256
500	0.088	0.115
1000	0.062	0.081

See the Appendix for an expanded table

Note that the listed values of r in Table 9.A do *not* depend upon the population size, N. The reliability of a correlation coefficient obtained with a sample of size $n = 20$ is just as great when the population contains 1,000,000 cases as when the population is small, with perhaps only 200 cases.

level, is related to the size of sample. It is not possible to interpret any correlation coefficient unless the value of n is known. Figure 9.13 illustrates the sampling distribution of r for samples of size 10 and of size 1000.

Table 9.A shows the critical values of r for several values of n, for the 5% and 1% levels of significance. Table 9.B shows some values for r_S. More extensive tables appear in the Appendix. To use these tables, compare your obtained value of r or r_S with the proper critical value in the table; if the obtained coefficient equals or exceeds the tabled value, the null hypothesis may be rejected at the level of risk chosen. We'll do this for our hypothetical experiment, in which $n = 50$ and $r = 0.69$. We find in Table 9.A that a relationship that strong (or stronger) will occur less than 1% of the time if H_0 is true, the critical value of r being 0.361, less than 0.69. So you can generalize that extroversion and popularity are related—to some unknown extent—in the population. It is not possible, however, to use this procedure to infer the *strength* of the population correlation; you can merely infer its *existence* (always with not-quite-absolute certainty).

Table 9.B
CRITICAL VALUES OF r_S FOR
5% AND 1% LEVELS OF SIGNIFICANCE

n	5%	1%
5	1.000	—
6	0.886	1.000
7	0.786	0.929
8	0.738	0.881
9	0.683	0.833
10	0.648	0.794

correlation and the
concept of reliability

Reliability (remember?) is the extent to which a measure is consistent in doing its job; a steel tape measure is reliable, whereas an elastic

WE'VE FINALLY FOUND A USE
FOR THAT RUBBER TAPE
MEASURE OF YOURS,
SHERWOOD

ELASTIC
BANDS

rubber one is not. Suppose that two people were to measure the same object with the steel measure separately; then suppose they each measured another object, then another, and so on until n objects had been measured. Bearing in mind that we are interested in *the measure itself,* not the things being measured, what prediction could we make about the correlation between person A's and person B's measurements? Give or take a few minor discrepancies, we should expect the measures to be almost identical for each object, and hence that the correlation between the two sets of measures would be very high: maybe $+0.989$. Now let's repeat the process using the rubber measure. The discrepancy between measures would be substantial, and hence the correlation would be very low, perhaps only $+0.156$. The low correlation may be seen as a consequence of the unreliability of the rubber measure. This example shows that a correlation coefficient can express reliability. (The same result would occur if one person took each measure twice and correlated the first measure obtained for each object with the second measure.) Note, however, that the question of *validity* is not involved here. Validity cannot be quantitatively expressed.

words to gnaw on and growl over

CORRELATION	PEARSON r
RELATIONSHIP	CROSS-PRODUCT
DIRECTION OF RELATIONSHIP	LINEARITY
POSITIVE CORRELATION	MONOTONIC
NEGATIVE CORRELATION	CURVILINEARITY
0 CORRELATION	POPULATION CORRELATION
SCATTERGRAM	SPEARMAN r_S
LINE OF BEST FIT	ATTENUATION
CRITICAL VALUE	

self-test on chapter 9

As tiresome as always, each question is full of sneaky tricks and the need for thought; any number of choices can be correct, and each such one should be lightly circled in pencil.

1 A relationship may be considered an instance of Pearson product-moment correlation when, among other things,

   a the X and Y variables are ordinally scaled.

   b the X and Y measures are quantified cardinally.

   c the scattergram shape justifies finding the mean value of all cross-multiplied z scores.

   d the relationship between X and Y falls into an egg-shaped scattergram.

   e it is only appropriate to calculate r_S.

   f it is strongly U-shaped.

2 Two hundred healthy children were measured for basal metabolic rate and attention span in class, and a monotonic positive relationship was found. The value of r turned out to be $+0.65$.

   a The correlation must have been linear.

   b The value of n was 400.

   c Children with high basal metabolic rates tended to have long attention spans.

   d One reason for poor attention span has been identified.

   e The total cross-product of z scores was 130.

   f r_S would have been a legitimate correlation measure.

   g We can conclude that an *abnormally* high basal metabolism rate should be associated with abnormally long attention span.

   h Four hundred scores were involved, but n was 200.

SCORING AND ANALYSIS of Self-Test items on the other side of this leaf. Align the left-hand edge of this scoring side with the encircled answers on the test side.

Answers to Question 1

a F Ordinal scaling is not enough for the Pearson r; scaling must be at
b T least interval, if not ratio, so cardinal scaling is required. This is because of the leverage effect of the cross-products of the scores.

c T The mean value of $z_X z_Y$ is the Pearson r; it is appropriate when the scattergram shape is not curvilinear.

d T Egg-shaped is linear—banana-shaped is not.

e F If it is *only* appropriate to use r_S, it must be the case that the relationship is monotonic and curvilinear, not linear. So Pearson r can't be used.

f F With a recursive (folding back on itself, or U-shaped) relationship, neither r nor r_S can be used.

Answers to Question 2

a F The correlation need not have been linear; you can't tell from the value of r alone, unless it is *very* close to 1 or -1.

b F n refers to the number of *pairs* of scores: 200 children, 200 scores.

c T As basal metabolic rate goes up, so does attention span. That's a positive correlation.

d F A correlation does not necessarily indicate causation!

e T Write out the z score formula for r, substitute into it, and you'll find that $r \times n = +0.65 \times 200 = 130$.

f T r_S is quite legitimate for any monotonic relationship.

g F The children were healthy, so the basal metabolism rate measure did not extend into the abnormal range; therefore we cannot conclude anything about the latter. Normal relationships often break down under extreme conditions.

h T Again, n is the number of *pairs* of scores.

3 In your research you obtain a correlation coefficient of $r = -0.70$. For various reasons you wish that your r was positive. How might you accomplish this?

  a There would not be any legitimate way.

  b You could find some reason for multiplying all the X (or Y) scores by -1.

  c You should multiply *both* sets of scores by -1.

  d Were you to do either (b) or (c), the absolute value of r would be unaffected.

  e You could use as data the reciprocals of the scores in either set, except that this would destroy any linear relationship present in the original data.

  f If you used reciprocals on the one set, you should also use reciprocals of the other set.

  g The value of r would be unaffected by (e) or (f).

4 In an experiment you find a Pearson correlation of $+0.60$. Later, on a bet, you try the Spearman r_s on the same data and get a correlation of $+0.95$. Your n is quite large. What can you conclude from these facts?

  a The relationship could not have been linear.

  b The relationship must have been arched or U-shaped.

  c The relationship must have been linear, but not very strong.

  d There is a very strong monotonic correlation, but the Pearson r was not appropriate.

  e There was a strong curvilinear relationship.

  f An appropriate transformation of the scores on one of your variables would possibly make a Pearson r correlation appropriate.

5 Sampling data from only a very narrow range of the X axis values

  a should not affect the size of the correlation between X and Y, provided that one is present in nature.

  b will reduce the value of r.

  c would give a valid r, but one whose implications are restricted within that narrow range.

  d would make both r and r_s smaller than they otherwise might be.

  e is likely to result in a spuriously high correlation.

  f is referred to as *attenuation*.

Answers to Question 3

a F   If some reason can be found, it is legitimate to multiply one set of
b T   scores by -1. The absolute value of r will be unchanged, but the sign
c F   will be different. If both sets of scores are multiplied by -1, you'll be
d T   back where you started.

e T   Reciprocal transformation on one set of scores also would be possible,
f F   since small scores have large reciprocals and large scores have small
g F   reciprocals. The correlation would then be positive. However, since
  the values of the scores would be changed, so would the actual value
  of r, and the relationship, if previously linear, would no longer be so.
  If the relationship was previously curvilinear and monotonic, a reciprocal transformation might be just the thing to give it linearity.
  The reciprocal transformation should be restricted to only *one* of the
  measures; otherwise, r will still work out to be negative.

Answers to Question 4

a T   An r_S of 0.95 is an almost perfect correlation. Since r was much
b F   smaller, the relationship could not have been linear. Recursive
c F   (arched or U-shaped) relationships give neither larger r values nor large
d T r_S values. Conclusion: the relationship must have been monotonic and
e T   curvilinear.

f T   A transformation of the scores along one axis could, if properly
  chosen, cause the curved relationship to straighten out, thus permitting use of the Pearson r.

Answers to Question 5

a F   Truncating the range of the scores on the X axis will probably reduce
b T f T   the value of r, possibly below its critical value.

c T   If your interest lies only within the narrow range (e.g., IQ and academic performance *only* among tenth-grade students) your obtained
  attenuated r would be valid, provided you did not generalize beyond
  your chosen range.

d T r and r_S would be equally reduced by truncation of the range. The
e F   correlation is not at all likely to be made higher.

Matching: Try to match the items on the right with the items on the left. Score one point for each correct match.

  6 Cigar-shaped   ( )
  7 Positive r ( )
  8 Scattergram   ( )
  9 Arched
      relationship   ( )
10 Cross-product   ( )
11 Egg-shaped   ( )
12 Negative r ( )
13 Curvilinear and
      monotonic   ( )
14 Linear   ( )
15 Cannot happen   ( )

a e.g., relationship of physical strength and age in unattenuated sample
b slope down from left to right
c slope generally up, left to right
d Pearson r
e a subject's X score multiplied by Y score
f $r = 0.9$
g two frequency distributions of ungrouped scores arranged at right angles
h $r = +1.50$
i Spearman r
j $r = 0.4$

16 Several values of r and r_s have been obtained from samples of various size. For each one, give the critical value of the statistic, and state whether or not you would reject the null hypothesis. Use the level of confidence shown.

  a $r = +0.30, n = 14, p = 0.01$
  b $r = 0.95, n = 4, p = 0.01$
  c $r = -0.19, n = 169, p = 0.05$
  d $r_s = 0.78, n = 6, p = 0.05$
  e $r_s = +0.90, n = 6, p = 0.01$

 6 f
 7 c
 8 g
 9 a
10 e
11 j
12 b
13 i
14 d
15 h
16 a 0.661, retain
   b 0.990, retain
   c 0.148, reject
   d 0.886, retain
   e 1.000, retain

4 What is the probability that a Pearson r of -0.50, from a sample of 12 subjects, could have been merely a long-shot accident?

5 Identify and describe three instances of the "causation fallacy" as it sometimes affects us in everyday life.

6 Inasmuch as it is possible to use the Spearman r any time it is legitimate to use the Pearson r, why is it considered wise to use the Pearson r whenever possible?

7 A lab technician performed a breathalyzer test twice in quick succession on each of 10 "clients." The pairs of readings were 0.21, 0.20; 0.18, 0.19; 0.10, 0.12; 0.06, 0.06; 0.05, 0.07; 0.19, 0.18; 0.23, 0.22; 0.12, 0.14; 0.04, 0.03; 0.02, 0.04. Using the *split-half technique* (every second score is compared with the one preceding), determine the *reliability coefficient* of the breathalyzer.

8 The next morning one of the clients, the one with the fifth pair of *breath*alyzer readings, appeared in court and was charged with having a *blood* alcohol level of over 0.06. Is there an assumption being made by the court officials, and, if so, what is it? (In technical terms, please.)

9 A student finds a correlation of -0.65 between duration of hospitalization and amount of initiative shown in ward activities, in a group of 50 mental hospital patients. Initiative was measured on a rating scale of the type shown below. What measure of correlation was used?

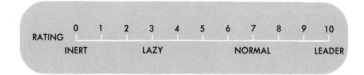

10 How could the student researcher of Question 9 establish whether prolonged hospitalization was the major *cause* of diminished initiative?

11 A statistical survey was undertaken in the Emerald Isle to ascertain, among other things, the degree of affinity between Irish humans and Irish leprechauns. The method chosen was ingenious and also had certain fringe benefits for the researcher. A list of Irish pubs was obtained, and outside each pub an observer made a count of the number of each species seen entering. At the same time, a measure was made of how much bourbon and bog water was consumed in each pub. The survey was continued over the weekend of March 17. Here are the data for County Kayo.

Pub	Humans	Leprechauns	Pints
The Horse & Onions	36	4	132
Tara's Harp	14	12	44
The Black and Tans	0	0	0
Paddy's Place	27	15	50
Boggy Hollow Motel	4	19	6
The Blarney Hilton	50	5	196
The Sow and Piglets	44	10	56
The Leprechauns' Rest	6	18	10
Harvey's Hole	30	10	31
Mr. O'Malley's	41	8	160
The Bent Shillelagh	18	18	11
The Tourist Hotel	51	11	203

a Draw a scattergram of each relationship that can be ascertained from the data.

b What would you do to determine the amount of affinity between humans and leprechauns?

c Go ahead and do it.

d What would you do to determine the relationship between human patronage and spirit consumption?

e Go ahead and work it out.

f Explain any difference between the techniques chosen in (b) and (d).

g What would be the critical value of r for making a generalization about the relationship between human patronage and leprechaun patronage?

h Can such a generalization be made? Explain.

12 Duplicate the graphs shown in Figure 9.13, and relabel the abscissa for the following values of n.

100 4 40

You may do this by eye. What general rule seems to apply here?

13 You are on the trail of a correlation that you suspect is "out there" but is basically rather weak. You figure that, at best, a 0.1 correlation might be found in any study you run. What should you do?

chapter 10
regression: measuring predictability

Let us return to the problem with which we started Chapter 9, the correlation between educational level and income. Suppose we know that the correlation is high, say +0.90, and want to consider another question: "Mr. Scoreshooter makes $24,000 per year; what educational level is he likely to have?"

You may reply that Mr. Scoreshooter *could* have *any* educational level, and this is of course true. But the problem is not to state his educational level with certainty; it is to make the best possible educated guess (known, as you will recall, as a *point estimate*). Discrepany will almost certainly exist between the guess and reality, but over the long run we want the "average guess" to correspond with reality.

First, suppose that you did not know Mr. Scoreshooter's income and merely had to guess at his educational level. What figure would you suggest for the amount of schooling possessed by this completely unknown person? With such total ignorance, the best guess you could make would be the mean educational level of all the men in the population. Figure 10.1 shows the distribution of such educational levels for 20 randomly selected men. Mr. Scoreshooter could be any one of them, so the population mean value μ would be the best guess, and the sample mean value $\bar{X}$, as an unbiased estimate of μ, would be the best *available* guess. Since we know of the +0.90 population correlation educational level and income, let us redraw Figure 10.1 so that the educational distribution on the Y axis is plotted against income distribution on the X axis. Such a cross-plot, or scattergram, is shown in Figure 10.2. Examining Figure 10.2, we see clearly that with a $24,000 income, Mr. Scoreshooter very probably has at least a B.A. degree. Prediction of educational level was rendered more exact by knowledge of

a the degree of correlation between educational level and income, and

b Mr. Scoreshooter's income.

How accurate can such predictions be? The answer depends partly upon the size and randomness of the sample that gave the inferences about population mean income and the population variance of income. The precision of estimating μ from $\bar{X}$ increases with larger n, and bias in the estimate is avoided with random sampling. Accuracy of prediction also depends upon the degree of correlation. Imagine a perfect correlation of +1.00, in which the plotted points lie along a straight line. Prediction in this case would be exact: a perpendicular

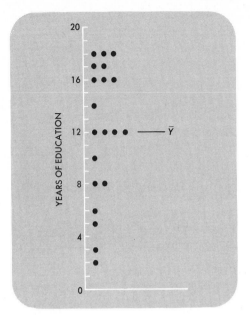

Figure 10.1
Distribution of educational levels of 20 randomly selected men. These data are the same as sown in Figure 7.1. As shown here, the data would justify a "best guess" prediction that Mr. Scoreshooter (a possible newcomer) should have 12 years of education.

from the known value on the X axis to the line, and another perpendicular from there to the Y axis would give an exact prediction. If we are using z scores, we would find that predicted z_Y = given z_X).

With a 0 correlation, no relationship between X and Y, the best prediction of any Y value would be the mean value of scores on the Y axis. With z scores, we would have predicted z_Y = mean z_Y = 0.

With a correlation coefficient between 0 and 1.00 and a linear relationship, the predicted z value of Y would be *between* 0 and a value of z_Y equal to the known z_X value. The reason can be considered as a kind of compromise. Obviously, the greater the value of r, the less compromise is necessary because the greater is the precision of prediction. Thus arises the principle of regression: the predicted value on the Y axis *regresses* toward the mean as the coefficient of correlation diminishes from 1.00 to 0, for any given value of X. (The term **regress** is from the Latin *regredi,* meaning "to go back"—in this case, back toward the mean.)

Provided that the raw scores on X and Y are converted to z scores, the regression principle is easily quantified as a formula:

$$\text{Predicted } z_Y = rz_X$$

As can be seen, the degree of regression depends entirely on the value of r, the Pearson coefficient. Figure 10.3 shows this relationship graphically for any desired given values of X (or z_X). The straight lines are termed **regression lines**.

calculating regression lines

In Figure 10.3 the slope or steepness of the regression lines is seen to be a function of the correlation coefficient. When $r = 1.00$ the line

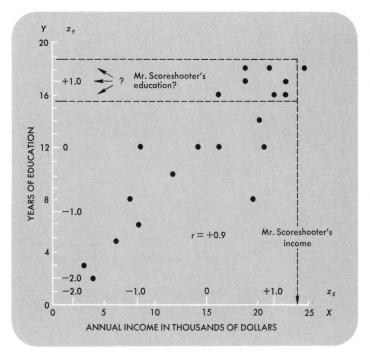

Figure 10.2
The Y scores of Figure 10.1 are now plotted against their corresponding X values, and the correlation is very high, $r = +0.9$. With an income of $24,000, Mr. Scoreshooter is likely to have 16 or more years of education. This is very probably a much more accurate guess (or estimate) than the one based on Figure 10.1.

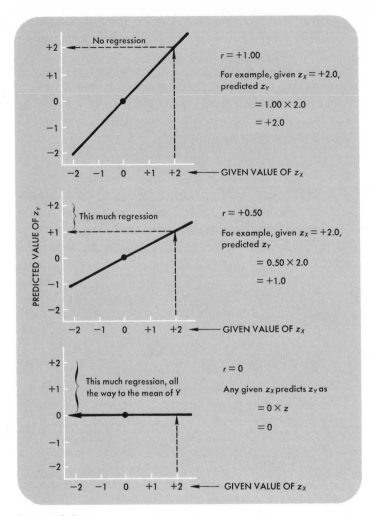

Figure 10.3
Regression lines for rs of 1.00, 0.50, and 0. The slope of the regression line is a function of r. The z scores on the coordinates refer to the X and Y values of the scattergram, but the latter is omitted for greater clarity.

slopes 45° from the X axis; when $r = 0.5$ it slopes 22.5°; when $r = 0$ the line slopes at 0°, i.e., not at all. In general, when the coordinates are scaled in z units of equal size on both axes, the angle the regression line makes with the X axis is a linear function of r.

$$\text{degrees of slope} = 45r$$

This principle may be used to construct actual regression lines, but there is an even easier way, based upon the fact that a straight line is determined by the positions of any two points on it. It's still a good idea to scale your X and Y axes so that they have identical z units.

One of the two points that will determine the regression line can be the intersection of the two means, $\bar{X}$ and $\bar{Y}$. Whether the coordinates are scaled in raw or z scores doesn't matter, provided only that the scaling along the coordinates properly indicates the cardinal spacing among the scores. The other point may be established as follows.

Arbitrarily select any score value X (but it should be reasonably far from $\bar{X}$). Then, using the regression formula provided below, calculate the predicted Y value, and plot it on the graph. Draw a straight line through the two points, and there's your regression line. If you wish to verify it, calculate a third point on the other side of $\bar{X}$ and check that the regression line runs through it.

calculating regression from raw scores

The z-score version of the regression formula is convenient to memorize and awkward to use. However, it is quite easily translated into a raw-score version that is awkward to memorize and easy to use:

$$Y' = \bar{Y} + rs_Y \frac{X - \bar{X}}{s_X}$$

The symbol Y' is read "Y prime" and stands for the **predicted value of Y**.

example of use

Among our old friends the leprechauns there is a reasonably high correlation between years of experience and amount of gold each one manages to scrounge on an average day from under the world's rainbows. Indeed, the value of r is $+0.90$. Leprechauns have a mean age $\bar{X} = 100$ years, with $s_X = 25.0$ years. (The wee folk are highly cooperative about furnishing convenient round numbers for their vital statistics.) The mean gold rip-off $\bar{Y} = 20$ beqa per day, with $s_Y = 4.0$

beqa. So in walks old Eamon O'Malley, aged 160 years, with a twinkle in his eye and a bet that you can't guess the amount of today's haul. Well, you can at least make an inspired and unbiased point estimate:

$$\text{estimate} = 20 + (+0.9)(4)\frac{(160 - 100)}{25}$$

$$= 20 + 8.64 = 28.64 \text{ beqa}$$

When you coyly suggest this amount to old Eamon, he's a bit annoyed, because it's a pretty close guess. Not exact of course, but close.

You could try working the same data using the z formula, and it would not be too difficult. Only with leprechaun data, however, can such smooth nonfractional z scores be counted on.

Try this one yourself: How much would you expect young Kevin O'Portly, age 20, to have made?

When calculating predicted values or considering regression in general, make it a rule always to predict from X to Y; let Y be the set of scores among which your predicted value will fall. It's equally possible to predict the other way, from Y to X, but why go to the trouble? One set of formulas is quite enough; with two sets it is too easy to get confused. Predict from X on the abscissa to Y on the ordinate: this is consistent with predicting from the independent variable on the abscissa to the dependent variable on the ordinate.

prediction and the regression line

The basic purpose of a regression line is *prediction*, giving a point estimate of the value of Y' that corresponds with any given value of X. It should not be thought of as a line of best fit to the data. In Figure 10.4, where $r = -0.6$, the line of best fit runs through the middle of the scattergram, at a 45° slope. It is the line that you might draw free hand to summarize the relationship. But it has no real use. The regression line, you will note, is flatter and closer to horizontal. It is *not* there to describe what the data look like; it is there merely to give unbiased point estimates of Y, given r and values of X. The two lines are the same of course, when $r = 1$, but when $r = 0$ the regression line is still very much there, in a horizontal position. However, the concept of line of best fit no longer makes any sense, since the scores are scattered all over the place.

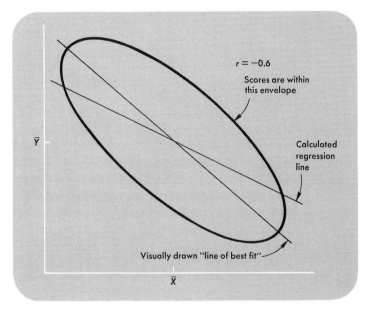

Figure 10.4
As the degree of correlation becomes less than perfect, the value of r
decreases, the scattergram becomes oval in shape, the line of best fit
becomes increasingly inappropriate for prediction, and the regression
line pivots toward the horizontal.

Although the regression line provides a best estimate of Y given X,
the actual value of Y (if it were known) is probably not exactly the
same as our predicted value, Y'. For any particular value of X, say
X_j, there is a sampling distribution of Y scores, as illustrated in Figure 10.5. The sampling distribution of Y scores corresponding to X_j is
the distribution of all the Y scores that you would find corresponding
to X_j if you looked at every possible sample. Let's now turn to this
idea.

"explained" and
"unexplained" variation

The concepts of explained and unexplained variation are difficult for
most students, probably because the terminology is a bit misleading.
In the interests of comprehension, if not tradition, the topics will now
be renamed.

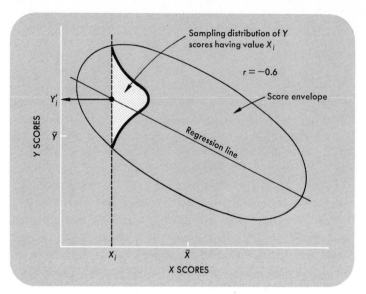

Figure 10.5
The regression line gives a point estimate of Y', but around this will be a sampling distribution of Y scores for any value of X, here shown as X_j. The sampling distribution (shaded here) is an up-ended Gaussian distribution. By treating it as a probability distribution, we can judge the probability of any Y_j score falling within any distance from Y'. The scores are not shown here, but they fall inside the oval envelope.

predicted and unpredicted variation

The first thing to bear in mind is that we are considering only the variation of the Y scores around $\bar{Y}$. Here we would like to think of the overall sampling distribution of Y around μ_Y, *but* must be content with $\bar{Y}$ as a close approximation of μ_Y. The distinction need not concern us here. Now let us begin by considering a case where two variables, X and Y, are perfectly correlated: $r = 1.00$. As for any distribution we know the values of $\bar{X}$, $\bar{Y}$, s_X, and s_Y. Concentrating upon $\bar{Y}$ and s_Y, we can see that most of the scores differ from the mean, since s_Y has a value greater than 0. There is *variation* in the values of Y. To what extent can such variation be predicted from the values of X, bearing in mind that $r = 1.00$? Since all the scores fall exactly along the regression line (because $r = 1.00$), their variation away from $\bar{Y}$ is 100% predicted. For any particular X score, there can be only one corresponding Y score. Thus, when a perfect correlation exists, we can say that all the variation in Y is necessarily predicted from the fact that the X scores vary. All the Y scores are right on the regression line, and being located on the regression line means that every value of Y will equal Y'. See Figure 10.6(a).

For this argument and the ones following to be valid, we must be able to assume that r has the value 1.0 (or whatever other value) by necessity rather than accident. A different sample would be expected to yield the same value of r; essentially, we assume that we are using the value of ρ rather than r. We also assume that our predictions from X to Y involve their sampling distributions, not just the particular samples that we have actually taken. If the sample was random (which we have assumed already), it will be a fair, if somewhat inaccurate, representation of the sampling distribution. These points need not concern us except in one way: since the closeness of r to ρ and also the similarity of the sample to the sampling distribution both depend upon n, efforts to predict Y from X should be undertaken only when r and n are jointly large enough to make r significant. The Table of Significant Values of r in the Appendix can be used to decide whether r is large enough to reject the null hypothesis that $\rho = 0$. Clearly, it would be illogical to use r as a basis for predicting Y from X, when you have no basis for believing that ρ is other than 0.

Let's now consider a second case, where $r = 0.5$ and we still know the values of $\bar{X}$, s_X, $\bar{Y}$, and s_Y. (See Figure 10.6(b)). In this example the scores will be in a slanted oval that is bisected by a horizontal perpendicular from $\bar{Y}$. The Y scores within the oval quite obviously vary from $\bar{Y}$, that is, from the horizontal line. The oval is also bisected by the regression line, which crosses the line from $\bar{Y}$ at some angle. The regression line indicates the extent to which the scores making up the oval vary predictably from $\bar{Y}$ as a result of their $+0.5$ correlation with X; the shaded areas between the two lines are analogous to the amount of variability in Y that is predicted on account of the correlation. The unshaded area is analogous to the amount of variability that is not so predicted.

When we look at a single Y score whose X value is shown, we can see that its total variation away from $\bar{Y}$ can be broken down into two portions, shown to the right of the scattergram. Part of the variation is unpredictable, but a certain amount *is* predictable because of the

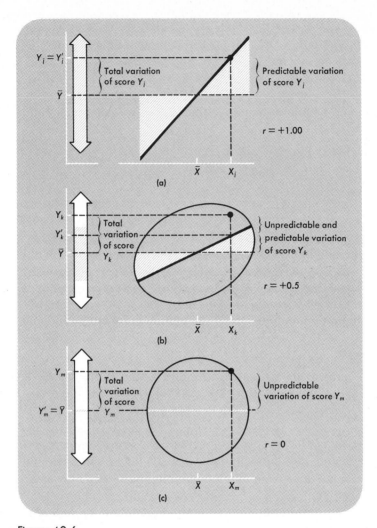

Figure 10.6

The amount of variation of any Y score from $\bar{Y}$ may be divided into two
components: variation *predicted* on account of the correlation of Y with
X, and variation *not predicted* from the correlation of Y with X. The
total variation of any one score, and hence of all scores taken together,
must be the sum of the predictable and unpredictable amounts of variation. (a) When $r = 1$, all the variation in Y (denoted by the vertical arrow
to the left) is accounted for by variation in X. A single instance is shown
for score Y_j. The predictable variation is shown as the shaded area.
(b) With $r = 0.5$, some of the variation of Y_k from $\bar{Y}$ is predictable, but
some is not. The white areas of the arrow and of the scattergram envelope symbolize the unpredictable amount. (c) With $r = 0$, none of the
variation of Y scores from their mean $\bar{Y}$ can be accounted for by corresponding values of X.

+0.5 correlation. The sum of the predictable amounts for all scores obviously makes up the shaded area as a whole.

Our third case, shown in Figure 10.6(c), is with 0 correlation. The scores lie within the circle, bisected by the line from Y, which is also the regression line. The shaded area has disappeared, and it is easily seen that the variability of any or all Y scores from $\bar{Y}$ is unpredictable from the X scores.

In any distribution there is a total amount of variance, s^2. As implied in the discussion above and in Figure 10.6, the total variance may be broken down into two components:

total variance = predicted amount + unpredicted amount

$$\frac{\sum (Y - \bar{Y})^2}{n} = \frac{\sum (Y' - \bar{Y})^2}{n} + \frac{\sum (Y - Y')^2}{n}$$

At this point, a word of comfort: don't worry about the prospect of a lot of arithmetic involved with this formula. It is just conceptual. In fact, we can simplify it in terms of percentage of total variance:

$$1.00 = r^2 + k^2$$

Here, you can see, the gods have come to your rescue, because the proportion of total variation in Y that is predictable from a given correlation with X is merely the **square of the correlation coefficient**, r^2. So if $r = 1.00$, 100% of the variation in Y is accounted for. If $r = 0.5$, 25% of the Y variation is predictable from X. If $r = 0$, none of the Y variation is predictable. Figure 10.6 presents a visual analogy to this argument, but should not be interpreted mathematically, since it does not portray the effects of squaring all the deviations between Y, Y', and $\bar{Y}$.

The name given to r^2 is the **coefficient of determination**. We symbolize the quantity $(1 - r^2)$ as k^2. This quantity is called the **coefficient of alienation**, referring to the degree to which the Y values are "alienated" from the X values, i.e., not predictable from or related to them. It is apparent that r^2 and k^2 must sum to 1.00, 100% of the variance of Y.

Let's now see how these two coefficients are useful.

The main virtue of r^2, the coefficient of determination, is that it cures unjustified optimism. Let's say you have found a correlation of +0.7 between two variables; this is quite a strong relationship. You think you have found something! But then you look at r^2 and realize that less than half the Y variance ($0.7 \times 0.7 = 0.49$) is predicted by—

or explained by—the correlation with X. Over half the Y variance must thus be predicted from things you know nothing about. Suddenly you feel less omniscient.

The major value of k^2, the coefficient of alienation, is found in the next section, which contains the pot of gold you deserve for having mastered (though in disguised terminology) the difficult topics of **explained and unexplained variation**. Please note that **explained** and **unexplained** do *not* mean *caused by* and *not caused by*. If you are not convinced that there is a difference, contemplate the startling fact that *everything* above could be looked at the other way around. We could just as well ask about predicting X from Y. Some of the variation in X is predictable from the variation Y. Obviously, though, causation cannot flow simultaneously in opposite directions. Of course, this parting shot need not be fired if your correlation involves *manipulation* of the X variable and *observation* of the Y. Then you are doing an experiment, in which predictability reflects causation, which does flow just the one way.

standard error of an estimate

We return to the patiently waiting Eamon O'Malley, whose gold-getting performance you so amazingly predicted, though with some slight error. It is now possible to relate that error to the coefficient of alienation.

Suppose you could persuade old Eamon to record his daily gleanings for the next several months (a short time for leprechauns). Provided that the statistical values given earlier in connection with this example remain valid, you can be sure that your original estimate of 28.64 beqa will remain valid too. This figure will thus turn out, in the long run, to be the mean value of what he delivers. But the actual values will vary around that mean; there will be unpredictable day-to-day variability in his performance, unpredictable variation that some version of a standard deviation might measure. Apart from going ahead and actually computing the standard deviation of Eamon's many hauls, is there any way in which that value of s might be obtained?

Why not use the standard deviation for the entire leprechaun popu-

lation? The answer is that some of that total variation (for all leprechauns) is explained or predicted by the known value of r, which is
$+0.9$, and the predicted portion should not be involved here; after all,
predicted variation is not the same thing as *variable* error, or unexplained variation, which is what we are talking about. We want the
unexplained portion of the total variance. This can be had easily
by multiplying s_Y^2 (*the total Y variance*) by $(1 - r^2)$, which is the unexplained fraction of the total. We call this quantity

$$s_{Y \cdot X}^2 = s_Y^2 (1 - r^2)$$

and it can also be given as the standard deviation,

$$s_{Y \cdot X} = s_Y \sqrt{1 - r^2}$$

The term $s_{Y \cdot X}$ (reads Y dot X) is the **standard error of the estimate
of Y, given the value of X**. The term *error* reflects the fact that
an explainable portion of the standard deviation has been removed,
and variable error is the only part left. Anyway, the standard deviation of Eamon's performance, and that of all leprechauns of any given
age, is

$$s_{Y \cdot X} = 4.0 \sqrt{1 - 0.90^2}$$

$$= 4.0 \sqrt{1 - 0.81}$$

$$= 4.0 \times 0.436$$

$$= 1.743 \text{ beqa}$$

What we've done is to gain precision in our estimate, as a result of
removing the explainable part of the variability in Y.

Looking at the up-ended frequency distribution shown shaded in
Figure 10.5, we might wonder about its shape. Yes, the shape is
Gaussian, or close to it. At least for our present purposes we assume so.
The value of $s_{Y \cdot X}$ is the variable (i.e., chance-determined) error remaining after the constant "error" due to r^2 has been removed. And
as we've seen, the distribution of chance-determined scores—as in an
infinitely large Pascal triangle—tends to be Gaussian. (This is an
oversimplification, but it will suffice for our purposes.)

We know the standard deviation of that Gaussian distribution: it
is the quantity we are calling $s_{Y \cdot X}$. We also know, from Chapter 7, that

68.26% of the area lies within one $s_{Y \cdot X}$ of the mean, the latter being the value of Y'. Or using the Table of Areas under the Gaussian curve in Appendix 1, we could find the probability of a Y value lying any distance from Y'. We know that 99.74% of the Y values corresponding to $X = 160$ years should fall between -3 and $+3$ standard deviations from Y' on that Gaussian curve. Translated into probabilities, we know (*if* our many assumptions about reliability, Gaussian shape, etc., are correct), that 99.74% of Eamon's ingot harvests will fall between -3 and $+3$ standard deviation units (Eamon's standard deviation units, 1.743 beqa) away from his estimated mean harvest. That is, they will fall between

$$28.64 - (3 \times 1.743) = 23.411 \text{ beqa}$$

and

$$28.64 + (3 \times 1.743) = 33.869 \text{ beqa}$$

If you can persuade old O'Malley to let you bet on a *range* of possible scores, rather than an exact *point*, you should find about 99% of his assays between these limits, and that could be useful to know, gold prices being what they are. But remember, too, probability theory *never* guarantees anything.

It is not always possible to make the assumption that only random factors, apart from the constant relationship with X, are causing variation of the Y scores around the regression line. If another systematic factor is at work unknown to us, it can produce some strange problems. There could be interference with a necessary assumption we made in guessing the amounts of Eamon's gold, namely, that the value of $s_{Y \cdot X}$ holds constant across all values of X. When it does hold constant, the overall relationship between X and Y is said to be **homoscedastic** (pronounced *homo-sked-ast-ik*). Figure 10.7(a) shows an instance of **homoscedasticity**, and parts (b) and (c) show instances of **heteroscedasticity**. In the latter cases, predicting the 68%, or 95%, or 99% confidence range for any Y value is going to be a chancy affair. The essence of being homoscedastic is for $s_{Y \cdot X}$, the standard deviation of the Y scores above the value X on the abscissa, to be the same for all possible values of X. If $s_{Y \cdot X}$ is smaller for some values of X and larger for others, as in Figure 10.7(b) and (c), you're up the creek. You can still make a point estimate of Y' given X, but no one value of $s_{Y \cdot X}$ will determine your confidence interval.

You may be wondering about one thing: We've been using the

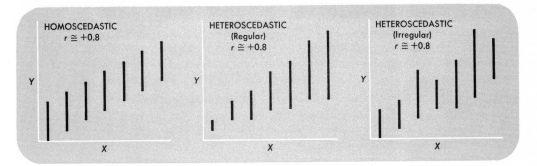

Figure 10.7
Imagine that the vertical lines represent the variability of the Y scores over each value of X. In (a) this variability is constant, so the correlation is homoscedastic. In (b) the variability of Y around its regression line gets greater as X increases, a situation quite often found in real-world correlations. In (c) the Y variability varies unpredictably for different values of X, a situation that can happen when not enough cases have gone into the correlation. Parts (b) and (c) show heteroscedastic correlations.

Pearson r as the bedrock for our discussion of regression, but what about the Spearman rank-order coefficient of correlation, r_S? Can it be used similarly? The answer is no. It should be intuitively clear that cardinally scaled scores are necessary in regression, and r_S is merely ordinal.

words the knowledge of which will correlate with grades

PRINCIPLE OF REGRESSION

REGRESSION LINE

SLOPE

Y' (PREDICTED VALUE OF Y)

LINE OF BEST FIT

EXPLAINED and UNEXPLAINED VARIATION

PREDICTED and UNPREDICTED VARIATION

COEFFICIENT OF DETERMINATION, r^2

COEFFICIENT OF ALIENATION, $(1 - r^2)$ or k^2

Vocabulary (*continued*)

STANDARD ERROR OF ESTIMATE, $s_{Y \cdot X}$

HOMOSCEDASTIC

HETEROSCEDASTIC

/elf-te/t on chapter 10

The same rules as for earlier similar efforts.

1 Any Irishman knows that the chief business of leprechauns is hunting for gold beneath rainbows. Not so well known is the fact that the necessary ability must be learned; it is certainly not innate to leprechauns. About 10,000 years ago they decided to set up a training academy. A problem became apparent immediately: The necessary training period was longer for the younger leprechauns; the value of r was 0.80 between age and duration of training. And leprechauns of all ages wanted to take the training. Chaos would have resulted if the wee folk had not discovered regression about 9950 years before people managed to. The mean leprechaun age is (and always has been) 100 years, with $s = 25$, and the mean training duration was 5 years, with $s = 1$. The variability of necessary training time was the same among all leprechaun age groups.

a The value of r was +0.80.
b r was −0.80.
c The coefficient of determination was −0.64.
d r^2 was +0.64, and k^2 was +0.36.
e Leprechauns aged 125 years would average a training period of about 50 to 51 months.
f The 125-year age group would average about 14 to 15 months of training.
g The correlation was heteroscedastic.
h Only about 1% of the trainees in the 125-year age group would not finish training somewhere in the range between 2.4 and 6.0 years.
i With the knowledge available as given above, the organizers would have been able to predict the necessary training time for 95% of its leprechauns, without being more than about 1.2 years off in either direction.

2 A relationship has the following peculiarity: it is linear and involves cardinal scaling, but the variability of the Y values at certain values of X is very small, while at other values of X the Y values vary very widely. What can we say in this connection?

a r would not be the appropriate coefficient.
b Prediction of Y' from X as a point estimate would be completely impossible.
c The confidence interval for any predicted value of Y would not be reliably known.

SCORING AND ANALYSIS of the Self-Test questions on the other side of this page.

Give yourself +3 points for each choice circled correctly, −2 for each wrong choice that was circled, and −1 for every correct one you missed.

Answers to Question 1

a F The younger the leprechaun, the longer the training. This is an in-
b T verse relationship, so we have a negative correlation.

c F If r is −0.8 (or +0.8 for that matter), the value of r^2 is +0.64. k^2 is
d T 1.00 − 0.64 = +0.36.

e T $Y' = \bar{Y} + rs_Y(X - \bar{X})/s_X = 5 + (-0.8)(1)(125 - 100)/25$
f F $= 4.2$ years

4.2 years is the same (even for leprechauns) as 50.4 months.

g F See the very last piece of information given in this question. This says that the relationship was homoscedastic.

h T 99% of cases fall between $\pm 3\, s_{Y \cdot X}$ on each side of the value of Y', which was 4.2 for this age group. $s_{Y \cdot X} = s_Y \sqrt{1 - r^2} = 1.0 \times 0.6 = 0.6$. And $3 \times 0.6 = 1.8$, so 99% of cases will fall between (4.2 ± 1.8) and $(4.2 + 1.8)$.

i T It's a homoscedastic relationship, so 95% of any age group will finish training between $(Y' - 1.2)$ and $(Y' + 1.2)$.

Answers to Question 2

a F If the line of best fit is really linear, then r is okay.

b F Not completely impossible. A mean predicted value (Y') would be
c T possible, but its range of confidence would not be accurately known because of the heteroscedasticity.

2 (*continued*)

   d r could not be ± 1.00.

   e The relationship is not homoscedastic.

   f A regression line would be meaningless.

   g The coefficient of alienation would, strictly speaking, have different values depending upon the value of X.

3 The fact that the regression line rotates around the intersection of $\bar{X}$ and $\bar{Y}$ as r changes

   a indicates that the mean value of X *always* predicts the mean value of Y, regardless of r.

   b reflects the fact that when $r = 0$, any X score predicts a Y score equal to $\bar{Y}$.

   c means that the phenomenon of regression does not affect average scores in a distribution.

   d makes it similar to the line of best fit.

   e depends upon the score clustering of the scattergram also rotating.

   f results in the slope of the line having a perfect positive correlation with r.

4 Explained variation in Y

   a may be caused by variation in X.

   b cannot be caused by variation in X.

   c is 0 when r is 0 and 1.00 when r is 1.00.

   d is 0.50 when r is 0.50.

Answers to Question 2 (*continued*)

d T An r of 1.00 requires *all* scores to lie along the regression line. Hence, there could be no variability around that line, so for any value of X there would be *no* variability of the corresponding Y scores.

e T It certainly isn't.

f F Regression lines refer to *mean* predicted Y values, which heteroscedasticity does not affect.

g T True. Wherever there is little variability of the Y scores at given locations along the X axis, it would seem that the coefficient of alienation should be very small at those locations. Where there is a lot of Y variability, there would be a huge coefficient of alienation at those locations. The official computed k^2 must therefore be an average of some kind.

Answers to Question 3

a T Look at Figure 10.3; the mean always predicts the mean.

b T Look at the bottom panel of Figure 10.3(c); the regression line has rotated to a horizontal position level with the value of $\bar{Y}$.

c T Regression is always toward the mean of Y. An average score is already at $\bar{Y}$; therefore, there is nowhere for it to regress to. Thus, it isn't affected by regression.

d F No, it is not at all the same thing as the line of best fit. The latter does not rotate as r changes; its significance merely fades as r diminishes from 1 toward 0.

e F The regression line does not describe the scores present in the scattergram; it merely predicts future values. It has no direct relationship with the scattergram.

f T Yes; slope $= 45r$. This is a linear relationship, so the correlation between slope and r would be $+1.00$.

Answers to Question 4

a T It *may* be caused by X, but it does not have to be. Remember, *not*

b F *necessarily* does not mean *necessarily not*.

c T True. Look at Figure 8.5, and ponder deeply.

d F Go on! r^2 (explained variation) does not equal r (correlation) unless both are 1.00!

4 (*continued*)

    e is -0.49 when r is -0.70.

    f necessarily involves explained variation in X.

5 The value of $s_{Y.X}$

    a can never be identical to s_Y.

    b will always be 0 when r is 0.

    c must always be larger than s_Y.

    d is s_Y minus the explained portion.

    e is inversely correlated with the value of r^2.

    f is reliable when the correlation is homoscedastic.

Here are some short-answer (completion) items. Put the answer in the space at the left of the question, in which the missing word, number, or short phrase as indicated by *. Two points each.

6 _____ When r is $+0.5$ and z_X is $+2.0$, then z'_Y is *.

7 _____ When $r = -0.5$, the unpredicted amount of variance in Y is * % of the total variance.

8 _____ When $r = -0.5$, $\bar{X} = 16.5$, $s_X = 4.0$, $\bar{Y} = 40$, and $s_Y = 8$, a value of $X = 8.5$ would predict Y' as *.

9 _____ For the previous question, $s_{Y.X}$ would be *.

EVALUATION OF YOUR PERFORMANCE on this insane self-quiz. Add $+3$, or -2, or -1, as per previous directions, and see what you get. The maximum possible is 65.

        45 to 65   You have found the pot of gold.

        30 to 44   You don't quite deserve it yet.

         5 to 29   A pot, maybe, but *gold*?? Do some more digging, fast.

        under 5   If you have a good singing voice, try for a chorus part in *A Midsummer Night's Dream*. It will take you away from all this. Alternatively, find out what's holding you back—it sure isn't the weight of any gold you dug out of Chapter 10.

Answers to Question 4 (*continued*)

e F Were you caught here? The values are correct, but the sign is wrong. r^2 cannot be *minus* 0.49; +0.49, yes.

f T Remember that variation along one coordinate explains variation along the other. We have chosen to stay with explaining Y in terms of X, but the other way around is equally okay.

Answers to Question 5

a F It can when r is 0, and none of s_Y is explained. Then all of s_Y must be $s_{Y \cdot X}$.

b F When r is 0, $s_{Y \cdot X}$ must be maximal, equal to s_Y.

c F $s_{Y \cdot X}$ can equal s_Y when r is 0, but it can never be larger than the amount of which it is a part.

d T True in a general sense. For arithmetic exactness, you would have to use the variance rather than s.

e T If the value of r^2 increases, the amount of explained variation is increasing, and thus the unexplained portion is decreasing. Increase in the one with decrease in the other means an inverse relationship.

f T Yes; it is reliable *only* when there's homoscedasticity. See Figure 10.7.

6 +1.00. $z_Y' = r z_X = (+0.5)(+2.00) = +1.00$

7 75%.   The predicted amount is 5^2 or 25%.

8 32.

$$Y' = \bar{Y} + r s_Y (X - \bar{X})/s_X$$
$$= 40 + (-0.5)\, 8(16.5 - 8.5)/4$$
$$= 40 - 8$$

9 ≈ 6.9.

$$s_{Y \cdot X} = s_Y \sqrt{1 - r^2}$$
$$= 8 \sqrt{1 - 0.25}$$
$$= 8 \sqrt{0.75}$$

*s*ome ordinary problem*s*

1 Draw a series of ten graphic figures, one for each integer value of r between 0 and 1.00 inclusive. Draw on each, in such a way as to show their interrelationships,

   a the "envelope" (ellipse, oval, or what have you) that would surround the scatter points in a typical homoscedastic relationship;

   b the line of best fit, drawn to a thickness proportional to its goodness of representation;

   c the regression line of Y on X, at its proper angle or slant;

   d the size of the value $s_{Y \cdot X}$.

You might even draw them on transparent plastic, for use on an overhead projector. The message comes through very clearly when the figures are superimposed one after the other.

2 A clever friend of yours got a score of 88 on the first statistics exam, on which the mean score was 64 and s was 8. Partly because he was so clever, he contrived to miss the second exam. On the second exam the rest of the class got a mean score of 45, and s was 5. Our friend was given a prorated score of 55 by an instructor who knew simply *everything* about regression. What was the correlation between the two exams?

3 Draw a graph (and perhaps keep it for future reference) showing the relationship between r and the amount of explained variation.

4 Starting with the data given, determine r, r^2, and so forth, enough to estimate the value of Y' for $X = 16$, and also to find the 95% confidence limits for Y'.

X	13	17	14	10	9	19	16	11	14	20	15
Y	7	9	8	10	3	10	11	7	8	12	10

   a List the statistical formulas that are needed, and define them by name.

   b Find the value of r.

   c Find the values of r^2 and k^2.

   d Find the values of s_X and s_Y.

   e What is Y' for $X = 16$?

   f Why is your Y' value for $X = 16$ different from the Y value of 11 associated above with $X = 16$?

   g What is the value of $s_{Y \cdot X}$?

   h Between what limits would 95% of the Y values associated with $X = 16$ fall?

5 A leprechaun, Sean Bogthumper, is 2 standard deviations younger than the mean age for the wee folk. How much daily gold (in z score units) would you expect him to bring in, if r equals

+1.00   +0.75   +0.50   +0.10   0   −0.60   −0.90

6 Are all values of Y' subjected to the same amount of regression toward $\bar{Y}$ when r is constant? Explain.

7 Draw a scattergram to show what would happen if a Y' were predicted from a value of X, and if X' were then predicted from a Y value equal to your obtained Y', and if then a further prediction of Y'' were made from a new X value equal to your X', and so on, back and forth. State in words what happens when this is done repeatedly.

chapter II
experiments: mathematics meets reality

At its simplest, an experiment is just a comparison. Essentially, you take two similar groups and do one thing to one group but not to the other, thus producing the **independent variable** of the experiment. You then compare the two groups for any difference in the characteristic in which you are specifically interested, the **dependent variable**. Since the two groups were similar before you manipulated one of them, any subsequent difference between the mean dependent-variables scores of the two groups may be attributed to your different treatments, your independent variable. In fact, such a difference *must* be attributed to your independent variable if your experimental procedures excluded other possible causes. However, this line of reasoning makes one assumption: that the difference noted between your two groups was not *accidental* in the first place.

Accidental? Such things as chance inclusion of generally more skilled subjects in one of the two groups, chance variations in your experimental procedures, and chance fluctuations in subjects' luck do happen. Any number of chance-determined differences *could* combine to push one group's scores slightly higher than the other group's, without the independent variable being involved at all. And yet, because we are concentrating on the independent variable and its parent theory, we are in danger of crediting it with being the cause of our experimental effect even when it was not. Amateur experimenters often fall prey to this error. Of course, there is also the opposite error: being so nervous about the possible role of chance errors that actual effects of the independent variable are discounted when they should not be.

How do we approach this problem?

We start with a very conservative assumption that *the independent variable has no effect*; equivalently (as you recall), *the null hypothesis is true*. The reason for this strange initial assumption will shortly become clear.

Now, to say that the null hypothesis is true means that the difference between the experimental groups is just the end product of a series of random accidents. How likely is such an accident? This depends upon how large the difference is. Remember, we are discussing differences between the *means* of the two groups, and (as you recall from Chapter 8) means of different samples from the same population should fall into a Gaussian distribution, whose standard deviation (the standard error of the mean, $\sigma_{\bar{x}}$) can be estimated. We actually have two of all those possible means. If H_0 is true, those two means should differ only by chance. If many more pairs of experimental and control groups were compared, such chance differences between their means should average out to 0. That's *if* the null hypothesis is true. There should be a *sampling distribution of differences* between means, centered on 0. This sampling distribution will have a parametric mean of $\mu = 0$ and will be Gaussian in shape. Like the sampling distribution of the mean (discussed in Chapter 8), it is the product of chance errors and is Gaussian for the same reason; it has a standard deviation, $\sigma_{\bar{D}}$, whose value it would be nice to know. All of this, remember, rests on the assumption that the null hypothesis is true. $\sigma_{\bar{D}}$ indicates the standard deviation of the difference between means. It is called the **standard error of the difference between two means**.

Figure 11.1 illustrates our discussion up to this point. Look at it now, and keep referring to it as you follow this discussion. Let's suppose that in just one experiment we find a difference between means, $\bar{D}$, as shown. The shaded portion of the distribution represents the probability of any such difference falling at least that distance from μ (whose value is 0). All we need is the Table of Areas under the Gaussian Curve to give us that probability exactly. Then, we could decide whether to retain or reject the null hypothesis.

In order to do this, however, we need the value of the standard error

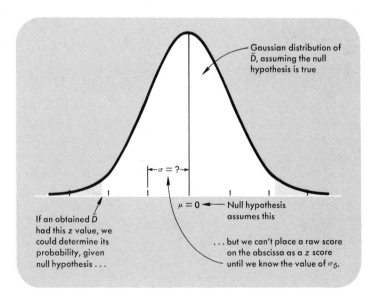

Figure 11.1

Sampling distribution of values of $\bar{D}$ (differences between sample means) assuming the null hypothesis. The value of $\sigma_{\bar{D}}$ is estimated from the sample variances.

of the difference between two means. Only then can we plot our value of $\bar{D}$ as a z score somewhere along the abscissa. Our table of areas applies to the Gaussian distribution of z scores, not raw scores.

the z test in experiments

What we are considering is a straightfoward use of the z test, as described in Chapter 8. Recall that it was used to find the probability that a given sample came from a given population. It's exactly the same here. We want the probability that a given difference between means, $\bar{D}$, is merely one of a population whose mean value μ is 0 and whose standard error we will now estimate.

In referring to the difference between two means, it's common to find the expression $\bar{X}_{\mathrm{E}} - \bar{X}_{\mathrm{C}}$, which is an almost unpronounceable mouthful. In this book we'll use instead the expression $\bar{D}$ (D-bar) which arithmetically works out to the same thing and is more pronounceable. All you need remember is that we are interested in the difference between two means, and the symbol $\bar{D}$ stands for such a difference, which the null hypothesis says should be 0.

the standard error of the difference between two means

The standard error of the difference between two means is quite a mouthful, so let's give it an abbreviation, fast! We'll refer to this parameter, shown in Figure 11.1, as $\sigma_{\bar{D}}$ (called "sigma difference," with the bar over the D to remind us that it's a difference between two *means*). We'll refer to the *best estimate* of the parameter according to our usual rule, using a circumflex: $\hat{\sigma}_{\bar{D}}$ (pronounced "estimated sigma difference"). This is an important concept to us, so begin training.

We might be happier if the *conceptual* basis of $\sigma_{\bar{D}}$ was reviewed first. Imagine that we have a population of scores, with any shape, and that we repeatedly take pairs of random samples, all of the same size n, from the population. Each sample should have a reasonable number of scores, say 30 or more. The means of the two samples are compared each time; sometimes the first will be larger, sometimes the second will be, but in the long run all these differences will balance out around 0, giving the distribution of Figure 11.1. The standard deviation of that distribution is $\sigma_{\bar{D}}$.

Assuming that sample sizes are constant within any sampling distribution, we can see in general that $\sigma_{\bar{D}}$ will be very small when the samples have a large n. The more scores in each sample, the more reliable their means. Remember that μ_{D}, the parametric difference between the two population means, should be 0; thus the values of $\bar{D}$ will be small and the distribution of $\bar{D}$ will cling close to 0, making $\sigma_{\bar{D}}$ small too. If n is relatively small, chance score fluctuations will more seriously affect each value of $\bar{X}$, so the difference between the two means in any given pair may be quite great. In all such possible pairs, these (probably large) differences will still be random, so the distribution will still be Gaussian, but its variability will be increased. The value of $\sigma_{\bar{D}}$ will thus be relatively large. Figure 11.2 illustrates the point.

We can estimate the value of $\sigma_{\bar{D}}$ from the standard deviations or variances of the two samples that make up one pair. The size of the samples is also taken into account, as you would hope after what has just been said about the effect of n on $\sigma_{\bar{D}}$. The estimation is as follows.

$$\sigma_{\bar{D}} \cong \acute{\sigma}_{\bar{D}} = \sqrt{\frac{\acute{\sigma}_{E}^{2}}{n_{E}} + \frac{\acute{\sigma}_{C}^{2}}{n_{C}}} \quad \text{or} \quad \sqrt{\frac{s_{E}^{2}}{n_{E} - 1} + \frac{s_{C}^{2}}{n_{C} - 1}}$$

where the subscripts E and C refer to experimental and control groups (which the null hypothesis says are from the same population anyway). Once we have our estimate of $\sigma_{\bar{D}}$, we can calibrate the abscissa of Figure 11.1 with z units.

To find the probability of a difference between means as large as the one shown in Figure 11.1, we merely perform a variation of the z test. Basically,

$$z = \frac{X - \bar{X}}{s}$$

Here we have

$$z_{\bar{D}} = \frac{\bar{D} - \mu}{\sigma_{\bar{D}}}$$

The $z_{\bar{D}}$ values thus obtained fall into the standard Gaussian (or normal) distribution, and their probabilities of occurrence can be judged from the Table of Areas under the Gaussian Curve, in the Appendix.

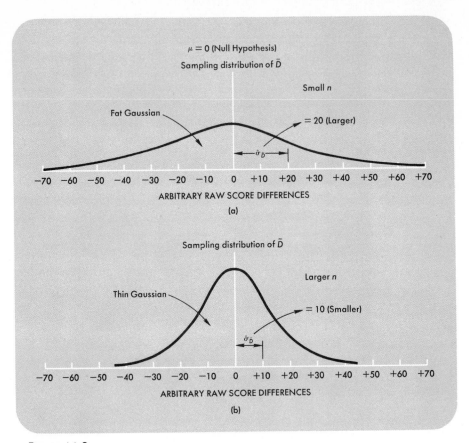

Figure 11.2
Sampling distributions of $\bar{D}$ when sample sizes are (a) small and (b) large. When n is small, the values of $\bar{X}$ will fluctuate more, so the differences between values of $\bar{X}$ will tend to be large.

That's if the null hypothesis is true. *If* our obtained $z_{\bar{D}}$ is too unlikely, we might prefer to reject the null hypothesis, inferring that the population of differences between means does not have $\mu = 0$. This, in turn, would signify that there are *two* populations of scores, one with a certain mean, μ_E, and the other with a different mean, μ_C.

Why should the two populations of scores have different mean values?

1 Well, let's not forget that they *might* still really have the same mean! We may have decided to reject the null hypothesis when it should not have been rejected. If our $z_{\bar{D}}$ value had been 5, for example, we would certainly have rejected H_0, but that $z_{\bar{D}}$ *could* have occurred as a very unlikely accident when H_0 was true. But the probability is low, so we put the possibility aside and look at some other possible reasons.

2 The independent variable might have systematically affected all the scores in the experimental group, thus making the mean of that group differ from the mean of the untreated control group. That would certainly have caused a substantial difference between $\bar{X}_E$ and $\bar{X}_C$. This is the explanation we would like to believe, that our experimental treatment had some effect. But unfortunately, there is another possible reason.

3 Some unsuspected variable might have been *confounded* with the independent variable. The former, not the latter, might be what changed the scores. Such a catastrophe can be prevented only by great care in setting up the experiment. If we manage to avoid (3), we can adopt (2) as our experimental conclusion, though with some slight risk of (1).

a worked example

The leprechauns of County Kayo are sampled at random to form two groups of 42 each. The experimental group is fed a special diet, mainly fermented prune nectar, on the hypothesis that this will sharpen their ability to sniff out gold. The control group gets treated exactly the same, but without the fermented prune juice. Two days of gold digging ensue, and then the gleanings of the two groups are compared.

Here are the data.

	Mean ($\bar{X}$)	Standard Deviation (s)	n
Prune Juicers	48 beqa	5 beqa	42
Controls	44 beqa	4 beqa	42

We see that $\bar{D} = 48 - 44 = 4$, and

$$\hat{\sigma}_{\bar{D}} = \sqrt{\frac{5^2 + 4^2}{42 - 1}} = \sqrt{\frac{41}{41}} = 1.0$$

We can now calculate $z_{\bar{D}}$ as

$$z_{\bar{D}} = \frac{\bar{D} - \mu}{\acute{\sigma}_{\bar{D}}} = \frac{4 - 0}{1} = 4.0$$

If we refer our calculated $z_{\bar{D}}$ to the Table of Areas under the Gaussian Curve, we find an exceedingly low probability of this value occurring by chance, assuming the null hypothesis. The probability is only 0.00006. We reject the null hypothesis, concluding that fermented prune juice does something for leprechauns' noses for gold.

The z test used here is somewhat different from the one we met in Chapter 8. In that chapter, the parameters μ and σ were known, and an exact probability could be expressed for any value of z. Here the parameters are not known. μ is assumed to be 0, and $\sigma_{\bar{D}}$ is estimated from the combined sample variances. Most of the time the estimate is reasonable, but sometimes it may not be.

what do we now mean by population?

The term "population" has taken on a subtle new meaning, not contrary to the former one, but needing a little extra thought. Previously, a statistical population was any defined potential collection of scores from which samples could be taken, or a population of people (or leprechauns), such as the population of County Kayo. *But* until the experiment was done, none of the leprechauns there had ever tasted fermented prune nectar! Yet, we claim that our experimental findings pertain to more than the sample of leprechauns; they must pertain (or why do the experiment at all?) to the FPN-swigging *population* of leprechauns, and such a population does not exist! How can this be?

The answer is that our findings would extend to any number of leprechauns *if* they were to indulge in FPN. The population as such is an abstraction. What we are really doing is making a generalization beyond our actual sample. We are saying that *if* leprechauns swig FPN before going to work, *then* they can expect, on the average, to smell out more gold than they would otherwise.

the concept of significance

The difference between two means $\bar{X}_E$ and $\bar{X}_C$ is called *significant* when its probability falls below some predetermined low value, customarily 0.05 or 0.01. This value, (whichever is chosen) is called *alpha* (α); it is the probability of the difference between the means being due to chance, if the null hypothesis is really true. When we say that a difference is *significant at the 0.01 level*, we are (a) inferring that the null hypothesis is false and that the experimental hypothesis is true, and (b) stating that this inference has a 0.01 chance of being mistaken, given that the null hypothesis is true. If it's false, then rejecting it has *no* probability of being wrong. But we don't ever *know* whether it is right or wrong, so a significant result is described not in a flat statement of fact, but in a statement of adequate probability.

In statistics, the term "significant" does not mean "large," "impressive," "important," or "of practical value." It means "extreme enough to reject the null hypothesis." And since it does not even imply certainty, we must be careful when using it. The null hypothesis is **rejected**, not disproven, and the experimental hypothesis is **supported**, not proven. Take careful note and display your professionalism whenever you can.

type I and type II errors

We are left with a feeling of unease about the probability of being mistaken in rejecting the null hypothesis. The value of α is always greater than 0, so we can always be wrong. Can the risk of error be minimized?

Yes. We only need to decrease α to a lower value such as 0.001 or 0.0001. Choice of α is perfectly arbitrary, and we can therefore decide exactly how much of a chance we are willing to take on rejecting the null hypothesis when it is in fact true. This chance is the value of α.

Why not play very safe and set α at a very low level, say 0.0000001? Because if we did, the value of z (or of any other critical ratio that we have yet to consider) would need to be *extremely* large. And then, if z were found to be very large but not quite large enough, we would run a big risk of retaining the null hypothesis when it was, in fact, false. And this type of error can be just as embarrassing as the first kind.

The first kind of error is called a **Type I error** or **alpha error**: rejecting the null hypothesis when it is true. The second kind of error is a **Type II** or **beta error**: retaining the null hypothesis when it is false. And as we reduce the probability of committing one, we increase the probability of the other. Where is the happy medium?

This question really goes beyond the modest limits of this book, but as a rule of thumb the 0.01 alpha level (associated with $z = 2.576$) will never be criticized. The 0.05 level is also commonly used, but ordinarily no greater risk of Type I error should be permitted. In science it's virtuous to be conservative. It keeps us from jumping to wrong conclusions too often.

How can we tell when a Type I or II error has occurred? We can't. Sometimes we will erroneously retain H_0 when it should have been rejected, and other times we will reject it when it should have been retained. It is always wise to replicate any experiment, re-doing it exactly and combining its results with the first experiment, if the conclusion is likely to have practical consequences. Replication greatly diminishes the likelihood of error, when the two sets of results point the same way.

Note very carefully that our discussion of errors has involved the *null hypothesis*, which is the opposite to the experimental hypothesis. Figure 11.3 provides a good mnemonic.

"I believe in $\alpha = 0.051$ . . ."

A horseplayer who lays a bet must do so *before* the nags have run and must rest with the odds (indicators of probability) stated when the bet was made. If you failed to be on the 20 to 1 long shot that won, you can't go and bet *after* the race.

Strangely, perfectly sane people try to do this all the time in statistics. Suppose you decided before collecting your data that you would reject H_0 if your z ratio reached an α of 0.05. So then you run the experiment, only to find that z did not quite reach the necessary value of 1.960, being only 1.959. Strictly speaking, you should retain H_0, but then, the figure is *so* close that you decide to reject it anyway. Maybe you justify this cleverly by rounding upwards, or retroactively deciding on $\alpha = 0.051$, or in some other crafty fashion. Alternatively, maybe you say that your critical ratio merely "approached significance," thus retaining your honesty along with H_0 but *acting* much as though you had rejected it.

This is rather like watching the end of a horse race and *then* trying to get a bet down on the winner.

The purpose of doing the experiment is to lay a bet with nature. You are saying: "If the evidence has at least such-and-such a degree of probability, I will believe it. Otherwise I will not believe it, but

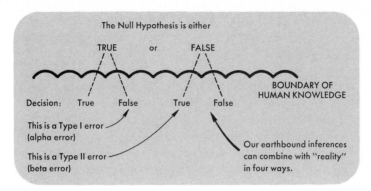

Figure 11.3

keep an open mind." Decisions to come or go, bet or stay, marry or stay single, and reject or retain H_0 are *either–or decisions*. This is a necessary return to two-valued logic, but it is a logic of *action*, not *belief*. While our beliefs will never be 100% positive, our actions must be. So it is not possible to hedge on an experimental conclusion. And if we resort to fudging on our evidence, changing $\alpha = 0.05$ to 0.051 or 0.052 or 0.060 just to suit our wishful thinking, then there is no reason to do experiments in the first place. Just believe anything you want to believe.

More seriously, however, we often do encounter z ratios (or other types of critical ratio) that *are* large without being quite large enough. To let the issue rest there is to run a strong risk of a Type II error. The next step is to repeat the experiment, using fresh subjects. As seen in Figure 11.4, two "nearly significant" probabilities can combine to give a significant probability.

It is wise to remember that while *action* is an either–or thing, *belief* is not. You can get results that have such α levels as 0.01, 0.05, 0.10, and 0.20, and these might be thought of, respectively, as the levels of overwhelming conviction, strong belief, cautious optimism, indecision, etc. From this point of view, you need not set your α level in advance. But if somebody asks you, "Well, *do* you believe it or *don't* you?" there must be a point at which you will say "yes." That would be your critical value for α.

We recognize that the 0.01 and 0.05 levels are arbitrary, doubtless chosen because we have five fingers on each hand. But any fixed probability would be equally arbitrary. If we had six fingers and used the 0.06 level, we could still make intelligent decisions about each null hypothesis we set up.

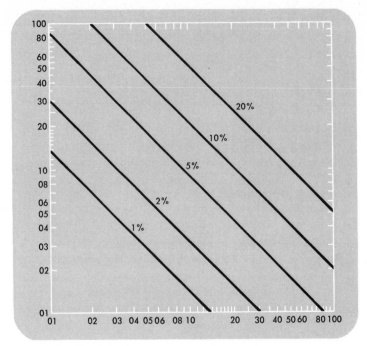

Figure 11.4

Finding the combined probability of two tests of significance. First locate
the value of one probability on the abscissa and the value of the other
probability on the ordinate. Then draw vertical and horizontal lines from
these points. Where the lines meet, draw a diagonal, which shows the
combined probability. The combined probability values shown here are
1%, 2%, 5%, 10%, and 20%. (Adapted from Figure 15.5 of F. J. McGuigan,
Experimental Psychology, Second Edition (1968), Englewood Cliffs, N.J.:
Prentice-Hall, Inc. Based on a communication from P. C. Baker to F. J.
McGuigan.)

controlling for type II error

While a detailed discussion of beta error is beyond our scope, it is
worth mentioning how we can avoid retaining the null hypothesis
when in fact the experimental hypothesis is true. The secret lies in
eliminating all possible sources of variable error in our experiments.
Suppose we have obtained a given difference between $\bar{X}_E$ and $\bar{X}_C$ that
actually reflects a real effect of the independent variable. If, however,
all sorts of sloppy variable errors have entered the picture, then when
we compute $z_{\bar{D}}$, the denominator value, $\hat{\sigma}_{\bar{D}}$, will be quite large (as a
result of the sample s's being unduly large). The larger the denomina-

tor $\hat{\sigma}_{\bar{D}}$, the smaller the value of $z_{\bar{D}}$, possibly to the point where the null hypothesis will be retained when it should not be. Moral: keep your sample values of s as small as possible by carefully controlling your experiment. You could also increase n to get $\hat{\sigma}_{\bar{D}}$ smaller. The art of research design is directed largely to the problem of controlling **error variance**, as $\hat{\sigma}_{\bar{D}}$ is often called. The two simplest prototypes of research design are discussed below.

simple randomized groups

The random-groups design involves two (or more) independently and randomly sampled groups of subjects from the same parent population. The groups are treated exactly the same except for the manipulated independent variable. This design has several advantages and disadvantages.

advantages

a Each subject belongs to just one of the groups. Therefore there is no possibility of the independent variable affecting the performance of a subject in the control group.

b Each subject need be tested, or give a score, only once, and thus the score will not be affected by experience or practice with the experimental conditions.

c Any number of subjects may be used in each group, giving one score apiece; n need not be the same for the several groups.

d This is an easy design to understand; in fact, it is the type considered with our fermented prune nectar example.

disadvantages

There is only one major disadvantage to the random-groups design. It relies on random sampling to ensure that the experimental and control groups are initially equivalent. And sometimes, as we have seen, atypical results can stem from random procedures. It is not hard to imagine situations in which the null hypothesis has been rejected, not because the independent variable was effective, and not because of poor experimental design, but because the two groups differed accidentally *at the very beginning*. (Suppose those 42 leprechauns treated with FPN were champion gold-getters to begin with?) This disadvantage can be minimized by using relatively large groups; with the z test, a minimum of 30 is required anyhow, but possibly groups of several hundred would be safer. For certain experiments, however, a large number of subjects may be unavailable, inconvenient, or impossible.

We have already exemplified random-groups design with our 42 leprechauns on fermented prune nectar, so we omit any further illustration here.

a caution

Whenever the z test is used, as with this kind of design, we assume that the sampling distribution of $\bar{D}$ is Gaussian. Now there are two ways in which we can ensure this in practice.

1 We can simply use samples of about 30 or more, if we know that the population distribution of scores (from which both groups are taken, assuming the null hypothesis) is approximately Gaussian, that is, reasonably bell-shaped and symmetrical.

2 If the population distribution is known or suspected to be otherwise, the z test can still be used, because the Central Limit Theorem ensures that the sampling distribution of means, or of differences between means, will be Gaussian *if the samples are large enough*. This could mean 50, 100, 200 cases per sample, or even more, depending upon how non-Gaussian the sampling distribution really is. (We rarely know for sure, but as a general principle, for this as well as many other reasons, it's good to have values of n as large as possible.)

the matched-subjects design

The major source of error in the random-groups design is that the subjects bring to the experiment intersubject differences which may add up to intergroup differences, adding a lot of "noise" to the data. Such "noise" ends up in the error variance, or error term, as $\sigma_{\bar{D}}$ is often called. As a result, a large n is required to bring $\sigma_{\bar{D}}$ down to a value that will allow rejection of the null hypothesis without great likelihood of a Type II error.

Another way out of this problem is to reduce the amount of noise that originally goes into the experiment along with the signal, which is the value of $\bar{D}$.[1] A major step toward this objective is to match the subjects in the two groups so that one member of each pair is in each of the groups. The matching procedure can vary: the requirement is merely that the members of each pair be similar in all ways that affect the dependent variable. Then the experimental treatment, applied only to the experimental group, will show up that much more clearly as the one thing producing a difference in the dependent variable.

Let's look at the matched-subjects design through a worked example.

sample problem

As we saw in a previous chapter, a major determiner of leprechaun gold-grabbing ability is age. The older the leprechaun, the bigger the wheelbarrow he needs to carry his day's haul. Now suppose that we are interested in whether *unfermented* prune nectar (UPN) helps a leprechaun to find gold. But we don't expect the unfermented product to be nearly as potent as the other stuff, so we must sharpen up our experimental design just a bit.

From that part of the County Kayo leprechaun population which received neither practice nor cirrhosis from the first experiment, we randomly sample 60 leprechauns of different ages. But in splitting them into the two groups, we don't use a random procedure as we did last time. Instead, we form them into **pairs**, of which both members have the same age. *Then* we assign, at random, one member of each pair to the E group and the other to the C group. (Flipping a coin for each pair will do it.) The unfermented prune nectar is gotten into the E group somehow, an inert concoction is given to the C group,[2] and the contest begins. Over a week's period, the following results pour in:

---

[1] "Signal" and "noise" are terms borrowed from the field of communications. Very apt words: the problem often is to reduce the noise enough to allow a weak signal to be heard. That's exactly analogous to statistical analysis.

[2] Remember, we want to make the groups similar in all ways that can affect the independent variable: we wouldn't want the control group to be handicapped by thirst!

Age of leprechauns in pair (years)	Hauls for UPN group (beqa)	Hauls for control group (beqa)	Difference (E − C)
400	150	145	5
375	148	149	−1
370	149	147	2
369	142	138	4
365	140	141	−1
364	145	143	2
364	144	144	0
359	139	137	2
356	139	140	−1
353	135	130	5
351	136	133	3
348	130	127	3
347	130	126	4
341	125	123	2
336	119	122	−3
335	120	118	2
331	115	110	5
329	114	110	4
325	112	110	2
321	110	108	2
320	109	108	1
314	101	97	4
310	97	97	0
306	95	90	5
305	96	94	2
301	93	95	−2
298	90	87	3
294	87	85	2
290	85	80	5
286	84	81	3

The first thing to be noticed in this table of data is that a clear correlation does exist between age and income. The oldest and wealthiest are at the top of the table; age decreases as you go down, and so does income. The variability in income is very largely accounted for by age.

But the comparison we wish to make is between the two groups, a sideways comparison in the table, not a vertical one. To ease this chore, the fourth column of differences within pairs is shown. Again, just glancing at the difference column, you will see a slight advantage for the UPN group: small positive differences are the rule. The edge for the UPN group is so slight, however, that it would be blanketed by the much greater age-related variability if we had not filtered that out by matching like-aged leprechauns in pairs.

To apply the z test, we use just the difference data. The first step is to obtain the mean difference $\bar{D}$, and then the standard deviation, s_D, of the differences.

$$\bar{D} = \sum D/n = 64/30 = 2.133$$

$$s_D = \sqrt{(\sum D^2/n) - \bar{X}_D^2} = \sqrt{(278/30 - 4.551}$$

$$= \sqrt{4.715} = 2.171$$

The z-test formula for matched subjects is a little different from the one already seen. It is

$$z_{\bar{D}} = \frac{\bar{D} - \mu}{\sqrt{s_D^2/(n_D - 1)}}$$

where n_D stands for the number of *pairs* of scores, not the total number. So $n = 30$ in this example. We find now that

$$z_{\bar{D}} = \frac{2.133 - 0}{\sqrt{4.713/29}} = 5.291$$

If we refer this value to the Table of Areas under the Gaussian Curve in the Appendix, we see that it is exceedingly unlikely to have happened by chance, assuming the null hypothesis: It is less than 0.00006 probable, although the table, stopping at $z = 4$, does not tell us just how much less. This is quite enough to lead us to reject the null hypothesis.

It is important to note that we very probably could not have rejected H_0 with exactly the same data, if we had not filtered out the strong relationship between income and age. In other words, we have used a strong experimental design and avoided a Type II error.

The matched-subjects design is hardly perfect; no one design is. You give a little to gain a little where you need it.

advantages

The matched-subjects design is great primarily when

a some matching variable, like age in our example, is available and subjects can be found who make enough complete pairs;

b relatively large values of n are difficult or impossible.

It is often possible to reject the null hypothesis with the matched-subject approach when it can be done in no other way. The matching need not be with human subjects. Populations of rats, farm acreages, food specimens, and urban areas can be sampled, and cases matched and then randomly assigned to either of the two groups.

disadvantages

1 One clear disadvantage is the need to have already, or find out in advance, a variable which is known to correlate highly with the dependent variable, which can be applied conveniently, and which will not spoil the subjects for the experiment proper. Matching age is no problem, but suppose you wanted to match aggression, or appetite, or genes: it would be very difficult.

2 You gain *and lose* your subjects in pairs. If *one* 400-year-old leprechaun had decided to go on vacation and not cooperate, you would have lost the other 400-year-old one too, unless you found another partner to match him with.

3 The power of the matching procedure is greatest when the correlation is highest between the dependent variable and the matching variable. If the correlation is too low, you are going to a lot of trouble for very little payoff.

the within-subjects or repeated-measures design

As a special case of the matched-subjects design, you can use the same subjects twice, once under the experimental conditions and once under the control conditions. The principle is that any subject is matched best by himself or herself. So a within-subjects design has all the benefits of the matched-subjects design, with the added advantage that you gain your subjects singly rather than in pairs, but with a couple of major disadvantages to slow you down a little. The first is that the experiment must not involve procedures whose effects last long enough to affect the subject's second session with the experiment. The E and C conditions (experimental and control) can rarely be given together; one must usually come first, and the first may well affect the second. The other problem is that *time*, and all the things which fill it, must come between the first and second sessions; and time itself may be a variable that affects the dependent variable.

There is no way out of the first problem. That is why some experiments cannot be done within subjects. The second problem is solved by counterbalancing the order of the experimental treatments. Half of the subjects would get E then C, the other half C then E. That way, any effects of time act equally both ways and should cancel each other out overall.

why the null hypothesis?

Students are often puzzled about why social and behavioral scientists use the rather circuitous logic of statistical experiments. Although they lean toward believing one hypothesis (the experimental hypothesis), they pretend to believe its opposite (the null hypothesis), even while hoping that the null hypothesis will turn out to be untenable so they can adopt the experimental hypothesis by process of elimination. Why go through that? Why not test the experimental hypothesis directly?

The reason can be shown most easily by an example. Let's imagine that a researcher has found a drug that may be an effective treatment for the common cold. Some people will believe, based on the casual evidence available to them and perhaps on a social, emotional, or cultural need to believe in the drug's effectiveness, that in fact it does cure colds. But suppose that these people want to test this belief, for the sake of fairness, and set out to look for evidence. Two things can happen: Positive evidence (strong or weak) may appear in favor of the hypothesis, in which case their belief would continue unchanged. Or evidence could fail to appear, in which case they might say, "Well, we still believe that the drug cures colds. Even if *some* colds aren't cured, the drug could still be useful against *other* colds." This is logically correct, but the evidence that has been acquired has not, and cannot, oblige the members of this group to change their minds. Positive evidence fits in with their belief; negative evidence has no effect because absence of proof is not the same as proof of absence. In fact, it is well-nigh impossible to prove the *absence* of anything.

Another group of people might begin by saying, "It *may* be true that the drug cures the common cold, but we can't be positive either way. Until we have some evidence that it does, we'll have to be neutral."

This state of neutrality is the null hypothesis. Now if the research shows no effect of the drug on the common cold, our second group would still have an open mind, because later evidence could still come in that would change their opinions. The null hypothesis would be retained in the meanwhile. But the evidence might show clearly that the common cold *was* improved by the drug. If the experimental design was good, the statistical analysis correct, and the level of α appropriate, such evidence would oblige this group to change its collective mind. Thus, only the second group is likely to be influenced by the evidence, whichever way it comes out.

Obviously, the difference between these two approaches would not justify any argument if evidence or lack of evidence were clearcut. The physical sciences frequently deal with questions for which evidence is either obvious or totally lacking. Sir Isaac Newton did not need the null hypothesis as one stage in discovering gravity. The social and biomedical sciences are, unfortunately, not like the physical sciences in this regard: evidence is usually less than clearcut, as we saw in our discussion of Type I and Type II errors. To the extent that experimental evidence is fuzzy, the null hypothesis is necessary as a starting point if we are to avoid adopting the experimental hypothesis when it is not true. The first of our two hypothetical groups above is certain to make Type I errors quite often, an unknown amount of the time. The second group would be able to specify exactly what risk of a Type I error it is willing to take, and therefore it would be in a far stronger position. Scientists are acutely aware of how seductive an interesting idea (such as the existence of a cure for the common cold or of extrasensory perception) can be, and they developed the null hypothesis approach in order to counteract their natural human tendency to get carried away too readily by fascinating (and unproven) ideas.

testing the difference between two correlation coefficients

Occasionally we find experiments in which the dependent variable is a correlation coefficient, r. For example, the correlation between age and success of treatment by Behavior Modification is $+0.08$ in female

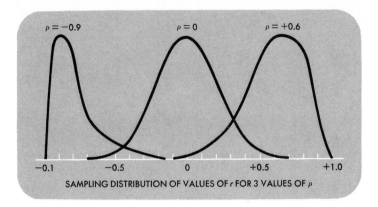

Figure 11.5
The sampling distribution of r when ρ has certain values. The skewness when $\rho \neq 0$ may be straightened out by transformation of r to Z, as in Figure 11.6.

leprechauns and -0.55 in male leprechauns. Question: Does the difference between the r values denote a *real* (population) difference between the sexes, or is it an artifact of sampling? A modification of the z test can help us to decide.

A requirement of the z test is, of course, that the relevant sampling distribution be Gaussian. Unfortunately, only when ρ (rho) is 0 is this requirement met. Otherwise, the distribution is sharply skewed, as shown in Figures 11.5 and 11.6; the degree of skewness depends upon the value of ρ, and the scaling of the abscissa depends upon n, the number of pairs in the samples whose probability distribution is portrayed.

What we need is a nonlinear transformation of r, a transformation that creates a Gaussian distribution. A *relatively* simple one, which gives an *approximately* Gaussian distribution, was proposed by Sir R. A. Fisher. The symbol Z is given to the values of this distribution, which are obtained from the values of r in the original skewed distribution.[3] The relationship between r and Z is

$$Z = 0.5 \log_e \left[\frac{1 + r}{1 - r} \right]$$

Possibly, if you are already a mathematician, you may understand this! But if not, imagine an invisible hand pushing against a skewed

---

[3] A suggestion: if you already call z *zee*, the American pronounciation, call Z *zed*, the British and Canadian pronunciation. Or if you wish, use *little zee* and *big zee*.

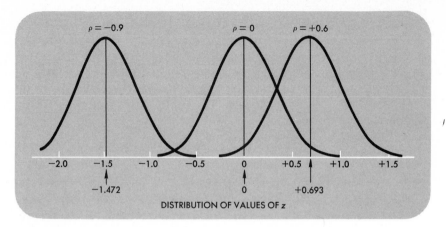

Figure 11.6
Distributions of Z for the values of ρ shown in Figure 11.5. The skewed distributions of r have been changed to Gaussian distributions of Z. The variance of the Z distribution depends upon the value of n; here, the abscissas are scaled for n = 5 pairs for each.

r distribution to make it stand up straight; you'll have the idea. Table 11.A gives Z values equivalent to some values of r, so you don't have to work them out. The sign attached to Z is the same as the sign of r. (A complete table of equivalent values of r and Z is included in the Appendix.)

To scale the abscissa of our newly Gaussianized distribution in z scores, we need only find $1/\sqrt{n - 3}$, where n is the number of pairs in the sample giving rise to our original r. The resulting value, the **standard error of Z**, is the standard deviation of the Gaussian distribution that is centered on the value of Z corresponding to ρ. The standard error of Z therefore has the formula

$$\sigma_Z \cong s_Z = \frac{1}{\sqrt{n - 3}}$$

Now let's move on another step. Remember, we are interested in deciding whether two correlations come from the same population of correlations. By the same reasoning that we used in the z test, we know that (when the null hypothesis is true) the differences between pairs of Zs will average out to 0. The standard deviation of such differences will be

$$\sigma_{Z_1 - Z_2} \cong \sqrt{s_{Z_1}^2 + s_{Z_2}^2} = \sqrt{\frac{1}{n_1 - 3} + \frac{1}{n_2 - 3}}$$

in which $\sigma_{Z_1 - Z_2}$ is called the **standard error of the difference between two Zs**.

The next step is to form a ratio between the obtained $Z_1 - Z_2$ difference and the value of $\sigma_{Z_1 - Z_2}$. The sampling distribution of such differences is Gaussian, so we can use Table 8.A or the Table of Areas under the Gaussian Curve in the Appendix to assess the likelihood of the two obtained r values occurring by chance from the same population with parameter ρ.

Table 11.A
TRANSFORMATION OF r TO Z

r	Z	r	Z
0.01	0.010	0.50	0.549
0.02	0.020	0.51	0.563
0.03	0.030	0.52	0.577
0.04	0.040	0.53	0.590
0.05	0.050	0.54	0.604
0.06	0.060	0.55	0.618
0.07	0.070	0.56	0.633
0.08	0.080	0.57	0.648
0.09	0.090	0.58	0.663
0.10	0.100	0.59	0.678

a worked example

We continue the above example. Among 67 females, there is a correlation of $+0.08$ between age and success of behavior modification. Among 52 males, the correlation is -0.55. From Table 11.A we find that the respective equivalent Z values are $+0.080$ and -0.618. Substituting into the formula, we find

$$z = \frac{Z_1 - Z_2}{\sqrt{1/(n_1 - 3) + 1/(n_2 - 3)}}$$

$$= \frac{+0.080 - (-0.618)}{\sqrt{1/(67 - 3) + 1/(52 - 3)}} = 3.677$$

which exceeds the critical value for the 0.01 level in Table 8.A. The null hypothesis is rejected, and we conclude that the relation between age and behavior modification is not the same in females and in males.

key words and concepts of this chapter

NULL HYPOTHESIS (again, please!)

EXPERIMENTAL and CONTROL GROUPS

STANDARD ERROR OF THE MEAN (yes, once more!)

STANDARD ERROR OF THE DIFFERENCE BETWEEN TWO MEANS, $\sigma_{\bar{D}}$, $s_{\bar{D}}$

POPULATION (again, but slightly different this time)

TYPE I or ALPHA ERROR

TYPE II or BETA ERROR

SIGNIFICANCE

ERROR VARIANCE

REPLICATION

r TO Z TRANSFORMATION

STANDARD ERROR OF Z, σ_Z, s_Z

STANDARD ERROR OF THE DIFFERENCE BETWEEN TWO Zs, $\sigma_{Z_1 - Z_2}$, $s_{Z_1 - Z_2}$

SIMPLE RANDOM GROUPS DESIGN

MATCHED-SUBJECTS DESIGN

COUNTERBALANCING

exercise for a rainy garden party

This is merely an exercise in observation and personal interaction. No darts, rubber bands, or coffee are required. However, it *can* still be dangerous. You may find that you are questioning and threatening people's beliefs. Keep away from politics, religion, race, and whether the Packers know how to play football. Stick to beliefs that are amenable to scientific investigation.

Infiltrate into conversation as opportunity permits, and after sufficient idle chatter, mention that you've "heard a lot of controversy about whether _____ is _____." (You fill in the blanks with opportune topics.) Keep mental notes. Gently press your partner for *reasons.* At the end of the afternoon, estimate the proportions of elicited beliefs that were (a) scientifically arrived at, (b) hearsay, (c) biased and need-fulfilling, (d) based upon biased selection of evidence, (e) illogical, (f) socially dangerous as well as wrong. Unless the data make you too depressed, think about this experience and perhaps use it as a basis for discussion.

self-test on chapter II

The same rules hold.
1 An experimenter wishes to determine the effect of a new tranquilizing drug upon driving skill and judgment. She has available a
driver-testing machine and a total of 100 subjects drawn from the
student body of a large and typical state college. She divides the
subjects into two groups of 50 each. One group is given a moderately high dosage of the drug, and each subject's performance is
measured on the testing machine. The other group is given a
placebo, a harmless inert substance that resembles the drug in the
form in which it was administered. The instructions to the two
groups were identical. The drug group achieved a mean score
$\bar{X}_D = 35$, with $s_D = 15$; the placebo group achieved $\bar{X}_P = 40$, $s_P = 10$.
Which of the following can we correctly say about this experiment?

a The value of $\hat{\sigma}_{\bar{D}}$ was 2.57.
b The value of $\hat{\sigma}_{\bar{D}}$ was 6.81.
c The value of $\hat{\sigma}_{\bar{D}}$ is impossible to estimate from the data so far
given.
d The independent variable was amount of drug given, and all
other relevant variables were held constant.
e The independent variable was whether the subject received a
drug dose or a placebo.
f Any possible conclusions would depend upon the method by
which the 100 subjects were formed into two groups of 50.
g Assuming that the experimenter did everything properly, we
know that his results are free from error.
h The critical value of z for $\alpha = 0.05$ is 1.945.
i The critical value of z for $\alpha = 0.05$ is 1.960.
j The results allow us (if only just barely) to reject the null hypothesis.
k It is possible from this experiment to conclude that *patients* receiving the drug will not be strongly affected by it in their driving ability.

SCORING AND ANALYSIS of self-test items overleaf. Scoring +3 for each correctly chosen answer, −2 for each incorrectly chosen, and −1 for each correct answer not chosen.

Answers to Question 1

a T $\hat{\sigma}_{\bar{D}}$, the standard error of the difference between two means, has the
b F formula

$$\hat{\sigma}_{\bar{D}} = \sqrt{s_{\mathrm{E}}^2/(n_{\mathrm{E}} - 1) + s_{\mathrm{C}}^2/(n_{\mathrm{C}} - 1)}$$

Substituting actual values, we then have

$$\hat{\sigma}_{\bar{D}} = \sqrt{\frac{15^2}{49} + \frac{10^2}{49}} = \sqrt{\frac{325}{49}} = 2.575$$

c F The independent variable must be one thing at a time: in this case,
d T **e** F the amount of active drug given, whether 0 or a larger amount. The placebo was given as a control measure, to prevent confounding between getting the *drug* and getting a *pill*. The placebo ensured that both groups got a pill, so taking a pill was not a variable. So the independent variable was *not* concerned with the placebo. Merely with amount of active drug.

f T Very true. The method should be random. If not, the good drivers might have been put into one of the two groups beforehand. Then the results would have seemed significant even if the independent variable were totally impotent.

g F We never know this. Type I or Type II errors can strike any time! But fortunately with low probability.

h F The critical value of *z* is the value that must be reached or exceeded
i T by the experimentally obtained one. And 1.96 is the critical value; see Table 8.A. The obtained value of *z* is

$$z = \frac{\bar{X}_{\mathrm{E}} - \bar{X}_{\mathrm{C}}}{\hat{\sigma}_{\bar{D}}} = \frac{40 - 35}{2.57} = 1.94$$

j F This obtained *z* does *not* permit rejection of the null hypothesis, because it is less than the critical value.

k F Patients are certainly not the population from which the student subjects were drawn. Thus no conclusions about them can be drawn, although speculation is permissible.

1 (*continued*)

    l The experiment would need to be repeated before any conclusions are drawn concerning the null hypothesis.

    m With the information so far given, an obtained z ratio of 3.0 would be seen as significant but not at all useful in any practical sense. (Note: $+3$ or -3 not specified!)

2 Suppose that an experiment was done to determine the influence of hunger on the responses given by subjects to certain stimulus words. A group of 100 subjects is deprived of food and water for 10 hours. The researcher finds that at the end of this time, 83% of the responses of the subjects to the stimulus words are food related and that 10% are thirst related. It is reasonable to conclude

    a that food responses increase under conditions of hunger.

    b that thirst responses do not increase under conditions of thirst.

    c that one-half of the necessary information has not been provided.

    d that hunger has a more powerful effect than thirst.

    e that if responses to the same stimulus words had been recorded for the same subjects previously, while they were not hungry and thirsty, valid inferences about the independent variable could be drawn.

    f that the "experiment," as reported here, does not fulfill the minimum requirements for being called an experiment.

    g that no basis has been provided for calculation of error variance.

Answers to Question 1 (*continued*)

l T   Yes. The obtained z is nearly large enough. You can only retain the null hypothesis, but a Type II error may have occurred.

m T   A z ratio of 3 is significant at the 0.01 level. The problem is, however, that no information has been given concerning whether the drug group was *better than* or *worse than* the placebo group. We only know that it was one or the other.

Answers to Question 2

a F   b F   The major thing wrong with this "experiment" is that there is no
d F   f T   comparison with a control group or condition. So (a), (b), (d), and (e) are all false, and (f) is certainly true.

c T   Before any conclusion can be drawn, we must have at least as much again in information, this time about a control group.

e F   The nondeprived base line taken *before*? Has a time-related change perhaps taken place since then? Who knows? Anyway, *time* would be confounded with the independent variable, and that's dangerous.

g T   The data as reported must refer to the mean percentage of food- and water-related responses in the experimental group. There is no mention of the standard deviation of either quantity. So even if a control group had been present, its data if reported as in this question would still not have permitted an error variance estimate.

3 A clinical psychologist desires to test whether frustration is a cause of aggression. He uses Rorschach "ink blot" cards as stimuli for aggressive responses and acquires the services of 65 subjects. He gives all the subjects the first five out of the ten cards and records their responses without comment. Then he attempts to frustrate them by berating, ridiculing, and sneering at the level of their imaginative responses to the ink blots. Then he administers the last five Rorschach cards, again recording the responses. The total number of aggressive responses is counted for each subject under each condition. The mean and standard deviation are calculated for the difference (second five minus first five) within each subject between the two conditions. $\bar{D}$ (mean difference) = 15, and s = 24. He infers that his technique demonstrated that frustration is a cause of aggression. What do you think?

a His inference is completely unacceptable.

b His inference is substantially correct.

c His inference would be acceptable if he had shown that the first five cards and the second five cards were equally evocative of aggression under normal circumstances.

d The value of z pertaining to this experiment is 40.

e The value of z is 5.

f The null hypothesis states that $\bar{D}$ should equal 0.

g The null hypothesis states that s should equal 0.

h The experimental design has permitted *two* different confoundings, either of which would invalidate any inference about frustration and aggression.

i The flaws present in this experiment would be overcome if the experiment were repeated exactly with new subjects.

4 A large value of z causes a researcher to infer support for his experimental hypothesis. His selected α value was 0.01. Subsequently, several other researchers replicate the experiment, but they never succeed in supporting the experimental hypothesis. What *may* have happened?

a The first researcher made a Type I error.

b The subsequent researchers made Type I errors.

c The first researcher's obtained z value may have been less than 2.57.

d The first researcher may have made no error, but the subsequent ones may have set their levels of α much too low.

Answers to Question 3

h T   There are two major problems in this "experiment." Either one is fatal to any possible conclusions. First, there is a confounding between the cards and the experimental condition: the first five cards were always associated with the first condition, and the second five with the second condition. It is possible (indeed, it is actually true) that the last five Rorschach cards are more evocative of aggression than the first five. So the difference between the two experimental conditions could be explained by this fact alone. Second, the two experimental conditions always occurred in the same order: relaxed conditions for administration of the first five and defensive conditions for administration of the second five cards. It is possible that subjects will become more and more aggressive as time passes in such exercises, and if so, the data are explainable even if frustration has no effect at all.

a T **b F**   Unacceptable for the reasons above.

     c F   This would repair half of the total problem, but the other problem remains. Generally, if an experiment has one thing wrong with it, other things are probably wrong too.

     d F
     e T

$$z = \frac{\bar{D}}{\hat{\sigma}_{\bar{D}}} = \frac{\bar{D}}{(s/\sqrt{n-1})} = \frac{15}{24/\sqrt{64}} = \frac{15}{3} = 5.0$$

     f F   The null hypothesis always refers to a 0 difference between popula-
     g F   tion means. It certainly does not imply that the sample mean must be 0 or that variable error around the mean (or mean difference) should not occur.

     i F   The flaws would still be present if the experiment were replicated exactly. It's the design that needs to be changed.

Answers to Question 4

     a T   Support was inferred for the experimental hypothesis, which means that the null hypothesis was rejected. If it was rejected erroneously, a Type I error occurred.

     b F   If, however, the first experiment came to the correct conclusion, lack of support from subsequent experiments must mean that they failed to reject the null hypothesis when it was false. This is a Type II error.

     c F   If the first experiment rejected H_0 with $\alpha = 0.01$, the obtained z value must have been 2.57 *or higher*.

     d T   It could have happened if their α had been too small, e.g., 0.000001.

4 (*continued*)

    e The first researcher made a Type II error.

    f If all researchers used the same α value, it may be that the first one made an honest mistake which would not occur 99% of the time when H_0 is true.

5 An experiment is being done to decide upon the effectiveness of a new treatment for cancer. The considerations are as follows: If the treatment is adopted on the grounds of erroneous experimental support, then money is lost on an ineffective treatment but the patients do not suffer, since they have nothing more to lose anyway. If, on the other hand, the treatment method is rejected by the medical profession as a result of erroneous lack of experimental support, then the patients will die when they might have been helped. Under these circumstances,

    a the value of α should be set extremely small, e.g., 0.0001.

    b as large an n as possible should be used in the experiment.

    c as much experimental control as possible should be employed.

    d it would be necessary for some of the experimental patients to be left untreated, at least for a time.

    e the value of α should be relatively large, perhaps 0.05 or even 0.10.

    f a Type II error is much less important than a Type I error.

    g a Type I error is preferable to a Type II error.

    h the probability of a Type I error is much more precisely known than that of a Type II error.

Answers to Question 4 (*continued*)

e F A Type II error can never occur when the null hypothesis is rejected, only when it is (erroneously) retained.

f T True enough. If α is 0.01, we have assurance that the null hypothesis will be rejected erroneously 1% of the times that it is really true. This may have been one of those times. An honest and unavoidable error.

Answers to Question 5

This question illustrates a common problem in practical research: since we must run some risk of making an error, which type of error is preferable?

a F
e T Certainly we do not want to retain the null hypothesis when it is not true; we don't want a Type II error. Setting α at 0.05 or even higher might therefore be wise.

b T
c T As big an n and as much control as possible are good for any experiment. Here, they will lower the probability of a Type II error, in conjunction with the somewhat relaxed criterion for α.

d T Unfortunately, this is true, otherwise there would be no control group. Usually, however, clinical research is done so as to provide the treatment for control group patients, but *after* they have served their control function.

f F
g T
h T In research of this kind, a Type I error costs nothing but time and money. A Type II error could cost lives and human suffering, since adoption of a successful treatment would be delayed or even prevented by such an error. The probability of a Type I error is equal to α and is therefore precisely chosen by the experimenter. The probability of a Type II error is less precisely ascertainable, since it depends upon many factors.

6 In the spaces below, enter the values requested, which pertain to the data given here.

	Experimental	*Control*
n	170	145
ΣX	3060	2175
ΣX^2	57,800	34,945

			Pronunciation Review
a	$\bar{X}_E$	= _____	X bar E (the mean of the E group)
b	$\bar{X}_C$	= _____	X bar C (the mean of the C group)
c	s_E	= _____	s sub E (the standard deviation of the E group)
d	s_C	= _____	s sub C (the standard deviation of the C group)
e	$\hat{\sigma}_{\bar{D}}$	= _____	estimated sigma difference (the standard error of the difference between two means)
f	$\hat{\sigma}_{\bar{X}_E}$	= _____	sigma sub X bar E (standard error of the mean, E group)
g	$\hat{\sigma}_{\bar{X}_C}$	= _____	sigma sub X bar C (standard error of the mean, C group)
h	z	= _____	

Answers to Question 6

Data	Experimental	Control
n	170	145
$\sum X$	3060	2175
$\sum X^2$	57,800	34,945

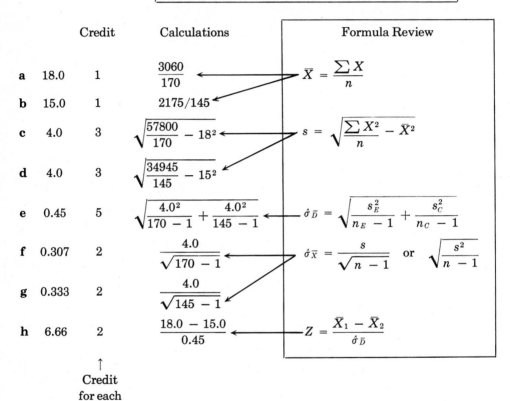

		Credit	Calculations	Formula Review
a	18.0	1	$\dfrac{3060}{170}$	$\bar{X} = \dfrac{\sum X}{n}$
b	15.0	1	$2175/145$	
c	4.0	3	$\sqrt{\dfrac{57800}{170} - 18^2}$	$s = \sqrt{\dfrac{\sum X^2}{n} - \bar{X}^2}$
d	4.0	3	$\sqrt{\dfrac{34945}{145} - 15^2}$	
e	0.45	5	$\sqrt{\dfrac{4.0^2}{170 - 1} + \dfrac{4.0^2}{145 - 1}}$	$\acute{\sigma}_{\bar{D}} = \sqrt{\dfrac{s_E^2}{n_E - 1} + \dfrac{s_C^2}{n_C - 1}}$
f	0.307	2	$\dfrac{4.0}{\sqrt{170 - 1}}$	$\acute{\sigma}_{\bar{X}} = \dfrac{s}{\sqrt{n - 1}}$ or $\sqrt{\dfrac{s^2}{n - 1}}$
g	0.333	2	$\dfrac{4.0}{\sqrt{145 - 1}}$	
h	6.66	2	$\dfrac{18.0 - 15.0}{0.45}$	$Z = \dfrac{\bar{X}_1 - \bar{X}_2}{\acute{\sigma}_{\bar{D}}}$

↑
Credit
for each
correct

7 An investigator finds an $r = +0.35$ between academic success and a measure of social awareness in a group of 50 black children. In a group of 64 white children, the correlation is $+0.12$. (These are hypothetical but plausible data.) In the spaces below, enter the values requested, which pertain to these data.

a $Z_{white} = $ _____	Zed-white (r to Z transformation)
b $Z_{black} = $ _____	Zed-black (r to Z transformation)
c $\sigma_{Z_w} = $ _____	sigma sub zed-white (standard error of Z)
d $\sigma_{Z_b} = $ _____	sigma sub zed-black (standard error of Z)
e $\sigma_{Z_1 - Z_2} = $ _____	sigma sub zed difference (standard error of the difference between two zeds)
f $z = $ _____	zee (the z test statistic)

g ( ) retain

$$H_0$$

( ) reject

h $\alpha = $ _____

EVALUATION

The maximum possible score is 94 points. The merit attached to the score you got without peeking (at anything other than Table 8.A and the transformation of r to Z table) is

80 to 94	Utterly meritorious
55 to 79	Quite meritorious
40 to 54	Pseudomeritorious
25 to 39	Negatively meritorious
12 to 24	Stupendously negatively meritorious
under 12	Awful

Answers to Question 7

DATA

$$r_{white} = +0.12, \ n = 64$$
$$r_{black} = +0.35, \ n = 50$$

		Credit		Formula Review
a	+0.121	1	$Z \leftarrow r = +0.12$	$Z = 0.5 \log_e \left[\dfrac{1 + r}{1 - r} \right]$
b	+0.365	1	$Z \leftarrow r = +0.35$	... see Table 11.A.
c	0.128	2	$\dfrac{1}{\sqrt{64 - 3}}$	$\sigma_Z = \dfrac{1}{\sqrt{n - 3}}$
d	0.145	2	$\dfrac{1}{\sqrt{50 - 3}}$	
e	0.194	2	$\sqrt{\dfrac{1}{64 - 3} + \dfrac{1}{50 - 3}}$	$\sigma_{Z_1 - Z_2} = \sqrt{\dfrac{1}{n_1 - 3} + \dfrac{1}{n_2 - 3}}$
f	1.258	2	$\dfrac{0.365 - 0.121}{0.194}$	$z = \dfrac{Z_1 - Z_2}{\sigma_{Z_1 - Z_2}}$
g	retain H_0	1	Critical value of	
h	$\alpha = 0.05$	1	z is 1.96 for $\alpha = 0.05$.	

↑
Credit
for each
correct

some ordinary problems

1 Define what is meant by *confounding*, and give two examples from hypothetical experiments.

2 Explain why all the arguments concerning probability in this chapter involve the Gaussian distribution of probability.

3 Two groups of students were given the same test in Introductory Biology. One group was allowed unlimited time, and the other was allowed only 40 minutes. Following is a summary of what happened:

n	Time Allowed	$\bar{X}$	s
401	Unlimited	112	48.0
401	40 minutes	117	36.0

a What type of experimental design was used?
b State the critical value of z needed to reject H_0 with $\alpha = 0.01$.
c Obtain the z from the data shown.
d Did you retain or reject the null hypothesis?
e What type of error *might* have been made in your decision?
f If α had been set at 0.05, would this have changed your conclusion?
g If α had been 0.05 and both values of s had been 36.0, would your conclusion have been the same or different?
h Name a general effect of reducing the values of s as much as possible in experiments.
i With the data otherwise as shown above, but with the values of n equal to 1801 for each group, what would your decision have been?
j Name a general effect of increasing the value of n as much as possible in experiments.

4 Show how it is possible, using the r-to-Z transformation, to decide whether a given sample r is significantly different from 0.

5 In a certain city, there is a correlation of $r = +0.69$ between income and scholastic success of children in school. This correlation was established on the basis of a sample of 1000 randomly sampled school children. The educators of a second city, aware of the significant correlation, decide to concentrate upon programs and evaluation procedures that are as free as possible from relationship with economic level. After a year of trying, they find $r = +0.58$ be-

tween income and scholastic success in a random sample of 950 children. Did the second city's program and evaluation policy differ from that of the first?

6 In the previous question, could the second city's educators claim that their policy had resulted in a less income-oriented evaluation of school pupils?

7 It is a fact that the r-to-Z transformation Gaussianizes the frequency distribution *only* when the correlation coefficient is a legitimate Pearson r. What precautions are necessary, therefore, in testing the differences between correlations?

8 Turn to the data given for our matched-subjects design example on page 318. Perform a random-groups style of z test, and compare your results with the matched-subjects results. What accounts for the difference?

chapter 12
the
t test

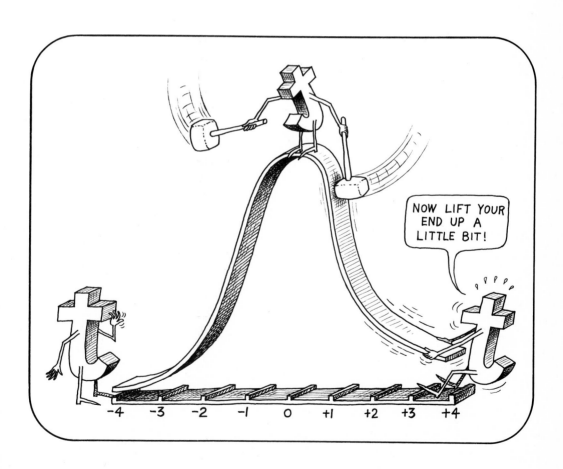

In Chapters 7, 8, and 11 you were lied to, just a little bit. It was a white lie, in the interests of your own grasp of why a frequency distribution of many means, coming from many random samples drawn from the same population, is Gaussian in shape.

If the latter argument is not thoroughly understood, stop here and go back over Chapters 7, 8, and 11 until you are firmly up to date.

Assuming that you are, we may still agree that a brief recapitulation is in order. Recall that if many pairs of random samples are drawn *from the same population*, their mean values will differ only by chance, so the mean (μ) of all the differences between means is 0, and the distribution of mean differences will be Gaussian. The standard deviation of this Gaussian distribution is closely estimated as the **standard error of the difference between two means**, $\hat{\sigma}_{\bar{D}}$. Such an estimate comes from the standard deviations of any two random samples of scores; in other words, two samples will give us all the information needed to make a conclusion about the parent population(s) of the experimental and control groups, whether similar or different.

So far, the validity of this assumption has depended upon two further assumptions: (a) that the sampling distribution of ($\bar{D} - \mu$), when translated into z scores, is Gaussian; and (b) that the denominator of that z score always has more or less the same reliability. Neither of these assumptions tends to be true, unless the sample sizes are substantial. *If* the population from which $\bar{X}_E$ and $\bar{X}_C$ are taken is Gaussian, then so are the distributions of $\bar{D}$ and of ($\bar{D} - \mu$)/$\hat{\sigma}_{\bar{D}}$, when n is about 30 or more. If the population is non-Gaussian, the n for each group must be much larger for the distribution of ($\bar{D} - \mu$)/$\hat{\sigma}_{\bar{D}}$ to be Gaussian (and thus for the areas under the Gaussian curve to be applicable).

The numerator of the z test gives no trouble. The null hypothesis always states that $\mu = 0$ and that the difference between two sample means, $\bar{D}$, will have a Gaussian distribution around that value of 0. The trouble comes with the estimation of $\sigma_{\bar{D}}$ from values of s taken from the E and C groups.

When n is very small, s has a high probability of being much smaller than σ, to an extent that cannot be accurately compensated for by the correction factor considered in Chapter 8. With the null hypothesis true, many very small samples would give values of s that fall in a skewed distribution; the result is that when s is used[1] in the

---

[1] As when calculating

$$\sigma_{\bar{D}} = \sqrt{\frac{s_E^2}{n_E - 1} + \frac{s_C^2}{n_C - 1}}$$

z-test denominator, the resulting quotients (value of $z_{\bar{D}}$) do *not* fall into a Gaussian sampling distribution. When n is made large enough, the problem diminishes; when n is less than 30, even with a Gaussian population distribution, the sampling distribution of $z_{\bar{D}}$ is leptokurtic, not Gaussian. This ruins the Table of Areas under the Gaussian Curve as a basis for rejecting H_0 when values of n are small, because in leptokurtic distributions the value of z is no longer tied to area in the tails as given in the Table.

In 1908, however, an English statistican named William S. Gosset published an article (under the psuedonym "Student") which showed that the entire statistic

$$\frac{\bar{D} - \mu}{s/\sqrt{n - 1}} = t$$

has a known sampling distribution, but a different one for every value of n. When n is large, the statistic has a Gaussian distribution, but when n is small, it has more the shape shown in Figure 12.1 Figure 12.1(a) shows the shape associated with $n = 2$, an extreme case. For larger values of n, the departure from the Gaussian shape is less extreme, as exemplified by Figure 12.1(b).

The statistic shown above, though similar to z, is called the **t statistic**, its sampling distribution is called the **t distribution**, and it is the basis of the **t test**, one of the most useful, sensitive, and powerful tests of the null hypothesis. Note that t is lower case; capital T has a different meaning, which you may recall from Chapter 6.

Gosset was an employee of a major British brewery, occupied with quality-control work. Employees were not permitted to publish under their own names, presumably in case the brewery should find itself dragged into something embarrassing. So the publication was signed simply *Student*. Altogether, he was a remarkable and brilliant man.

characteristics of the *t* distribution

As with the Gaussian distribution, the t distribution is based on a variety of z score. The mean of the distribution is 0, equal to μ_D, which the null hypothesis says is the value of $\mu_E - \mu_C$. The distribution is symmetrical. There are two ways in which the t distribution is *very* different from the Gaussian.

1 The abscissa is scaled differently. The Gaussian abscissa is scaled in terms of its own standard deviation, $\sigma_{\bar{D}}$, which is estimated as

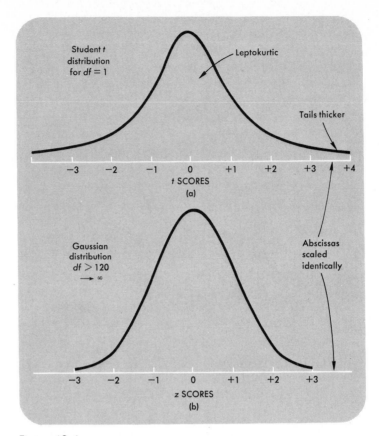

Figure 12.1
Comparison of Student *t* distribution with Gaussian distribution. The
t distribution shown here is an extreme example; usually, the shape is
closer to that of the Gaussian curve.

$\hat{\sigma}_{\bar{D}}$. The *t* abscissa is scaled by means of the identical estimate, even
though this estimate pertains to a different distribution! Both *z*
and *t* values utilize the same denominator, $s/\sqrt{n-1}$, which for the
z score is close to $\sigma/\sqrt{n}$. This is not true for the *t* distribution, however. Because for small values of *n*, *s* is likely to be unduly small in
comparison with σ, the *t* denominator is likely to be unduly small
in comparison with the actual standard deviation of the *t* distribution. No matter: that is how *t* is defined. But this does lead to our
second big difference.

2 Very large values of *t* may occur relatively frequently under the *t* distribution, even when the null hypothesis is true. Note in Figure 12.1(a) how thick the tails are; thus there is a reasonable probability that *t* values can arise, even given H_0, that would make us reject H_0 if they were found during a *z* test. This is most common when *n* is very small; remember, with large *n* the *t* test becomes almost identical with the *z* test. In Figure 12.1(a) we would need to go out to *t* values of ± 63.657 before 99% of the area was encompassed. Compare this with $z = \pm 2.57$ being the values that enclose 99% of the area in Fig. 12.1(b).

the *t* test of the null hypothesis

If you are already familiar with the *z* test, you are also already familiar with the *t* test, except for two minor refinements. The procedures for gathering the data, finding the values of $\bar{X}$ and *s*, and finding the difference between the two means to give $\bar{D}$ are all the same.

The first refinement is in the denominator of the *t* formula.

In principle, the denominator of the formula for *t* is similar to the one for the *z* test,

$$\hat{\sigma}_{\bar{D}} = \sqrt{\frac{s_E^2}{n_E - 1} + \frac{s_C^2}{n_C - 1}}$$

This formula ought to be generally understandable as one that combines the two standard errors of the two samples whose difference between means, $\bar{D}$ appears in the numerator. But it is appropriate, unfortunately, only when $n_E = n_C$ and when the variances, σ_E^2 and of σ_C^2, of the two populations (experimental and control) are identical. The problem is with the *weighting* of the two terms under the square-root sign; as shown here, it is evident that the two standard error estimates $\sqrt{s_E^2/(n_E - 1)}$ and $\sqrt{s_C^2/(n_C - 1)}$ have equal weight in determining $\hat{\sigma}_{\bar{D}}$. It should be apparent that if σ_E is much greater than σ_C, or if n_E is much greater than n_C, the two *s* values or the two *n* values should *not* be treated equally. The bigger one should be more important in determining the value of $\hat{\sigma}_{\bar{D}}$.

The problems of unequal population variances and unequal sample sizes are really separate problems, and we shall treat them separately. The easiest way to handle the variance problem is to assume that the variances of the control population and the experimental population are equal. This assumption is legitimate if we assume that the experimental treatment affects each score in the treated population by a constant amount, or by a random amount. But if already potentially low scores are selectively made even lower by the treatment, and potentially high scores are selectively made even higher, then clearly the independent variable will increase the variance of its sample and population (or the variance could be lowered). But it is usually reasonable to assume that this does not happen. The assumption of equal population variance is customarily made.

The problem of equal sample sizes is not so easy. For one thing, it's quite obvious that sample sizes do sometimes differ. You may start with equal numbers of subjects in the two groups, but some will drop out, or move away—whether they are white rats or people. You can never count on winding up with the same *n* you started with. So the best way to compensate for unequal sample sizes is to weight the denominator terms so as to take the effect of the inequality into account. This is done by adopting the following formula for the denominator.

$$\sqrt{\frac{n_{\mathrm{E}}s_{\mathrm{E}}^2 + n_{\mathrm{C}}s_{\mathrm{C}}^2}{n_{\mathrm{E}} + n_{\mathrm{C}} - 2}\left(\frac{1}{n_{\mathrm{E}}} + \frac{1}{n_{\mathrm{C}}}\right)}$$

If $n_{\mathrm{E}} = n_{\mathrm{C}}$, this formula reduces to the one given above; if not, it is accurate, so this one, although harder to memorize, is the one to use.

The second refinement involved in the *t* test is the interpretation of the *t* value once it is obtained. Remember that the shape of the *t* distribution varies with the number of subjects *n*, or $(n_{\mathrm{E}} + n_{\mathrm{C}})$, in the experiment. Therefore, *many* sampling distributions, not just the one Gaussian distribution, are necessary to give accurate rejection regions, each set off by a different critical value of *t*, for every value of α.

In Table 12.A we have a partial list of critical values of *t*. (A fuller table appears in the Appendix.) Across the top of the table appear various values of α, our selected probabilities of rejecting H_0 when it

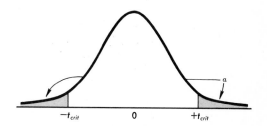

Table 12.A
CRITICAL VALUES OF *t*

df	Level of significance for two-tailed test				
	0.20	0.10	0.05	0.02	0.01
1	3.078	6.314	12.706	31.821	63.657
2	1.886	2.920	4.303	6.965	9.925
3	1.638	2.353	3.182	4.541	5.841
4	1.533	2.132	2.776	3.747	4.604
5	1.476	2.015	2.571	3.365	4.032
6	1.440	1.943	2.447	3.143	3.707
7	1.415	1.895	2.365	2.998	3.499
8	1.397	1.860	2.306	2.896	3.355
9	1.383	1.833	2.262	2.821	3.250
10	1.372	1.812	2.228	2.764	3.169

should be retained. Usually, we adopt 0.05 or 0.01. Down the left side of the table are values known as **degrees of freedom**, or **df**, which are related to *n* and which we'll consider shortly. Within the table are critical values of *t* that must be reached or exceeded by our obtained *t* if we are to reject the null hypothesis at the risk selected. Once we know our *df* and our value of α, we can find the critical value at the intersection of the appropriate row and column in the table.

Before we consider the topic of degrees of freedom, let's look at a worked example of our argument to this point.

a worked example: random groups design

A leprechaun specialist wonders whether equipping leprechauns with prospecting kits affects the amount of gold they can find. He outfits one random sample of the wee folk with pick, shovel, pan, and sourdough; another random sample is left to shift for itself in the time-honored way. There are five leprechauns in each sample. In due course, the data come in looking like this.

	$\bar{X}$	*s*	*n*
Outfitted	41	8	5
Control	20	9	5

If these data are substituted into the formula above, we get

$$t = \frac{41 - 20}{\sqrt{(5 \times 8^2 + 5 \times 9^2/5 + 5 - 2)(\frac{1}{5} + \frac{1}{5})}} = \frac{21}{6.02} = 3.488$$

We now must compare this obtained *t* with the critical value of *t*, and to do this we must now consider the topic of **degrees of freedom**.

degrees of freedom

The concept of degrees of freedom is rather difficult, and for our present purposes simplification is necessary. Basically, the argument is this: Every statistical distribution involves variation among scores. The variation may be due to chance, or it may not. The amount of variation clearly depends upon *the opportunity the data have to vary*. For instance, a sample made up of two scores has far less scope for variation than one made up of 20, or 200, or 2000. The more scores, the more freedom they have to vary.

So degrees of freedom are associated with the size of n. This part is easy. But there is a complication.

Suppose we have a sample of size $n = 10$. The mean value of the scores has been calculated and is thus known. Notwithstanding, the ten scores themselves could take on different values without constraint, without forcing any change in the given value of their mean, $\bar{X}$. Or to be more exact, *all but one* of the scores could have different values without affecting $\bar{X}$. One score's value, however, would be determined by the remaining nine perfectly arbitrary values, because of the necessity that their ten scores add to a total of $n\bar{X}$. So the value of one score out of the ten would *not* be free to vary. The number of degrees of freedom would be 1 less than n, or $(n - 1)$.

A more general way of stating this is that degrees of freedom are equal to the number of sampled scores in a distribution minus the number whose values can be ascertained from the remainder of the information.

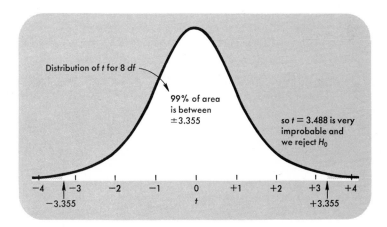

Figure 12.2
An obtained t of 3.488 will cause us to reject H_0 when the critical value is 3.355.

For example, suppose we know that $\bar{X} = 6, X_1 = 4, X_2 = 8$, and $n = 3$. Can we determine the value of X_3? If not, it is free; if so, it is not free. Of course, X_3 can have only one possible value, 6, so it is not free. With $n = 3$, $df = (3 - 1) = 2$. Now we can establish the critical value of t for our last example, the well-equipped leprechauns. With $n = 5$ and a mean value $\bar{X}$ known for each group, we have

$$n_E - 1 + n_C - 1 = n_E + n_C - 2 = 8 \ df$$

Since we have calculated 2 means on 10 leprechauns, we have $10 - 2 = 8 \ df$. From Table 12.A, we see that for $8 \ df$ and $\alpha = 0.01$, the critical value of t is 3.355. Look it up now, to make sure you grasp the principle of the table. The obtained t value was 3.488, which is larger than the critical value. Therefore, we reject the null hypothesis and conclude that prospecting kits *do* help leprechauns find gold. This logic is illustrated in Figure 12.2.

determining *df* for
matched-subjects experiments

In a matched-subjects design, the presence of a subject in one group is tied to the presence of a similar one in the other group. The two are interdependent, being matched in some way. The calculations are performed exactly as for the example in the previous chapter; since n_E and n_C must be equal, the denominator of the z test formula is appropriate. The only change from the z test is the use of the Table of Critical Values of t rather than the Table of Areas under the Gaussian Curve. The rule is simple; there is one *df* for each *pair* of subjects, minus one for calculation of the mean difference between pairs. (It happens that the difference between two means is equal to the mean of the differences.) Therefore, in the matched-subjects experiment, there are $n - 1 \ df$, where n is the number of *pairs* rather than *scores*. Or, if you like, *df* could be the number of *difference scores* minus one.

You might ask, why not use n instead of *df*, and adjust the table of t accordingly? There are two reasons.

1 Two types of t test exist, for random groups and for matched subjects, and they could not share a table of t whose rows were determined by n rather than *df*.

2 *df* is a concept that will become more important later in the book,

so we might as well grapple with it now and get to understand it while it's still relatively simple.

Another question: Is it not rather silly to argue backward, as it were, calculating the value of $\bar{X}$ and then saying that it forces the removal of a degree of freedom from the sample? Possibly it seems quite silly, but we do it because the t test involves certain assumptions, called *parametric assumptions*, which must be met by the data. These are listed in the next section, and the first is the one that applies to the present discussion.

another worked example: repeated-measures design

As you may recall, the repeated-measures design is a special variation of the matched-subjects design, one in which the same subjects are used in both of the experimental conditions, in the hope that variability between groups will then be filtered out of the variance in the denominator of the test. But as usual, to get something, you give up something. In the repeated-measures experiment, when first one measure is taken of the dependent variable, and then another under the other level of the independent variable, the effect of *practice* must be removed from the experimental comparison.

Pretend that you are an educational researcher interested in the growth of verbal comprehension in children between the ages of three and five. Your basic intention is to use a standard test and administer it to a group of, say, ten children at the time of their third birthdays. You would then retest them on their fifth birthdays and note the difference.

From a statistical point of view, all we need do is find the difference (D) between each pair of scores, always substracting in the same direction (age 5 scores minus age 3 scores). This would give data such as those in Table 12.B.

By this time, we are all accustomed to summing score values, and also to squaring them and adding the squares. So this has been done, as shown. Note that if any of the values of D had been negative, the value of ΣD would have been appropriately affected.

The formula for the mean difference, $\bar{D}$, is $\Sigma D/n$, giving $46/10 = 4.6$. As it happens, this mean difference is identical with the difference be-

Table 12.B

Child	Score at Age 3	Score at Age 5	Difference	D^2
1	5	11	6	36
2	6	12	6	36
3	4	10	6	36
4	5	13	8	64
5	6	12	6	36
6	7	10	3	9
7	7	11	4	16
8	8	9	1	1
9	6	10	4	16
10	7	9	2	4
			$\Sigma D = 46$	$\Sigma D^2 = 254$

tween the two means $\bar{X}_5$ and $\bar{X}_3$, and you can work it out either way. Since we want the values of D^2, the values of D are required anyway; they enable us to calculate only one mean rather than two.

The standard deviation of the differences is, of course,

$$s_D = \sqrt{\frac{\Sigma D^2}{n} - \bar{D}^2}$$

and this is identical with the working formula already known, except for the substitution of D for X to remind us that we're using *difference* scores. Substituting the data into the formula, we get

$$s_D = \sqrt{\frac{254}{10} - 4.6^2} = 2.06$$

Now we can move directly to the formula for the repeated-measures t, which is

$$t = \frac{\bar{D}}{s_D/\sqrt{n-1}} = \frac{4.6}{2.06/3} = 6.699$$

With *df* equal to $n - 1 = 10 - 1 = 9$, we find that our obtained value of t exceeds any of the critical values of t in Table 12.A, so we reject the null hypothesis. Our obtained t is significant. Five-year-old children apparently differ from three-year-old children in verbal comprehension.

The repeated-measures t test is very easy, but there are a couple of things to watch out for. One is the problem of silly arithmetic errors, encountered when too many different formulas are involved in a calculation. Here we have used two when one would suffice: the denominator involved the calculation of s_D (the standard deviation) and then of $s_D/\sqrt{n-1}$ (the standard error of the mean difference). Two square roots had to be calculated, and they might have been quite difficult. It will be better, then, to use still another formula, giving t from the raw score values:

$$t = \frac{\bar{D}}{\sqrt{\dfrac{n \sum D^2 - (\sum D)^2}{n^2(n-1)}}}$$

This formula can be derived from the formulas you already know. Using this new formula, recalculate the t ratio for our example above.

The other thing to watch out for is that the repeated-measures t test is easy at the expense of complications in the research design. Here we have ignored these complications for the sake of simplicity. But consider: the significant t *could* have been due to the effect of practice rather than the effect of maturing. In an actual experiment, then, two equivalent versions of the test should have been used, so that no child would get the same test twice. In other words, the *confounding* between practice and maturing would be eliminated. But then another problem occurs: if Version A is given first to all the children, then Version B given two years later, the difference between the two groups might be attributable to the two versions being somewhat different, not to the effects of growth. So it would be necessary to *counterbalance* the order of the two versions, half of the children getting first A and then B, the other half getting first B and then A. This would clean up the experimental design. But then, complications would arise with the statistical analysis when we tried to filter out the effects of *different versions* and the *order of their administration* from what is supposed to be strictly random error in the denominator. These two procedural effects are systematic, not random; hence any effects they have will be constant errors which, if mixed in with the variable error of the denominator, might enlarge the latter to such an extent that a Type II error occurs.

We won't worry about such problems here; in Chapter 14 we begin to discover how such unwanted systematic variables can be turned into added independent variables, all handled within one experiment.

parametric statistical tests

The *t* test and others to be considered in the following chapters are powerful research techniques whose potency is focused by some very real constraints or assumptions that must be met. Statistical tests which meet them are called **parametric tests**; certain other tests of hypotheses are available which do not require such assumptions, and they are termed **nonparametric tests**. The latter are generally less potent than the former. The parametric assumptions are

1 All observations must be independent of one another. Each score must be sampled separately from every other score, and the mean itself is considered a score. Hence, when we are considering the *sample*, we must give up one degree of freedom even though we got the mean and the standard deviation *from* that sample. (But the *df* that we gave up for the mean also went for *s*.)

2 The scores must have been drawn from a Gaussian distribution. This assumption is necessitated by the fact that population mean μ and population σ must be independent of each other, and this is true only of Gaussian populations. (If it is not Gaussian in shape, it is not randomly caused. If something systematic is making it non-Gaussian, then this something must be affecting some scores more than others; hence both mean and standard deviation are being jointly affected and cannot be independent.)

3 The population or populations from which the scores are sampled must have the same standard deviations as each other.

4 The population or populations from which the scores are sampled must have the same means (μ) as each other. This assumption is the null hypothesis, which can be rejected if the previous three assumptions hold true and if the data are too improbable under all four of these assumptions.

Since the *z* test is now seen to be a *t* test for infinite *df*, we shall no longer consider the *z* test a separate entity. It has now completed its duty as a conceptual stepping-stone to statistical inference from small samples.

on reporting *t* values

When an experiment has been performed and a *t* value found, it must be reported as informatively as possible. Three bits of information should be included:

1 the appropriate *df*,
2 the value of *t*,
3 the probability of a Type I error, or the *alpha* value associated with the obtained *t*. They are strung together thus.

$$t(19) = 2.978, \quad p < 0.01$$

This is read as "*t* for 19 degrees of freedom equals 2.978, *p* less than 0.01." The implication is that the *t* value is significant, in the sense already discussed. If *t* had not reached the critical alpha value, say $\alpha = 0.05$, then the probability would be written as $p > 0.05$.

two-tailed and one-tailed tests

In Figure 12.2 you will have noticed and perhaps wondered about the use of *both* tails, negative and positive, of the *t* distribution. Surely, a *t* ratio cannot head off in both directions at once? Why use two tails for our odds, when our obtained *t* cannot fall in more than one of them?

If we review the *t* formula, we'll note that the sign of the value $\bar{D}$ in the numerator determines the sign of the obtained *t* ratio (because by convention the square root in the denominator is taken as positive). So whether the *t* value goes off to the left or to the right in the graph is determined simply by which way we subtract in the prospector-kit *versus* control comparison. Since we carefully (but sneakily) hypothesized that the two conditions would result merely in *different* mean scores, we didn't specify which way the subtractions should be done, and hence we didn't specify the expected sign of $\bar{D}$ and *t*. *Any* difference would have been fine so long as it was significant. Therefore, to keep the betting honest, we had to use both tails of the *t* distribution. Naturally enough, then, the statistical test that we used and that let us reject the null hypothesis is called a **two-tailed test**, and the experimental hypothesis is called a nondirectional hypothesis.

We could have done things differently. We could have predicted that the prospector outfits should *increase* (not merely change) leprechauns' ability to find gold. Naturally, we wouldn't expect the outfits to handicap discovery. Thus, if we subtract the control scores from the outfitted scores, we should expect a positive value of $\bar{D}$, and hence we should expect to find t only to the right of 0 in the t distribution. So we are really interested in only one tail, the right-hand one.

Suppose we were to make such a **directional hypothesis**. We would expect to reject the null hypothesis if our obtained t ratio was in the shaded **rejection region** in the tails of the t distribution (as it was). But why bother to incorporate a rejection region in the left-hand tail, if our hypothesis predicts the right-hand tail as the only meaningful one? So we might as well remove the left-tail rejection region as immaterial to our purpose. Nevertheless, we must maintain a total rejection region whose proportion of the whole distribution is equal to α, if we want the probability of a Type I error to be α. Of course, we could be content with the right-tail region and have an α that is half of the original. If we predict the *direction* as well as a *difference* in means, a given t value is only half as likely to reject H_0 erroneously. This can be useful, but we also have the option of keeping our original value of α and shifting the shaded area in the left-hand tail over to the right-hand tail. If this is done, we can reject H_0 at the original level of α, but with a smaller necessary value of t. This option is useful too. The argument is schematized in Figures 12.3 and

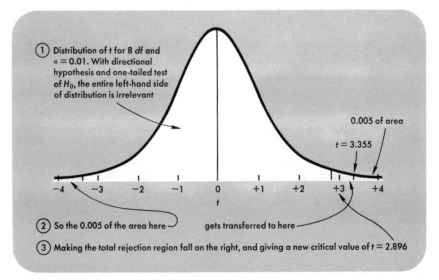

Figure 12.3
One-tailed and two-tailed tests of the null hypothesis.

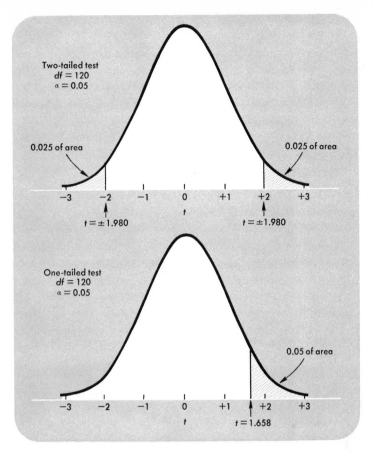

Figure 12.4
Note that for any given *alpha*, the critical value of t is smaller in the one-tailed test than in the two-tailed test.

12.4. Logically enough, this one-tailed procedure is called a **one-tailed test** of the null hypothesis.

Look carefully at Figures 12.2, 12.3, and 12.4 and relate them to the proper places in the Table of Critical Values of t. It is quite important for you to do this in order to understand the table fully.

The headings in the Table of Critical Values of t show α values for two-tailed tests. If critical values are desired for a one-tailed test, the probability values (α) across the top need only be halved. You may wish to do this on the page itself. When finding the critical value for t, remember to look in the correct column for the type of test being used.

how to know how many tails you have

As we've found, a one-tailed test allows rejection of the null hypothesis by smaller obtained values of *t*, which (we know) are easier to get. In other words, our experiments will "work out" better with one-tailed tests. We will be able to reject the null hypothesis with a *t* value that would not quite make it with a two-tailed test. Under some circumstances, that is.

The rule is this: the direction of difference between the experimental and control groups must be *predicted and stated before* the data are collected. We must also accept a risk that the obtained difference *might* be in the wrong direction. If a huge negative *t* value is achieved when we are doing a one-tailed test in the right-hand tail, we can conclude *nothing* even if it would have been significant with a two-tailed test. You can conclude that it just isn't one of your days, if you like, but nothing else.

There is sharp difference of opinion about the propriety of using one-tailed tests at all. Some people claim, probably quite correctly, that if you obtain an enormous *t* value in the wrong direction, you will strongly doubt that the null hypothesis is true anyway: the probability would be too small. Deep down you would wind up not *really* retaining H_0. At the very least, you would wonder what was happening. The only way to prevent this is *always* to use a two-tailed test, and some people recommend this.

Others say that a one-tailed test is necessary whenever your hypothesis is naturally and strongly directional; otherwise you'll be using α values that are not really valid. (In our example of the growth of verbal comprehension in children between the ages of three and five, a directional test would actually have been quite natural, because we certainly do not expect verbal comprehension to *decrease* between those ages.) This argument has a good point too. Perhaps the fairest way to leave the decision is *always* to use a two-tailed test, *except* when there is a strong, documented, obvious case (aside from your own mere preference) for a one-tailed test.

degrees of significance?

Often, when a *t* ratio comes out with a very high value, we hear the experimenter crow gleefully that the *t* was "*highly* significant." As we know, saying that a t value is significant merely indicates that the experimenter has decided to reject H_0 on the basis of the obtained evidence and arbitrary criteria. Whether it is rejected by a very large value of t or a marginally adequate one makes no difference. You reject H_0 if you reject it. You don't "highly reject" or "barely reject" it. Remember that the only reason for doing a t test is to find a rational basis for a decision. The *decision* cannot be qualified in terms of mere enthusiasm. Of course, strength of belief is another matter, which *can* depend on obtained t values. But a decision is for one choice *or* the other, with nothing in between.

We have seen the reverse argument with the notion of "approaching significance." If t only "approaches" significance, then it is not significant. Let's not substitute wishful thinking for statistical inference.

Most frequently, an experimenter will adopt a prior level of α, often 0.05, above which an obtained t will be significant. Subsequently the t values will be reported with the α levels at which they *are* significant. For example, if α is 0.05 but $t(10) = 3.50$, then the latter is significant at the 0.01 level as well as (obviously) the 0.05 level. This practice of upgrading the α level above the pre-set value is acceptable, but downgrading it is not.

what's the good word? here are some to mumble in your sleep

STUDENT t DISTRIBUTION

TABLE OF t VALUES

DEGREES OF FREEDOM

PARAMETRIC STATISTICAL TESTS

COUNTERBALANCING

TWO-TAILED TEST

ONE-TAILED TEST

REJECTION REGION

CRITICAL VALUE

DIRECTIONAL HYPOTHESIS

NONDIRECTIONAL HYPOTHESIS

/elf-te/t on chapter 12

Attempt these problems only after you have thoroughly read and understood this chapter. It will be necessary to use the table of critical values of t in the Appendix, and possibly some of the figures for conceptual clarification, but don't peek otherwise. The usual instructions apply.

1 A would-be experimenter wishes to employ a matched-subjects design, which would succeed in rejecting the null hypothesis if t worked out to be as little as 3.00. The experimenter would need a minimum of how many pairs?

   a 120, using a two-tailed test and $\alpha = 0.001$.
   b 14, using a one-tailed test and $\alpha = 0.01$.
   c 15, using a one-tailed test and $\alpha = 0.01$.
   d 15, using a two-tailed test and $\alpha = 0.01$.
   e 8, using a one-tailed test and $\alpha = 0.01$.
   f 4, using a two-tailed test and $\alpha = 0.05$.
   g 15, using a one-tailed test and $\alpha = 0.005$.

2 In a t test, certain factors combine to make it easier to reject the null hypothesis correctly, but certain factors make it harder. Identify them.

   a Easier: decreasing n and increasing s.
   b Harder: decreasing n and increasing s.
   c Easier: increasing $\bar{D}$ and decreasing s.
   d Easier: increasing s and decreasing $\bar{D}$.
   e Harder: increasing α.
   f Easier: using a one-tailed test rather than a two-tailed one.
   g Easier: increasing $\bar{D}$, decreasing s, increasing α, increasing n, and using a one-tailed test.

SCORING AND ANALYSIS of self-test questions. Give yourself 3 points for each correct answer. Subtract 2 points for each incorrect choice and 1 point for each correct one you failed to discover.

Answers to Question 1

a F
b F
c F
d T
e T
f T
g T

This question pertains to critical values for *t*. The first step is to discover the correct column in the table of critical values of *t*. This will establish α for the one- or two-tailed test, as required. Then, run down the column to the first value that exceeds 3.00. Next, run across to the column of *df* and note the value there. Since in a matched-subjects design, $df + 1 = n$, add 1 to the *df* value and you have the minimum required *n*.

   example: with a two-tailed test and $\alpha = 0.01$, we are in the column headed 0.01. Running down it, we find 2.977 at $df = 14$. So the minimum *n* is 15.

Answers to Question 2

a F
b T
c T
d F
e F
f T
g T

Rephrased, the question is, "What factors tend to make the value of *t* as large as possible?" The formula for *t* is

$$t = \frac{\bar{D}}{\sqrt{\dfrac{n \sum D^2 - (\sum D)^2}{n^2(n-1)}}}$$

for matched subjects and repeated measures, or

$$t = \frac{\bar{D}}{\sqrt{\dfrac{n_E s_E^2 + n_C s_C^2}{n_E + n_C - 2}\left(\dfrac{1}{n_E} + \dfrac{1}{n_C}\right)}}$$

for random groups

So the ratio will be made larger by increasing the numerator or by decreasing the denominator. The values of *s* should thus be as small as possible, and *n* should be as large as possible. Increasing *n* also permits a smaller critical value of *t*, since *df* is tied to *n*. Using a one-tailed test also reduces the critical value of *t*. In general, anything that causes the critical value of *t* to be as small as possible and the obtained value of *t* to be as large as possible will allow the null hypothesis to be rejected more easily. But if *too* easily, we may be victims of a Type I error!

3 A comparison of the characteristics of Gaussian and t distributions show that

   a t values are more variable than z values.

   b the distributions are approximately similar when n is larger than 1000.

   c the important areas of difference are in the tails.

   d the variances are always the same.

   e one-tailed tests can be used only with t.

   f there are as many t distributions as there are Gaussian distributions.

   g the t test is a special variety of z test.

4 You are doing an experiment to test a one-tailed hypothesis. However, at the last minute you decide to be conservative and use a two-tailed test. The result is that you did not reject the null hypothesis when you could have done so. Any comments?

   a You have invited a Type I error and have doubtless found it.

   b This is a Type II error.

   c The probability of a Type II error is increased under these circumstances.

   d Your chosen value of α is not valid.

   e This was equivalent to halving your α level.

   f This procedure would have been justified if a Type I error were to involve a serious practical risk of some sort.

Answers to Question 3

a T True. The tails of the *t* distribution contain relatively more values than the Gaussian tails.

b T Yes, but they are practically identical for much smaller *n* than 1000.

c T True. The tails are the sites of the rejection regions.

d F The variances are different, the more so when *n* is smaller.

e F One-tailed tests may be used equally well with *z* tests as with *t* tests. The number of tails pertains to the directionality of hypothesis, not the *df* value, which distinguishes between *t* and *z* tests.

f F The Gaussian distribution has a given mathematical shape. The *t* distributions are shaped differently for each value of *df*.

g T *t* certainly can be thought of as a *z* score:

$$z = \frac{X - \bar{X}}{s} \cong \frac{\bar{D} - \mu}{s/\sqrt{n-1}}$$

Answers to Question 4

a F Remember, rejecting the null hypothesis is a *personal* decision, merely
b F based upon certain criteria and evidence. Any decision to *retain H_0*
c T cannot lead to a Type I error, but it may lead to a Type II error. We never know. The probability of Type II error is increased, however, when α is made very small.

d T It would be mistaken to say that you had retained H_0 at the originally
e T adopted α level, because in effect you have halved its value and set a higher criterion for rejection.

f T Your procedure might be justified if risk of Type I error had serious consequences, but you chose a strange way to take precautions. Better to start out at a smaller value of α, e.g., 0.005 rather than 0.01, and keep the one-tailed test.

5 If an infinite number of samples of size $n = 2$ were drawn randomly and the values of $\bar{X}$ were subtracted from $\mu = 0$, within what abscissa values would 99.9% of all t values fall? Assume that the abscissa is sealed in t-units.

a ±3.00
b ±3.291
c ±2.326
d ±636.619
e ±31.598
f ±9.925

6 Assume that we are using a two-tailed test, 20 df, and $\alpha = 0.01$. The rejection region would run

a from 2.845 to 2.861, positive and negative.
b toward the tails on both sides, beginning with an abscissa value of ±2.845.
c differently between random groups and repeated measures experiments.
d from 2.528 toward the tail on both sides.
e from +2.528 toward the tail on the right side only.

7 The purpose of counterbalancing is to

a diminish the probability of Type I error.
b ensure that only the independent variable is allowed to affect systematically the dependent variable.
c reduce (but not totally eliminate) confounding.
d handle the problem of practice in repeated measures.
e ensure that the experiment pertains to its actual hypothesis.

Answers to Question 5

a F This question pertains to the characteristics of the *t* distribution

b F when $df = n - 1 = 2 - 1 = 1$. If we want the *t* values *between which*

c F 99.9% of the area falls, then we are looking at the values *beyond which*

d T the remaining 0.1%, or 0.001, of the area falls. In other words, an α

e F value of 0.001 with $df = 1$. Referring to the Table of Critical Values

f F of *t* in the Appendix, we see that the *t* value in question is 636.619.

If you picked any value other than the correct one, you may have a serious problem of comprehension. It would be a good idea to review Chapters 11 and 12 very carefully.

Answers to Question 6

a F (a) refers only to the critical values for $df = 20$ and $df = 19$ and the area lying between them.

b T Another question to ensure that you can use the Table of Critical Values of *t*. The only correct choice is (b).

c F (c) would be correct were we talking about *n* rather than *df*, because repeated measure designs have $df = n - 1$, while random groups have $df = N - 2$.

d F (d) and (e) are in the wrong column, for $\alpha = 0.02$ rather than 0.01.

e F

Answers to Question 7

a F The role of counterbalancing is to prevent an incidental (but possibly

b T potent) variable from being systematically associated with the inde-

c F pendent variable. Without counterbalancing, such an incidental vari-

e T able could have a strong effect on the data, falsely attributed to the independent variable, which may or may not be potent in its own right. Systematic association between two variables is *confounding*; the latter is prevented totally by correct counterbalancing procedures.

d T In repeated measures designs, the two measures cannot be made simultaneously, so time intervenes and may have an effect. Practice, insight, or learning may occur within this interim time and may be taken into account by counterbalancing.

8 You test a directional hypothesis, expecting a positive $\bar{D}$, and find a t of -9.888 for 3 df. What should you do?

  a Reject the null hypothesis.
  b Check all calculations.
  c Retain the null hypothesis.
  d Reconsider your grounds for the experimental hypothesis.
  e Retroactively adopt a two-tailed test.
  f Report a significant value of t.
  g Repeat the experiment with fresh subjects.

EVALUATION

The maximum possible score is 75. Since this self-test refers to one of the more important topics in statistics, please try to have done well. If not, review would be very wise.

65 to 75	Excellent; you have it made.
45 to 64	Try to discover where you ran aground.
25 to 44	Still not very bad, but you should be concerned. Review heartily.
under 25	You have problems. Stick your head into some cold water and begin again.

Answers to Question 8

a F   The main thing to note in this question is that the difference between
b T   means was in the wrong direction.
c T   It's a big *t*, but you can't reject H_0, unless, of course, you discover
d T   upon checking that you goofed in your calculations and the obtained
e F   *t* was really positive. If not, you'd better rethink your hypothesis and
f F   start over again with fresh subjects. And if you retroactively adopt a
g T   two-tailed test and report a significant *t*, you will have guilty dreams
      for nine years.

ſome ordinary problemſ

1 Using the data in the Table of Critical Values of *t*, draw a graph of
the positive tails of the *t* distributions having 2, 10, 20, and 120 *df*.
The tails should begin at the point where 10% of cases are cut off
from the distribution.

2 Study the graphs constructed in the previous question. What
characteristic is evident that shows the importance of *df* in ascertaining α for any value of *t*?

3 Give an example of a situation in which a Type I error would have
serious consequences and one in which a Type II error would.
Then discuss the implications for practical and theoretical research questions.

4 In a way, it can be claimed that any experiment (of the type sus-
ceptible to a *z* or *t* test) can involve *four* possible alternative hypotheses. What are they, and how do they relate?

5 As we know, the *z* scores in any regular distribution are tied to
the width of the distribution itself, so *z* scores beyond ± 3 are very
rare indeed. Explain why *t* scores (which are a form of *z* score) can
vary so much more than that, even when the null hypothesis is
true.

6 A somewhat naive experimenter goes through the motions of an
experiment, which actually is done properly. However, having
calculated the final critical ratio, the experimenter uses the *z* test,
not knowing about *t*. There are only six subjects in this repeated-
measures experiment, but it is possible (but only just) to reject H_0
at $\alpha = 0.01$. What type of error is being invited, and approximately what is its probability?

7 You get very ambitious in an experimental term project and wind up with eight experimental groups to compare. You then find that (to be fair) you must compare between all possible pairs. Using an α of 0.05, you find that two of the t tests are significant. In your conclusions you build a case upon the latter. Comment.

8 An independent-groups experiment is done properly and yields the following information.

	$\overline{X}$	s	n
Experimental	15.6	2.8	21
Control	17.2	2.2	25

The hypothesis was nondirectional. What should be concluded?

9 Make sure that you can define the following:

a rejection region
b degrees of freedom (df)
c significance

10 In Table 12.C is shown a population of scores upon which some experimental treatment has been performed. We don't know whether the treatment was effective or not. Initially, the Table 12.B population was the same as the Table 8.B population. Using a sampling procedure, test the null hypothesis using

a groups of $n = 4$ each.
b groups of $n = 8$ each.
c groups of $n = 16$ each.

Each group should be freshly sampled, by the method indicated in Table 8.B. Draw conclusions with $\alpha = 0.05$.

11 If Question 10 was a class-assigned question, compare your own conclusions with those of others. Specifically, if the class contains 100 students or more, compare the proportion of those who did not reject H_0 with the proportion of those who did for each of the three experiments (df values).

12 What is the general effect of increasing n?

Table 12.C
APPROXIMATELY GAUSSIAN POPULATION

53	70	36	53	96	42	48	73	42	78	73	53	56	102	68	51	63	70	28	88
64	62	60	78	61	78	71	44	44	42	80	53	58	63	52	38	58	44	74	56
44	86	73	42	70	36	88	63	70	32	48	61	50	96	51	68	54	82	56	64
56	22	63	68	58	74	46	40	84	73	58	40	68	44	92	51	80	73	40	48
61	66	71	68	56	52	63	62	62	50	78	51	80	54	78	68	61	51	68	50
44	62	42	73	58	61	73	76	58	78	71	62	63	62	68	96	82	44	78	46
70	63	52	61	54	68	51	96	88	88	60	61	28	53	74	50	60	96	51	110
62	72	48	80	25	32	73	44	60	63	80	62	52	73	40	51	25	56	62	66
73	51	74	68	50	60	61	72	38	60	70	46	36	72	38	46	66	54	61	92
38	58	54	78	82	50	46	44	48	48	64	53	73	56	66	105	28	36	64	62
71	58	50	58	52	84	73	74	52	51	28	80	66	61	62	61	51	84	78	76
66	71	53	73	86	96	63	46	68	74	80	68	58	61	66	66	64	96	74	44
74	96	58	40	61	46	70	52	64	73	62	82	51	63	82	46	38	58	86	48
42	58	84	52	54	50	80	72	74	40	70	80	92	54	38	82	46	78	71	52
64	62	19	70	70	53	66	68	48	66	63	84	80	84	40	53	73	72	72	32
92	32	54	52	76	48	42	71	66	32	38	58	70	86	80	73	60	66	60	32
82	13	88	25	51	73	92	66	74	28	73	74	84	62	36	52	53	53	61	56
61	56	71	80	70	53	70	70	61	56	46	42	60	44	62	64	72	60	63	56
58	44	51	105	28	78	56	54	60	56	48	66	58	52	48	22	66	51	60	92
48	60	63	64	74	46	64	64	102	71	64	68	88	42	42	62	74	40	76	62
61	70	66	50	72	36	63	52	88	74	44	54	82	62	32	74	76	25	60	58
82	46	46	32	56	52	62	53	71	52	54	72	72	63	62	65	70	56	42	70
36	78	66	46	70	92	60	54	54	84	56	76	76	71	80	48	56	50	64	56
86	68	28	48	73	68	64	44	61	63	73	40	72	52	51	52	60	84	61	51
54	48	46	71	58	62	63	53	64	70	40	60	62	44	92	61	82	42	38	53
61	73	88	62	84	32	40	46	50	63	66	86	62	32	86	76	46	84	53	74
38	78	72	36	78	71	28	76	50	64	58	53	38	68	58	48	60	51	61	60
72	36	70	88	84	70	38	84	54	64	42	40	53	62	82	80	71	62	64	63
44	66	80	66	82	56	61	38	38	62	62	68	40	84	64	63	56	40	80	82
78	76	40	64	56	36	36	72	99	76	51	82	38	56	46	80	52	46	70	60
58	66	78	60	76	64	44	80	52	48	44	63	78	88	50	71	54	71	58	54
80	60	78	60	68	38	42	71	51	76	54	58	25	74	53	68	84	60	52	96
80	64	51	50	64	80	54	92	62	71	63	66	99	84	76	44	52	64	74	86
74	36	82	63	80	70	60	72	48	72	92	62	72	73	48	68	74	71	80	66
64	36	63	78	61	48	63	61	76	61	71	68	82	71	74	61	71	68	58	66
54	68	76	80	76	40	64	46	73	64	52	46	92	63	78	42	54	82	64	50
86	73	96	48	62	53	40	86	73	52	68	92	61	78	71	48	48	63	78	52
61	66	53	68	78	58	71	61	51	42	82	56	51	54	50	38	86	82	76	50
72	74	74	44	73	58	96	54	86	58	61	42	82	44	64	63	61	46	96	46
48	53	71	52	51	70	76	70	52	64	56	46	72	74	68	53	50	56	52	70
42	62	88	54	84	84	72	60	28	63	32	71	102	61	64	60	76	61	92	22
56	51	62	76	60	64	28	50	82	60	71	74	38	82	84	73	74	68	53	62
42	66	28	54	62	56	52	53	32	58	99	66	74	63	62	88	64	36	50	76
72	86	61	66	51	99	46	61	71	54	82	61	40	58	78	68	68	54	36	66
76	36	76	28	38	99	50	42	44	58	48	72	58	40	68	72	86	60	70	61
88	71	86	86	56	62	84	62	92	58	82	73	73	53	58	56	53	60	51	66
92	78	50	42	40	71	32	50	46	78	82	80	64	68	52	50	62	38	88	50
54	88	42	36	58	72	63	54	64	42	51	32	44	48	52	60	72	71	70	72
50	76	86	70	78	44	42	88	54	66	88	72	72	56	62	78	32	72	53	48
66	54	72	76	44	63	60	84	74	51	64	63	46	63	42	68	44	50	56	40
76	38	36	86	66	54	70	74	46	61	53	48	86	60	86	80	88	53	50	63

chapter 13
analysis of variance and the *F* test

One of the troubles with research is that it gets complicated more easily than it gets simpler. Remember the experiment with fermented prune nectar? It involved two groups, one with nectar and one without. The experiment showed that nectar improved ability to find gold. Now the chief leprechaun, Mulligan Muldoon, wants to do a more elaborate study.

He wants to use *four* random groups, not just two, each to get one of the following:

1 Fermenting prune nectar that has not quite begun to bubble and froth;
2 fermenting prune nectar that has bubbled and frothed for one day;
3 fermenting prune nectar that has bubbled and frothed for one week;
4 fermenting prune nectar that has bubbled and frothed until it stopped.

Somehow, Mulligan has the idea that the secret ingredient has something to do with the amount of bubbling and frothing, and, like all wise experimenters, he wants to decide upon his method of statistical analysis beforehand. So far, the *t* test is the limit of his expertise, so he turns to us for advice.

We know that the *t* test is unexcelled for comparing two groups, that is, two experimental treatment levels. But frequently we wish to compare three, four, or more experimental groups. How can we do this?

An obvious answer would be to use the *t* test between all possible pairs of groups. Suppose, for example, that our experiment involved seven groups; Table 13.A shows the possible combinations, each

Table 13.A

NUMBER OF *t*-TEST COMPARISONS AMONG SEVEN GROUPS

	1	2	3	4	5	6	7
1		*	*	*	*	*	*
2			*	*	*	*	*
3				*	*	*	*
4					*	*	*
5						*	*
6							*
7							

asterisk denoting one of them. Evidently 21 such comparisons would be necessary.

There are two disadvantages to such a *t*-test strategy, one trivial and one tremendous. The trivial is that a lot of work would be necessary: 21 complete *t*-test calculations just for one experiment. The tremendous—and devastating—disadvantage is that no conclusion could be made concerning the null hypothesis anyway, after all the work was completed!

Why not? Suppose that we were using the 0.05 level for the rejection of the null hypothesis, meaning that we would decide to reject it on the basis of a *t* ratio (or ratios) that could occur by chance five times in 100, or once in 20 when H_0 is actually true. With 21 comparisons among seven groups, by chance alone we might *expect* at least one "significant" *t* ratio. Thus the probability values attached to the various critical values of *t* would be meaningless, and we could not arrive at any equivalent probability values (which are the whole reason for doing *t* tests in the first place).

What to do? There are only two possible solutions: simplify your experiments to only two groups, or use some other statistical test. The first alternative has considerable merit. We must remember that experiments are not supposed to be motivated by the personal need for a Ph.D or academic prestige, but by a real desire to know how nature is organized. But when the means—experimentation—becomes an end in itself, there is a tendency for it to become elaborate and unnecessarily complicated, often more than the research question would call for, if honed to a sharp edge and divested of all padding. The ideal, seldom reached, is for an important discovery to arise from a beautifully simple experiment that culminates in one *t* test.

Unfortunately, in the social sciences even relatively simple research projects go beyond the capability of the *t* test, so we are oblized to seek an alternative. Such an alternative, which permits comparison among many group means, not just two, is available. It is called **analysis of variance**.

If you turn back to Chapter 9 and look at Figure 9.11(b), you can get an intuitive idea of how analysis of variance works. The independent variable has five levels, not just two. The independent variable has a strong effect on the dependent variable. In the lower right-hand panel the same picture is given, except that the independent variable has a very weak effect.

Let's now try to get a basic idea of what analysis of variance is and how it works, from a mathematical point of view.

First, we should review the basic concept of variance. We know that a sample has variance, as measured by the formula

$$s^2 = \frac{\Sigma (X - \bar{X})^2}{n}$$

and that a population variance is estimated by

$$\hat{\sigma}^2 = \frac{\Sigma (X - \bar{X})^2}{n - 1}$$

We shall be concerned with the second of these formulas, the estimate of population variance.

Now we must direct our attention to the numerator of the variance formula, $\Sigma (X - \bar{X})^2$. Let's expand it fully for an n of (for instance) 3. We would have

$$\frac{(X_1 - \bar{X})^2 + (X_2 - \bar{X})^2 + (X_3 - \bar{X})^2}{df} = \hat{\sigma}^2$$

remembering that df is equal to $n - 1$, one df being lost by calculation of $\bar{X}$ from the values of X.

This principle applies to any number of scores. The closeness of the obtained $\hat{\sigma}^2$ to σ^2 would, of course, increase with the value of n.

In Figure 13.1 a distribution of six scores is shown along with its variance. Let's suppose that this sample of scores comes from the control group of an experiment; say it's the "unbubbled FPN" control group of leprechauns being run by Mulligan Muldoon.

Now, suppose that one of the experimental groups is added to the picture *and* that the experimental treatment has no effect—that the

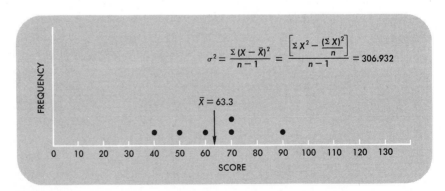

Figure 13.1
Random sample of scores having $\bar{X} = 63.3$, $n = 6$, and $\hat{\sigma}^2 = 306.932$. The latter is an unbiased estimate of population variance.

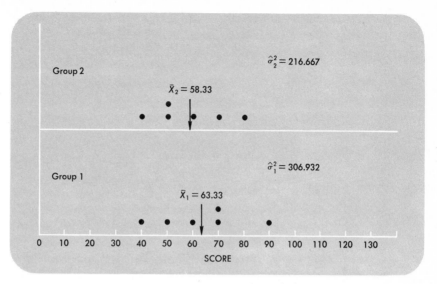

Figure 13.2
A second random sample is added. The statistics are somewhat different. We are assuming that the null hypothesis is true, so the parameters (whatever they are) are the same, and the statistics differ by chance. The combined within-group variances estimate the population variance, and the variance between the means can do the same.

null hypothesis is true. Let's look at the data of the two groups combined in Figure 13.2.

In Figure 13.2 we see two groups of scores, each a random sample from (according to the null hypothesis) the same population. The values of $\bar{X}$ are slightly different, as we would expect by chance, and the values of $\hat{\sigma}^2$ are different, too, again presumably because of chance fluctuations in the values of X sampled in each group.

Let us now consider the problem of population variance estimation again. An estimate of σ^2 can be obtained from *each* of the two samples shown, or from any number of samples. Or, alternatively, we could find just *one* variance, that of all the scores in all the samples together, working from the grand mean $\bar{\bar{X}}$ (X bar bar) of all the scores without regard for which sample they came from. The principle is that any time you have a bunch of scores you can find their variance: very easy.

While this is being digested, let's simplify our terminology. We have not been calling the numerator of our variance equation by any special name, but it is now becoming important, so $\Sigma\,(X - \bar{X})^2$ will now be called the **sum of squares**. That is, the sum of the squared deviation scores of X from $\bar{X}$. An easier formula for calculation is

$$ss = \hat{\sigma}^2 = \sum X^2 - \frac{(\sum X)^2}{n}$$

The abbreviation *ss* traditionally denotes sum of squares. A moment of thought will reveal that our estimate of population variance, $\hat{\sigma}^2$, now can have the formula

$$\hat{\sigma}^2 = \frac{ss}{df}$$

which is the same as

$$\frac{\sum (X - \bar{X})^2}{n - 1}$$

Let's now go back to Figure 13.2 and consider how we might estimate population variance from our two (or, in principle, any number) combined groups.

Clearly, as we've seen, we can estimate from the two sets of scores combined, the deviation of each score being measured from its *group* mean. This would have the formula

$$\hat{\sigma}^2 = \frac{\sum (X_1 - \bar{X}_1)^2 + \sum (X_2 - \bar{X}_2)^2}{N - 2} = \frac{ss_w}{df} = ms_w$$

where the symbol N means the total of all values of n. If we had done it just the same way but with the *grand* mean, the result should be the same, except for the possibility that the group means might (by accident) have differed somewhat from the grand mean. So we use each value of $\bar{X}$, not $\bar{\bar{X}}$, for this calculation.

The sum of squares obtained as shown is called the **within-groups sum of squares**, or ss_w. When divided by its *df*, which is $n_1 - 1 + n_2 - 1$, or N minus the number of groups, we have an estimate of population variance. It is usually called the **mean square**, or *ms*. In this case, it is called ms_w, the **within-groups** mean square, and it is an estimate of population variance.

So far, we have not looked at the mean values for the two groups, except to observe that they are slightly different (presumably due to chance, since we are assuming the null hypothesis). The fact that the means are different must be caused by population variance. (If there were *no* population variance, the means would be exactly the same, right?) So there is the possibility that the difference or variance among the means may also provide an estimate of population variance.

Recall from Chapter 11 that the variance of the sampling distribution of means is

$$\hat{\sigma}_{\bar{X}}^{2} = \frac{\hat{\sigma}^{2}}{n}$$

We can turn this formula around and estimate σ^2 from $\hat{\sigma}_{\bar{X}}{}^2$:

$$\hat{\sigma}^{2} = n\hat{\sigma}_{\bar{X}}^{2}$$

Let's work out our estimate of σ^2 directly, using the values of $\bar{X}$ that we have (two of them in our sample, but any number would be fine). Let $\bar{\bar{X}}$ be the grand mean, or mean of all the scores in all groups. And in this case, *df* is one less than the number of means.

$$\hat{\sigma}^{2} = \frac{n_1(\bar{X}_1 - \bar{\bar{X}})^2 + n^2(\bar{X}_2 - \bar{\bar{X}})^2}{2 - 1} = \frac{ss_{\mathrm{b}}}{df_{\mathrm{b}}} = ms_{\mathrm{b}}$$

The numerator gives us the **sum of squares between groups**, ss_{b}, and when placed over its degree of freedom, it gives us the **mean square between groups**, ms_{b}, which is also an estimate of population variance.

We now have two independently obtained estimates of population variance. They are independent because, as we saw in Chapter 5, the mean of a sample (on which we based ss_{b}) is independent of the standard deviation of the sample (on which we based ss_{w}).

Why should we want two independent estimates of population variance, $\hat{\sigma}^2$? If the null hypothesis is true, as we've been assuming, the two estimates should be about the same, except for chance. If a ratio is formed between the two estimates, it *should* be about 1.0, but it will vary around that value as a result of chance. The values will fall into a probability distribution having a particular known shape, the *F* **distribution**.

the *F* distribution

A typical *F* distribution is pictured in Figure 13.3. It is very distinctive in contrast to the other distributions we have considered thus far. Obviously, it is sharply skewed, since its mean value will be 1.0 and the ratios that form it can sometimes be quite high but never less than 0.

The relatively high values in the distribution are the ones that interest us. Just as with the Gaussian and *t* distributions, the area under

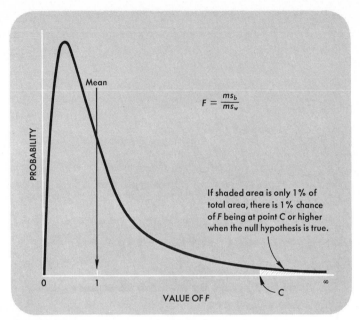

Figure 13.3
A typical F distribution, with essential features shown. Since F is the ratio between two separate estimates of the same population variance, its mean is 1. It can't go below 0; it can, however, go quite high by chance, once in a while. A very high F value, if obtained, would cast doubt on the truth of the null hypothesis.

the F distribution stands for probability. As shown in Figure 13.3, if 1% of the area lies beyond point C in the shaded zone, then there is only a 1% chance of an F value occurring high enough to be in that zone. This provides a basis for testing the null hypothesis that *any number* of means (not just two of them) have been sampled from one population having a given mean of μ. Such a test of the null hypothesis is referred to as an **F test.** (The symbol F was chosen to honor Sir Ronald A. Fisher, the British statistician whose work led to analysis of variance and the F distribution as a basis for the F test. He was active in the 1920s and 1930s.)

the rationale of the *F* test

As outlined above, when the ms_b and ms_w have been independently obtained from any number of experimental groups, their ratio will fall into the sampling distribution of F whenever the null hypothesis is true. Very large values of F, as in the shaded portion of Figure 13.3, will be rare, and the value of C in Figure 13.3 (the critical value of F) can be set to cut off any desired percentage of the F distribution, to give any desired value of α.

When the null hypothesis is false, however, the independent variable affects the different experimental groups so as to change their means (values of $\bar{X}$) and (by generalization) their population means, μ. When the values of $\bar{X}$ differ more than would usually happen by chance alone, this will increase the variance between these mean values, and thus greatly increase the value of ss_b, and thus of ms_b. As a result, the ratio ms_b/ms_w would become quite large, larger than would very often happen if the null hypothesis is true. So if an ob-

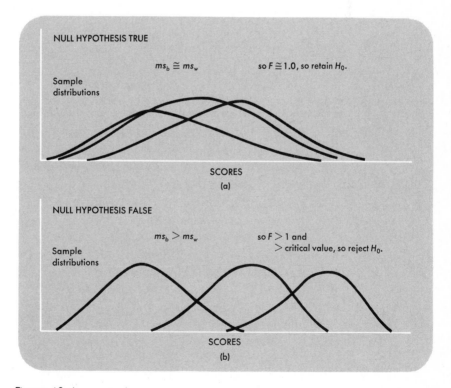

NULL HYPOTHESIS TRUE

$ms_b \cong ms_w$ so $F \cong 1.0$, so retain H_0.

Sample distributions

SCORES

(a)

NULL HYPOTHESIS FALSE

$ms_b > ms_w$ so $F > 1$ and
                  $>$ critical value, so reject H_0.

Sample distributions

SCORES

(b)

Figure 13.4
(a) The null hypothesis is true, and the three groups differ only by chance. They overlap closely, so their ms_b is about the same as their ms_w. (b) The means are far apart, relative to the within-group variances. Therefore, ms_b will be huge compared to ms_w, and the null hypothesis would be rejected.

tained value of F falls beyond the critical value C, one of two things must have happened. Either

1 The null hypothesis is true, and a very improbable ($p < 0.01, 0.05$, or whatever value of α was chosen) event has occurred,

or

2 The null hypothesis is false, and the μ values of the several experimental populations are different, presumably because of the action of the independent variable—if your experiment was properly controlled, that is. The whole argument is presented in Figure 13.4.

the F distribution, again

Of course, life is such that things that are simple in principle become quite complicated in practice. The F distribution obeys this splendid old tradition, in that its shape depends upon the separate df values of numerator and denominator.

$$F = \frac{ms_b}{ms_w} = \frac{ss_b/df_b}{ss_w/df_w}$$

As you can see on the right of this formula, the df values must be obtained anyway, enroute to the F value. The only problems are how to determine df values properly and how to lay out the critical values of F in a table that takes both df values into account.

The solution is to use a "two-dimensional" table, in which the df_b values range as 1, 2, 3, etc., across the top, and the df_w values range from 1 to infinity down the side. Very few experiments will involve df_b as high as 12 (that's 13 different experimental groups), so it is safe enough to stop at that value for df_b. But because experiments can involve quite a few subjects, perhaps hundreds or thousands, a much larger range of df_w values must be provided. Table 13.B gives the critical values of F for a few values of df. A more extensive F table appears in the Appendix.

Unlike the t table, the F table has no room for a large selection of α levels, if we want to keep it compact. So just the 0.01 and 0.05 levels are provided. To use the table, determine α and your df values, find your experimental value of F, and compare it with the critical value at the intersection of the proper row and column. If your obtained F is at least as large as the critical value in the table, reject the null

Table 13.B

CRITICAL VALUES OF F

	Degrees of freedom for numerator			
	1	2	3	4
7	5.59	4.74	4.35	4.12
	12.25	**9.55**	**8.45**	**7.85**
8	5.32	4.46	4.07	3.84
	11.26	**8.65**	**7.59**	**7.01**
9	5.12	4.26	3.86	3.63
	10.56	**8.02**	**6.99**	**6.42**
10	4.96	4.10	3.71	3.48
	10.04	**7.56**	**6.55**	**5.99**
11	4.84	3.98	3.59	3.36
	9.65	**7.20**	**6.22**	**5.67**

Degrees of freedom for denominator

Light type: $\alpha = 0.05$
Bold type: $\alpha = 0.01$

hypothesis. If either of your actual *df*s are not in the table, use the next lowest ones listed.

one tail or two?

With the F test and the analysis of variance, which leads up to it, you are doing a two-tailed test. You might as well make no exceptions here.

The alert student may, however, detect a paradox: the F distribution obviously has just one important tail, the one to the right. How can this be a basis for a two-tailed test? To answer the question for yourself, recall that the procedures leading to F involve squaring scores, and squaring makes all differences positive, thus eliminating the implication of direction in any difference. So the F test is sensitive only to the *magnitude* of any differences among means, not to the direction of their differences.

Again, you may notice that the F distribution does have a tail of sorts, if a rather short one, on the left. Does it have any significance? Mentally working out how F value could arrive in this short tail, we find that it would happen when inter-group mean differences are exceedingly *small* as compared to the within-group differences. As you move left in the F distribution, the probability of the null hypothesis being true increases. But no actual probability figure can be ascertained, on this basis, of H_0 being true.

how to determine values of *df*

The previously stated principle for df still holds. Your df is your number of independently sampled scores or unrelated values minus the number of restrictions you've put on their freedom to vary without disturbing anything. If you have n subjects per group, you have $n - 1$ df for that group, because the mean must be calculated. If you have several groups of size n, the df_w is the total of the degrees of freedom, $(n - 1)$, for each of the groups. The total of $(n - 1)$ values over k groups will give $(N - k)$. If you have k different groups, $df_b = k - 1$, since the grand mean is the weighted average of the k group means, and 1 df is lost in calculating $\bar{\bar{X}}$.

When the experiment is considered as a whole, it is obvious that the df_b and the df_w are separate aspects of the df for the entire experiment. Their relationship is additive; that is, we can add them to find the total df. With a total of N subjects or scores, the grand mean can be calculated directly, losing only one df in the process; so the total df, df_t, is $N - 1$. But it is also true that

$$\begin{array}{r} k - 1 \ (df_b) \\ + N - k \ (df_w) \\ \hline N - 1 \end{array}$$

because

$$N - \cancel{k} + \cancel{k} - 1 = N - 1$$

So

$$df_b + df_w = df_t$$

In our two-group example as developed so far in this chapter, where there are two groups of $n = 6$ each, df_w is 10. This can be calculated either as $n_E - 1 + n_C - 1$ or as $N - k$; it works equally well either way. The df_b is 1. The total df, when calculated as the total of df_b and df_w, is $10 + 1 = 11$. The total as calculated directly is $N - 1 = 12 - 1 = 11$.

(Remember that N is the total of all the values of n.) Try entering these numerical values into the addition shown above.

the additivity of sums of squares

Just as df is additive, so are the sums of squares. Looking at the experiment as a whole, we see a total sum of squares, ss_t, which is the sum of ss_b and ss_w:

$$\sum (X - \bar{X})^2 + \sum n(\bar{X} - \bar{\bar{X}})^2 = \sum (X - \bar{\bar{X}})^2$$

$$ss_w \qquad + \qquad ss_b \qquad = \qquad ss_t$$

While you may be the type of person who glances only briefly at formulas, look for a long moment at this one. It is mostly symbolic. Note that the sum of squared differences between every score and the grand mean is subdivided into two components: the difference between the score and its group mean, and the difference between that group mean and the grand mean. This is just like dividing the distance between Chicago and New York into two components: from Chicago to Cleveland and from Cleveland to New York.

So the term **analysis of variance**, ANOVA for short, is appropriate. The total variance of an experiment's data is analyzed into its various components.

At this point it might be wise to stop reading for a while and go back and review from the beginning of this chapter. If you are steaming, relax: Remembering that no arithmetic (other than the mention of additivity) is involved in the *concepts*. Get them *as concepts* first. When you are ready, move on to the next section, which is easy for a change.

ANOVA calculations made easy

Let us imagine that Mulligan Muldoon, while waiting not so patiently for us to get back to him, has swigged most of the independent variable, leaving only enough to provide for three experimental groups and only enough for a short day's work at gold digging. This has the advantage of providing very small numerical values for our data. In this section, we work through the data by means of a systematic approach that not only gives the correct answers, but is also instructive in the basic principles of ANOVA, in case you're still not on good terms with them. We will use the letters A, B, and C to represent the three experimental groups, who partake of fermented prune nectar that is, respectively, slightly, moderately, or madly fizzing.

The major breakthrough in ANOVA is to organize the data into a proper layout. Once this is done, not even a formula is needed to get you working properly.

The first principle of organization is to lay out *all* the data so that all the scores from one **treatment** are enclosed in one **cell.** These are new terms but not new ideas. A treatment is an experimental manipulation; if the independent variable involves three experimental groups, then accordingly there are three treatments (three strengths of prune nectar in this case). A cell is merely the hypothetical "space" in which all the scores from one treatment fall. Each cell has three types of information noted within: (1) the actual values of the raw scores (X), (2) the *cell n*, or number of scores within the cell, available merely by counting, and (3) the sum of all scores within the cell, ΣX. This is how the layout should look at this point.

Cells
[Treatment]

A		B		C	
1	4	2	3	5	5
1		4		7	
3		4		4	
0				6	
				4	
5		10		26	

A   Cell designation

1	4	Cell n
1	Values of X	
3		
0		
5	ΣX	

Possibly this whole exercise will strike you as being absurdly simple, but there is excellent reason for attention to layout, as we shall see soon enough. Bear with us.

Our next objective is to lay out very clearly a summary table, which we can fill in as we go. We are looking for the various sources of variance that make up the total, so such a table will be set up as a *summation* both of *ss* and of *df*, which should add up to the total *ss* and total *df*. In addition, it should have provision for noting values of the *ms*, and finally a place for entering the value of *F* and its significance level. Such a **summary table** is begun here.

SUMMARY TABLE

Source of variance	*df*	*ss*	*ms*	*F*
between				
within				
Total				

Now, the first thing we must have is the total *df* and the cell *df* values. Total *df* equals $N - 1$ (where N is the total number of scores in the experiment). The number of degrees of freedom between cells (experimental groups) is the number of such groups (or cells) minus 1. There are two ways of finding the *within-cells df*: since df_t must equal $df_b + df_w$, we could merely subtract df_b from the total and use the remainder. Or else we could take the total number of scores, N, and subtract the number of cells, since 1 *df* is lost for each cell. If there are three cells, 3 *df* are lost. Hence, $df_w = N - C$ (where C denotes the

SUMMARY TABLE

Source of variance	*df*	*ss*	*ms*	*F*
between	2			
within	9			
Total	11			

number of cells). These df values are entered in the summary table.

Next we need the total sum of squares. The easiest working formula is

$$ss_t = \Sigma X^2 - (\Sigma\Sigma X)^2/N$$

which comes from

$$ss_t = \Sigma (X - \bar{X})^2$$

(Read $\Sigma\Sigma X$ as "the sum of the sums of X." Each cell gives a value of ΣX, so $\Sigma\Sigma X$, the sum of the ΣX values from all cells, is the total of all scores.) To get the first term, ΣX^2, every value of X in the experiment is squared, and the total is shown and labeled. To get the term $(\Sigma\Sigma X)^2/N$ we merely add the values of ΣX, square the total, and divide by N. It is clear that $N = \Sigma n$. These procedures are shown, and the value of ss_t is the first value, ΣX^2, minus the second, $(\Sigma\Sigma X)^2/N$; it is entered in the summary table below.

$$ss_t = \Sigma X^2 - \frac{[\Sigma\Sigma X]^2}{N}$$

$$N = 4 + 3 + 5 = 12$$
$$\Sigma X^2 = 1^2 + 1^2 + 3^2 + \cdots + 4^2 + 6^2 + 4^2 = 189$$
$$\Sigma (\Sigma X) = 5 + 10 + 26 = 41$$
$$[\Sigma (\Sigma X)]^2 = 41^2 = 1681$$
$$\frac{[\Sigma\Sigma X]^2}{N} = \frac{1681}{12} = 140.08$$

So

$$ss_t = 189 - 140.08 = 48.92 \quad \Leftarrow$$

SUMMARY TABLE

Source of variance	df	ss	ms	F
between	2			
within	9			
Total	11	48.92		

The value $[\Sigma \Sigma X]^2/N$ is sometimes called the **correction factor** in calculations of sums of squares. Failure to subtract this correction factor from every summation of squared X values would result in the summation of X^2 rather than of $(X - \overline{X})^2$.

Calculation of ss_b is next. The conceptual formula is

$$ss_b = \Sigma (\overline{X} - \overline{\overline{X}})^2 n \qquad (\overline{\overline{X}} \text{ is the overall mean, while}$$
$$\overline{X} \text{ is the mean of one cell)}$$

but the easier working formula is

$$ss_b = \Sigma \left[\frac{(\Sigma X)^2}{n} \right] - \frac{(\Sigma \Sigma X)^2}{N}$$

Within the framework of the cell layout, the formula is easily worked: square each of the values of ΣX, and divide each by its n. Add the three resulting quotients, and subtract the value of $(\Sigma \Sigma X)^2/N$, which is already available from the previous step.

A		B		C		Total
1	4	2	3	5	5	$N = 12$
1		4		7		
3		4		4		
0				6		
0				4		
5		10		26		41

correction
factor

$$ss_b = \frac{5^2}{4} \quad + \quad \frac{10^2}{3} \quad + \quad \frac{26^2}{5} \quad - \quad \frac{41^2}{12}$$
$$= 6.25 \quad + \quad 33.33 \quad + \quad 135.20 \quad - \quad 140.08$$
$$= 34.70$$

We can most easily get the value of ss_w by subtracting ss_b from ss_t.

$$ss_w = ss_t - ss_b$$
$$= 48.92 - 34.70 = 14.22$$

Enter the values of ss_b and ss_w in the summary table.

SUMMARY TABLE

Source of variance	df	ss	ms	F
between	2	34.70		
within	9	14.22		
Total	11	48.92		

The values for *ms* are easily calculated by dividing each *ss* by its attendant *df*. (Recall that the formula for *ms* is $ms = ss/df$.)

SUMMARY TABLE

Source	df	ss	ms	F
between	2	34.70	17.35	10.981
within	9	14.22	1.58	
Total	11	48.92		

The value of F is merely the ratio ms_b/ms_w. The latter, ms_w, is generally called the **error term** of the *F* ratio. *F* is entered in its space above.

The probability of the obtained *F* value being due to chance (when the null hypothesis is true) is found in the Table of *F*; we look for *df* values 2 and 9, which are easily seen in the summary table.

From Table 13.B we can see that the critical values for $df = 2$ and 9 are 4.26 for the 0.05 level and 8.02 for the 0.01 level. Make sure that you actually find these in the table! Our obtained *F* of 10.981 exceeds these critical values, so we reject the null hypothesis at both the 0.05 level and the 0.01 level. If we were reporting the results, we would write "$F(2,9) = 10.981, \quad p < 0.01$." This combines brevity with sufficient information for anyone to check your obtained *F* value if, as should be the case, your data were adequately summarized as well.

It would be instructive for you to work out the ss_w directly and check the value against those obtained as the difference between ss_t

and ss_b. Work with the formula

$$ss_w = \sum\left[\sum X^2 - \frac{(\sum X)^2}{n}\right]$$

where the contents of the brackets are the ss within any one cell. This result should equal the indirectly obtained result within the margin of rounding-off errors.

The calculations in this chapter are not complicated; the worst aspect is merely some very large interim values in the calculations. It is very important that such large values *not* be rounded. As a rule,

DESIGN / DATA MATRIX

LEVELS OF INDEPENDENT VARIABLE

Key		A		B		C		
X_1	n	1	4	2	3	5	5	$12 = N$
X_2		1		4		7		
X_3		3		4		4		
etc.		0				6		
						4		
$\sum X$		5		10		26		$41 = \sum\sum X$

SUMMARY TABLE

Source	df	ss	ms	F
between cells	$C - 1$ $= 3 - 1$ $= 2$	$\frac{5^2}{4} + \frac{10^2}{3} + \frac{26^2}{5} - \frac{41^2}{12}$ $= 34.70$	$\frac{34.70}{2}$ $= 17.35$	$\frac{17.35}{1.58}$ $= 10.98$
within cells	$N - C$ $= 12 - 3$ $= 9$	$ss_w = ss_t - ss_b$ $= 48.92 - 34.70$ $= 14.22$	$\frac{14.22}{9}$ $= 1.58$	
Total	$N - 1$ $= 12 - 1$ $= 11$	$1^2 + 1^2 + 3^2 + 0^2 + 2^2$ $+ 4^2 + 4^2 + 5^2 + 7^2$ $+ 4^2 + 6^2 + 4^2 - \frac{41^2}{12}$ $= 48.92$		

calculations should go to the number of significant figures found in the *F* table being used: two, in the present case.

It would be a good idea to linger over this worked example long enough to see the logic of the procedures. There is a definite logic there. The table on p. 389 summarizes the whole procedure.

What conclusion can we draw? The obtained *F* value was significant, so we can send Mulligan on his way, full of prune nectar and happy in the knowledge that it does make a difference to how many beqa of gold can be dug in a short working day.

general comparisons and planned comparisons

When significant results are found with three or more experimental groups, a question arises: between which of the groups is the experimental effect concentrated? A significant *F* test merely says that a difference among values of μ exists *somewhere* in the data. In our present experiment, it could be between A and B, or A and C, or B and C, or between any two of these pairs, or equally within all three pairs. How do we find out exactly where the difference exists?

At the beginning of this chapter you were warned that multiple *t* tests between all possible pairs of means were not appropriate. They still aren't, *unless* an ANOVA has revealed a significant effect *somewhere* in the picture. When that occurs, it may be necessary to discover more precisely where that effect is. (If, before you got your data, you had a particular hypothesis about a difference between two given experimental treatments, then it is all right to do a *t* test directly as a previously **planned comparison**. That's not the same as previously unplanned *t* testing on the basis of a generally significant *F* ratio.) Two kinds of question are possible.

a Before the data are known: Is the null hypothesis true or false overall?

b Knowing that a significant *F* has been found, will there be a significant difference between any given pair of means?

The second question is quite loaded regarded H_0, since H_0 has already been rejected, although we don't know where. Your odds on a horse race bet would certainly change if you knew that your horse would be one of the first three, even though you don't know which one of the first three.

We need an *a posteriori* (from the Latin for "after the fact") test for significance between any two mean values, once we have an overall significant F. There are several choices. You could use a t test. Another would be **Tukey's Test**, which is very easy. To compare two means *a posteriori*, multiply the critical value of F (not your obtained value!) by $\sqrt{ms_w/n}$ where n is the number of scores in each of the two groups being compared (the sample sizes must be equal); ms_w is available from the Summary Table, and F_{crit} is found in the Table of Critical Values of F in the Appendix. If this resulting value is *less than* the difference between the two means, the difference between them is significant at the α level of F_{crit}.

Let's do an example.

Among five experimental groups, F is significant at $\alpha = 0.01$. We wonder whether the difference is significant between a given pair of these means, whose values are 15.5 and 18.5. For each of the groups, $n = 10$, and the overall ms_w, from the summary table, is 4.0. Are these two means significantly different?

The value of F_{crit} is at the junction of $k - 1 = 5 - 1 = 4$ df and $N - k = 50 - 5 = 45$ df in the Table of Critical Values of F. We find F_{crit} is 3.78. (Please look it up yourself, too.) So we find

$$F_{crit} \sqrt{ms_w/n} = 3.78 \sqrt{4/10} = 2.391$$

The difference between means is $18.5 - 15.5 = 3.0$, larger than 2.391. Therefore, we can conclude that 3.0 is an "honestly significant difference" between the two means. (Tukey refers to this simple test as the **HSD test**, for "honestly significant difference.") To use this test, the overall F value must be significant first.

parametric assumptions

The parametric assumptions considered in Chapter 12 for the t test apply equally to the F test. Since they are identical, they won't be repeated here. Go back and look at them on p. 354 if you don't remember them.

The F test and ANOVA are highly sensitive only to the differences among the means, the values of μ. Good evidence exists that a non-Gaussian population distribution does not affect F very much unless the skewness is very extreme and in opposite directions for different experimental populations. (In that case, H_0 would be false, since the parameters that reflect the shapes of the populations would be different. But we don't want to blame the wrong parameter: the means are the parameters we are interested in.)

Equality of the variances in the populations (note the plural: same comment as immediately preceding) is another assumption whose failure does not much matter, provided that the variances are not *extremely* different. The fashion nowadays is to test for *homogeneity of variance* before going ahead with the ANOVA. Such a move is necessary more for mathematicians than for experimenters. Although marked heterogeneity of variance has a real effect on the obtained value of F, the effect is small, tending mainly to alter the alpha probabilities slightly. If you thought you were rejecting H_0 at $\alpha = 0.01$, you might really have a 0.02 chance of a Type I error, if H_0 is really true. Unless variances are extremely different, you can adjust by setting α just a bit lower than you had planned. One of the several tests for homogeneity of variance is not usually necessary.

another coefficient of correlation!

Leprechauns, and sometimes humans, are practical folk. Mulligan, between forays to the Used Prune Nectar Department, would like to know the *strength* of the relationship between the dependent variable and the independent variable of the experiment that we analyzed earlier. People can ask similar questions. "So you have a significant F. But how *strong* is the relationship? How much of the dependent variable's variance is accounted for by the independent variable?"

There is a very simple correlation coefficient, called η (eta, pronounced as in "*ate a* horse"). It is defined as

$$\eta = \sqrt{\frac{ss_b}{ss_t}}$$

or the square root of the proportion that ss_b is of ss_t. Let's illustrate with some examples.

example 1. 0 relationship

If there is no effect of the independent variable, the means of the groups will not vary, so all variance will be within groups. The numerator, ss_b, will be 0, and the denominator, ss_t, will be the same as ss_w, so the ratio and its square root will both be 0. So we get a correlation coefficient of 0, denoting 0 correlation.

example 2. perfect relationship

If a perfect relationship exists, *all* variability in the dependent variable must be due to the independent variable. A dependent variable score must then be completely determined by the value of the independent variable, so all scores within a given cell (level of the independent variable) must be the same. With no room for within-groups variance, the total variance would equal the between-groups variance. The numerator and denominator would be equal, the ratio would be 1.00, and so would the square root of the ratio. A perfect correlation has a coefficient of 1.00.

Eta may be used to determine the amount of total variance that is "explained" by the independent variable. The argument parallels that given in Chapter 10, which you may wish to review. Just as r^2 is the coefficient of determination in linear regression, η^2 gives the proportion of total variance accounted for by the independent variable.

example 3. the prune nectar relationship

Let's find out how much of the variance in the dependent variable of the prune nectar experiment was accounted for by the independent variable. From the summary table on p. 389 we find that

$$ss_b = 34.70$$
$$ss_t = 48.92$$

So eta equals $\sqrt{\dfrac{34.70}{48.92}} = \sqrt{0.709} = 0.842$

The correlation is 0.842, quite high; we see also that about 71% of the variance was accounted for by the independent variable.

Eta is different from the Pearson *r* in two ways.

1 It is nondirectional. It indicates strength of relationship but no direction, positive or negative.

2 It doesn't need to assume a linear or monotonic relationship. Thus we have a correlation method for the relations left over in Chapter 9.

There is just one problem with η. The values of n in the different experimental groups should not be too small, and they should be as nearly equal as possible. The value of η gets pretty unreliable if the values of n are less than 10. One more incentive to use samples that are as large as possible.

more words to yell in your sleep

ADDITIVITY (of df and of ss)

SUM OF SQUARES, ss

MEAN SQUARE, ms

POPULATION VARIANCE ESTIMATE, $\hat{\sigma}^2$, $ms_{between}$, (ms_b) ms_{within} (ms_w)

F DISTRIBUTION

PARAMETRIC ASSUMPTIONS

CELL

CELL n and ΣX

SUMMARY TABLE

MULTIPLE t TESTS

HONESTLY SIGNIFICANT DIFFERENCE, HSD

ETA

self-test on chapter 13

From now on, the format of self-tests will be different, to emphasize the value of statistics as a tool. To help prepare you for the real world of practical statistics, self-tests will combine procedures to be learned, insights to be gained thereby, and instant analysis. Not *too* instant, though: Please, always do *all* the self-test items without peeking!

1 Peform all the steps of the analysis of variance on the following data. Set up a summary table, a data matrix, and values of *df, ss, ms*, and *F* for all sources of variance; also calculate η. With $\alpha = 0.01$, retain or reject the null hypothesis.

Scores

Group A	1	1	1	1	1	1	1	1	1
Group B	2	2	2	2	2	2	2	2	2
Group C	4	4	4	4	4	4	4	4	4

2 Repeat the previous question with these data.

Scores

Group D	1	1	1	2	2	2	4	4	4
Group E	1	1	1	2	2	2	4	4	4
Group F	1	1	1	2	2	2	4	4	4

3 Make a short statement relating what is intuitively obvious about the two preceding sets of data to the outcomes of your procedures.

4 Repeat the steps of Question 1 with these data.

Scores

Group J	1	1	1	1	2	2	2	2	4
Group K	1	1	1	2	2	2	2	4	4
Group L	1	1	2	4	4	4	4	4	4

ANSWERS TO PROBLEMS 1–4

The first step is to construct a design/data matrix and enter the values of X, n, and $\bar{X}$.

Group

A		B		C		
1	9	2	9	4	9	$N = 27$
1	1	2	2	4	4	
1	1	2	2	4	4	
1	1	2	2	4	4	
1	1	2	2	4	4	
9		18		36		$\Sigma\Sigma X = 63$

D		E		F		
1	9	1	9	1	9	$N = 27$
1		1		1		
1		1		1		
2	4	2	4	2	4	
2	4	2	4	2	4	
2	4	2	4	2	4	
21		21		21		$\Sigma\Sigma X = 63$

J		K		L		
1	9	1	9	1	9	$N = 27$
1	2	1	2	1	4	
1	2	1	2	2	4	
1	2	2	4	4	4	
2	4	2	4	4	4	
16		19		28		$\Sigma\Sigma X = 63$

Next, construct the summary table and its *df* values. They are identical for all three questions; total $df = N - 1 = 26$; $df_b = C - 1 = 2$ (since there are three cells). $df_w = N - C$, since 1 *df* is lost from each cell, so there are $27 - 3 = 24$ *df* within cells. These values and the Summary Tables look like this.

SUMMARY TABLE

Source	*df*	*ss*	*ms*	*F*
between cells	2			
within cells	24			
Total	26			

Now calculate the values of *ss* and enter them in the summary tables. From here onwards, the values diverge for the three questions. The calculations for Question 1 are as follows.

$$ss_t = 1^2 + 1^2 + 1^2 + 1^2 + \ldots + 4^2 + 4^2 - \frac{63^2}{27}$$

$$= 189 - 147 = 42$$

$$ss_b = \frac{9^2}{9} + \frac{18^2}{9} + \frac{36^2}{9} - \frac{63^2}{27}$$

$$= 189 - 147 = 42$$

$$ss_w = ss_t - ss_b$$

$$= 42 - 42 = 0$$

Source	*df*	*ss*	*ms*	*F*
between	2	42	21	∞
within	24	0	0	
Total	26	42		

∞ = infinite (or indefinitely large)

As is obvious, there is no variance within cells, and the total variance arises from that between cells. Therefore the value of *F* is very large.

The calculations for Question 2 are as follows.

$$ss_t = 1^2 + 1^2 + 1^2 + 2^2 \ldots + 4^2 + 4^2 - \frac{63^2}{27}$$

$$= 189 - 147 = 42$$

$$ss_b = \frac{21^2}{9} + \frac{21^2}{9} + \frac{21^2}{9} - \frac{63^2}{27}$$

$$= 147 - 147 = 0$$

$$ss_w = ss_t - ss_b$$

$$= 42 - 0 = 42$$

Source	df	ss	ms	F
between	2	0	0	0
within	24	42	1.75	
Total	26	42		

As is again obvious, the variance is all within the cells, and there is no variability between cells. Consequently, the *F* ratio is 0.

The general statement requested in Question 3 should be to the effect that the scores for Groups A, B, and C have no variability within the cells, but they do have variability between cells, as is obvious by inspection. Hence, the obtained mean squares and *F* are not surprising. Similarly, the D, E, and F groups show variability within themselves but none between groups; therefore, the mean squares and *F* as obtained are quite reasonable.

The calculations for Question 4 are as follows.

$$ss_t = 1^2 + 1^2 + 1^2 + 1^2 + 2^2 + \ldots + 4^2 - \frac{63^2}{27}$$

$$= 189 - 147 = 42$$

$$ss_b = \frac{16^2}{9} + \frac{19^2}{9} + \frac{28^2}{9} - \frac{63^2}{27}$$

$$= 155.666 - 147 = 8.666$$

$$ss_w = ss_t - ss_b$$

$$= 42 - 8.666 = 33.334$$

Source	df	ss	ms	F
between	2	8.666	4.333	3.121
within	24	33.334	1.388	
Total	26	42		

FOLLOW-UP

Scoring seems a bit ridiculous in the real world. If you had made any mistake other than a rounding error, none of the things done properly would have been worth anything. A million-dollar decision (or a decision concerning a chance for extended life on the part of a dying patient denied a new drug because of a Type II error based upon bad arithmetic) could have been involved. From now on, consider that a passing grade is 100%. A failing grade is 99% or less.

some ordinary problems

1 An experiment resulted in the following data.

	Group 1	Group 2	Group 3
$\bar{X}$	18	12	10
s	5	4	4
n	10	10	10

Calculate the necessary values for ANOVA, and then perform it, testing the significance of F for $\alpha = 0.01$.

2 What proportion of the variance is accounted for by the independent variable in Question 1?

3 An experiment involved six experimental groups of respective sizes 6, 8, 9, 8, 8, and 9. Set up the "source of variance" column and the df column of the summary table.

4 Randomly take two samples of size $n = 16$ from Table 8.B, and then another two samples of 16 each from Table 12.B. Label these samples A, B, C, and D, respectively. Assume that the scores result from four levels of an independent variable, and conduct an ANOVA, working at $\alpha = 0.05$. Follow up with individual *t* tests, if indicated, and HSD tests as well. What conclusions are proper?

5 If Question 4 was a class assignment, tabulate the proportions of different conclusions. A and B should have been the same, with neither *t* nor HSD both significant; C and D are also not significant. The differences between A and C, A and D, B and C, and B and D should have been significant. How do the proportions check with the chosen value of α? How frequent were Type II errors?

6 List in proper order the steps taken in ANOVA procedures.

7 A traffic researcher desires to find out whether pedestrian con-formity to WALK signals is affected by various rates at which the WALK and DON'T WALK signs flash. He uses three rates of flashing plus a nonflashing control. He gets his data by observing pedes-trians at a busy intersection, scoring 0 if the pedestrian obeys the letter of the law, 1 if the pedestrian starts from the curb on an amber light but waits for green, 2 if the pedestrian starts on amber and tries to make it across, and 3 if the pedestrian tries to get across against a red light when traffic is not heavy. The score of 4 is given to any pedestrian who tries to cross against a red light in total disregard of heavy traffic. This researcher is con-templating the use of ANOVA in analyzing his data, and turns to you for advice. What would your advice be, and based upon what arguments?

8 What are the critical values of *F*, with $\alpha = 0.01$, for the following experiments?

   a $df = 4, 40$
   b 6 groups of $n = 5, 8, 7, 8, 9,$ and 8
   c 3 groups, $N = 60$, equally distributed among the groups.

9 A scientist whose arithmetic is always perfect is shown an ANOVA summary table. After merely glancing at the table, the scientist says: "I see that the population variance is 36.5." Was this statement likely to be correct, or reasonably so? If so, upon what was it based?

10 Using an *N* no *greater* than 60, design an experiment to test whether three patent-medicine cold "cures" actually reduce the duration of common colds. Have proper regard for control in this experiment. Set up design/data matrix with values of *n* in-cluded; then set up the Source and *df* columns of the summary table.

11 We can tell whether any value of r is significant by referring it to a Table. How do we know when a value of η is significant?

chapter 14
how to do two experiments at once

There was once a college athletic coach who believed in getting things done: *His* athletes were going to qualify for, and win, a few international meets, or else. He had two trainers, and one of them decided to find out the effect of increasing the already high motivation of his athletes. He hoped that this would further increase their level of performance. So he ran an experiment. Having some knowledge of experimental methods, he decided to take two random samples from all the athletes, use one as a control group, and subject the other to pre-meet pep talks, large financial incentives, and heavy penalties for failure. At the end of the year he knew that these measures had had some effect, because several students had complained to the coach about the pressure. So he staged a miniature meet between his two groups, to determine whether the experimental group performed better than the control group.

The results were disappointing, even shocking. There was very little difference between the two groups—certainly not a statistically significant one Figure 14.1 summarizes what he found.

This result came as a surprise to the other trainer, whose experience indicated that hypermotivated athletes sometimes *did* do much better. This trainer also had some knowledge of experimental psychology. Since it is known that increased motivation affects the performance of simple tasks and of complex tasks quite differently, tending to improve the former while interfering with the latter, he decided to run a further study.

This time, two types of athletes were sampled separately: those whose specialty was primarily a matter of muscle, motivation, and stamina, such as milers, and those whose specialty was very much a matter of sensorimotor discrimination and finesse, such as gymnasts. For each of these two groups he took two samples at random, one as a control group and the other as the experimental group. The experimental groups received the motivating procedures of the preceding experiment; the control groups did not.

For clarity, the experimental plan may be summarized as in Figure 14.2.

In Figure 14.2, you'll note that the design/data matrix is ruled with straight lines that set off the top and side headings (representing the independent variables) and form the cells into which the dependent variable data will be classified. *Unless* a complex experimental plan can be tabulated and ruled clearly, it is *probably* a fuzzy, poorly thought out experiment—or else a very complex one indeed, which in its way is undesirable too, unless no simpler alternative exists.

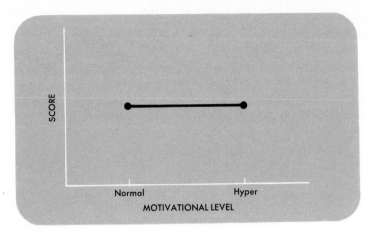

Figure 14.1
Athletic score as a function of motivational level; hypothetical data from
a hypothetical college.

Another advantage of Figure 14.2 is that it forms the framework for
our analysis of variance once the data are in.

EXPERIMENTAL CONDITION (MOTIVATION)

		Hyper	Normal
	Simple		
	Complex		

TYPE OF SKILL

Figure 14.2
Design/data matrix for two-way experiment. Each independent variable is counterbal-
anced over the other, so each can be studied separately.

Suppose the experimental results look like Figure 14.3. This picture
is rather more complicated than the previous one. There are now *two*
independent variables: type of athletic skill and level of motivation.
The first variable, shown within the body of the graph via separate
curves, is sometimes called the **parameter** of the graph.[1] The second
independent variable is scaled along the abscissa in the usual way.
They could change places arbitrarily, in which case the graph would
look like Figure 14.4. The important thing to note in either figure is

[1] This is a second meaning of the word "parameter"; the other is a population value for
the mean, variance, etc.

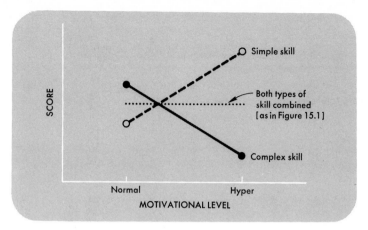

Figure 14.3
Aha! Figure 14.1 was wrong. Motivational level *does* make a difference.

that no adequate conclusion can be drawn about the effects of either independent variable without reference to the other one. Can we say that in general increased motivation improves performance? No. Or that in general it degrades performance? No. We would be in danger of making the first trainer's error and concluding that motivation has *no* effect, were we not to run the two experiments at once. The two experiments are, of course, (1) the effect of motivation on performance, and (2) the effect of the type of athletic skill on performance.

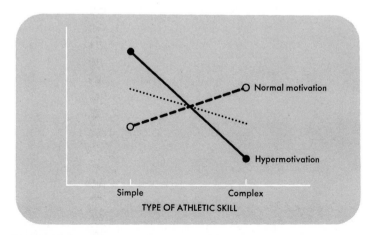

Figure 14.4
The same information as in Figure 14.3. Compare the two pictures carefully.

And in a way there is a third experiment, too: (3) the effects of motivation and type of skill upon each other, or how they interact when athletic skill is being measured.

In summary, the two independent variables have a *joint* effect upon performance; if they are not disentangled, an incorrect conclusion will result. When such a mutual entanglement occurs between two variables, they are said to **interact**.

interaction

Variables can interact to varying degrees. Looking at Figure 14.2, we can see that four cells are involved in the experiment; in Figures 14.3 and 14.4 we see four plotted points, corresponding to the mean measures from each cell. Now, it is clear that the effectiveness of an independent variable can be seen as the amount of slope in the lines connecting these plots. Study Figures 14.3 and 14.4 carefully. In the former, a dotted line shows what would be the case if normal and very high motivation were studied without regard to skill complexity. The line is almost horizontal, but not quite. The overall effect (known as the **main effect**) of motivation is a *slight* overall superiority under very high motivation. In Figure 14.4 there is a similar dotted line, showing the main effect of skill complexity without regard for motivational state. Here there is a moderate favoring of performance on simple skills. Neither of these two main effects expresses the quantitative (and sometimes qualitative) degrees of relationship between either of the independent variables and the dependent variable that we see when the other independent variable is taken into account. Since most of the variance between cells is not due to main effects, it must be made up primarily from interaction. Actually, in any two-factor experiment each main effect and the interaction can involve any proportion of the between-cells variance, as long (obviously) as the total is 100%.

The concept of interaction is most important in experimental design and statistics. The need to eliminate it from the comparisons between main effects is the major reason why complex experiments tend to outnumber simple ones. Experiments are usually labeled as to type by the number of levels assumed by each of their factors. The experiment above involved two levels of the complexity factor and two levels of the motivation factor. Hence it would be called a 2×2 (two by two) design. If we recall that the principle of *df* and *ss* additivity may pertain to *any* number of cells with *any* number of degrees of freedom, we can understand that there need be no limitation on the number of factors or the number of levels within each factor. Thus, a 5×8 or $299 \times 877 \times 56$ design would be just as possible, in principle, as the 2×2 design discussed above, and all within the context of analysis of variance.

but beware

Analysis of variance is clearly a very powerful statistical technique. With it, you can design very large, complex, grandiose experiments. Typically the neophyte does just this, putting in all sorts of "variables" just because they look good and there is no mathematical reason not to. There are, however, two excellent human reasons not to. For background, see Figure 14.5.

The first difficulty is in the mechanics of the technique. While they are simple in principle, they are awesome in practice. Every time a factor is added, the amount of complexity more than doubles. The use of computers makes a lot of difference here, but it doesn't help at all with the second difficulty.

In a one-factor experiment (as in Chapter 13) there are no interactions to worry about. In a two-factor experiment there is one. If the experiment has three factors, there are four interactions; with four and five factors there are, respectively, 11 and 26 interactions. The problem arises as to how such interactions may be explained. Suppose that in a five-factor experiment there is a significant interaction involving four out of the five. You would be left trying to say that "The effect of A depends upon the level you're at in B, but this depends in turn on where you are in C—but all this so far assumes that you are at such-and-such a level of D (at another level of D, the whole thing changes); fortunately, nothing relies on E being constant at a given level, because it isn't." That's not really a very useful conclusion, is it? Assuming that *you* understand it (which is doubtful), nobody else will. Anyway, such a high-order (multifactor) interaction may not mean anything at all. It might have been accidental to begin with; with five main effects and 26 interactions, you'll have 31 *F* values to interpret, and at the 0.05 level there are better than even odds that at least one is significant only by accident.

This entire nightmare can be avoided very easily. Don't include *any* factors in the experiment except those essential to your hypothesis or (cautiously) to control *major* sources of undesired variance which,

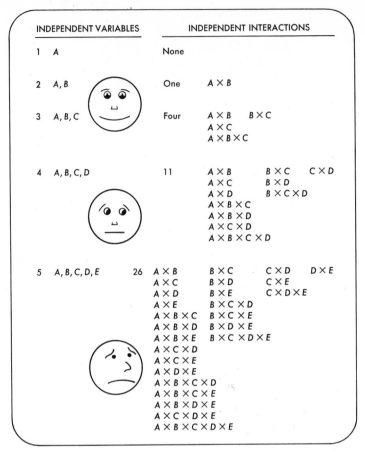

INDEPENDENT VARIABLES	INDEPENDENT INTERACTIONS			
1 A	None			
2 A, B	One	A × B		
3 A, B, C	Four	A × B B × C		
		A × C		
		A × B × C		
4 A, B, C, D	11	A × B B × C C × D		
		A × C B × D		
		A × D B × C × D		
		A × B × C		
		A × B × D		
		A × C × D		
		A × B × C × D		
5 A, B, C, D, E	26	A × B B × C C × D D × E
		A × C B × D C × E
		A × D B × E C × D × E
		A × E B × C × D
		A × B × C B × C × E
		A × B × D B × D × E
		A × B × E B × C × D × E
		A × C × D
		A × C × E
		A × D × E
		A × B × C × D
		A × B × C × E
		A × B × D × E
		A × C × D × E
		A × B × C × D × E

Figure 14.5
As you increase the number of independent variables in an experiment,
you increase the number of interactions much faster.

left uncontrolled, might lead to a Type I error. When terribly complicated experiments are necessary, there are legitimate ways of approaching them, but they are beyond the scope of this book.

So far we have not done more than discuss principles and problems. Now let us follow the practice of the previous chapter, by developing the computations in close step with the logic underlying them.

The first step is to enter the raw data scores into each cell of the table (a reproduction of Figure 14.2), put the cell n into the upper right-hand corner of each cell, and find ΣX within each cell. Our work so far is shown here.

	hyper		normal	
simple	5 7 7 6	4 $\Sigma X = 25$	2 1 3 2	4 $\Sigma X = 8$
complex	0 1 0 2	4 $\Sigma X = 3$	2 4 3 5	4 $\Sigma X = 14$

It is convenient to adopt shorthand labeling for the table. We denote the cells C, the motivational variable M, and the athletic complexity variable A. Thus, we relabel the table (from Figure 14.2) as shown.

M

	m_1		m_2		
a_1	5 7 7 6	4 $\Sigma X = 25$	2 1 3 2	4 $\Sigma X = 8$	2 levels of A, so 1 df for A
a_2	0 1 0 2	4 $\Sigma X = 3$	2 4 3 5	4 $\Sigma X = 14$	

A

2 levels of M, so
1 df for M

4 cells overall, so
3 df for cells

Next the summary table must be set up. As for the simple analysis of variance in Chapter 13, there are two overall components to the total variance: within cells and between cells. However, the latter may now be subdivided into three subcomponents: the main effect of M, the main effect of A, and the interaction $M \times A$. As before, the df for cells is $C - 1$ (the number of cells minus 1). The df for motivation is $M - 1$, and for complexity it is $A - 1$. Finally, the df for the interaction is $(M - 1)(A - 1)$. We find the very elegant fact that $df_c =$

$df_M + df_A + df_{M \times A}$, just as the additivity rule requires. There are four cells, so $df_C = 3$. There are two levels of M, giving 1 df, two levels of A, giving 1 df, and the interaction df is $(2 - 1)(2 - 1) = 1$.

The dfs for the total variance and within-cells variance are found as in Chapter 13. The total is $N - 1$, and the within-cells df is best found as the difference between the total df and the cells df. The summary table as constructed so far is shown here.

SUMMARY TABLE

Source	df	
Cells	$C - 1 = 3$	
M	$M - 1 = 1$	
A	$A - 1 = 1$	
$M \times A$	$(M - 1)(A - 1) = 1$	
within cells	$N - C = 12$	
Total	$N - 1 = 15$	

Our next step is to calculate the ss values, starting with the total ss. Every raw score is squared, and the squared values are summed. There is no need to distinguish among the various cells. From the total of squared scores is subtracted the correction factor $(\Sigma\Sigma\Sigma X)^2/N$. The result is entered into the summary table.

$$N = 4 + 4 + 4 + 4 = 16$$
$$\sum X^2 = 25 + 49 + 49 + 36 + 0 + 1 + 0 + 4 + 4 + 1 + 9 + 4 + 4$$
$$+ 16 + 9 + 25$$
$$= 236$$

$$\sum\sum\sum X = 5 + 7 + 7 + 6 + 0 + 1 + 0 + 2 + 2 + 1 + 3 + 2 + 2$$
$$+ 4 + 3 + 5$$
$$= 50$$

$$ss_t = \sum X^2 - \frac{(\sum\sum\sum X)^2}{N} = 236 - \frac{50^2}{16} = 79.75 \blacktriangleleft$$

SUMMARY TABLE

Source	df	ss	
Cells	3		
$\quad A$	1		
$\quad M$	1		
$\quad M \times A$	1		
within cells	12		
Total	15	79.75	$\blacktriangleleft$

Remember that a sum of squares is the sum of *deviation scores* squared, not raw scores squared. That is,

$$ss = \Sigma (X - \bar{X})^2 \quad \text{or} \quad ss = \Sigma x^2$$

not ΣX^2

However, when the correction factor is subtracted from ΣX^2, we get ss. Check back to Chapter 5 to review this point if necessary.

I'VE BEEN GIVING THE MATTER SOME THOUGHT, AND I'M CONVINCED THAT USING $\alpha = 100\%$ IS WHY WE'RE STILL HERE

Next, the ss_{cells} is found. The scores within each cell are summed, the sums are squared, the squares are divided by the cell n, and the resulting values are summed. Finally, the correction factor is subtracted. The resulting ss value is then entered into the summary table.

M

	m_1		m_2	
a_1	5 7 7 6	4 **25**	2 1 3 2	4 **8**
a_2	0 1 0 2	4 **3**	2 4 3 5	4 **14**

A

down columns
across rows
within cells

$\sum\sum\sum X = 50$
$N = 16$

$$ss_c = \sum \frac{(\sum X)^2}{n} - \frac{(\sum\sum\sum X)^2}{N}$$

$$= \frac{25^2}{4} + \frac{3^2}{4} + \frac{8^2}{4} + \frac{14^2}{4} - \frac{50^2}{16}$$

$$= 67.25$$

According to the principle that all sums of squares within the summary table must sum to ss_t, we can find the value of ss_w by subtracting ss_c from ss_t. But the within-cells sum of squares can also be calculated directly as shown below on the left

$$5^2 + 7^2 + 7^2 + 6^2 - \frac{25^2}{4} = 159 - 156.25 = 2.75$$

$$0^2 + 1^2 + 0^2 + 2^2 - \frac{3^2}{4} = +5 - 2.25 = 2.75$$

$$2^2 + 1^2 + 3^2 + 2^2 - \frac{8^2}{4} = +18 - 16.00 = 2.00$$

$$2^2 + 4^2 + 3^2 + 5^2 - \frac{14^2}{4} = +54 - 49.0 = 5.00$$

$$\overline{12.50}$$

79.75
-67.25
$\overline{12.50}$

You'll see that the answer is identical to the ss_w calculated the other way, as shown on the right. Doing the calculation directly serves to check your accuracy.

SUMMARY TABLE

Source	df	ss
Cells	3	67.25
A	1	
M	1	
$M \times A$	1	
within cells	12	12.50
Total	15	79.75

Our approach continues with the calculation of the sum of squares for independent variable A, type of athletic skill. As seen here, the other independent variable, M (motivation) is ignored, treated as though it did not exist. In our design/data matrix we find the row marginal totals of n and ΣX, which are treated exactly as in previous steps. The summed values of ΣX are here shown as $\Sigma\Sigma X$ to denote the double summing (within cells and across the two levels of M). They are squared and divided by their ns; then the resulting values are summed and the correction factor subtracted. The resulting value is placed in the summary table. Note that the marginal values of n are the totals of the cell ns, shown in the small boxes.

Row Marginal Totals

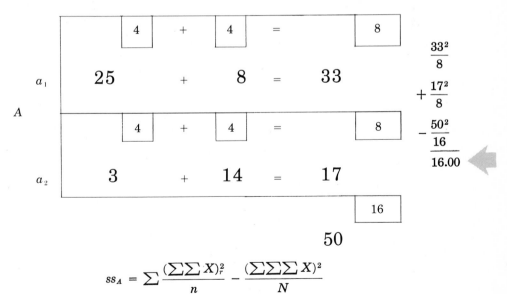

$$SS_A = \Sigma \frac{(\Sigma\Sigma X)_r^2}{n} - \frac{(\Sigma\Sigma\Sigma X)^2}{N}$$

SUMMARY TABLE

Source	df	ss
Cells	3	67.25
A	1	16.00
M	1	
M × A	1	
within cells	12	12.50
Total	15	79.75

The sum of squares for *M*, motivational level, is found along the same lines as previous sums of squares. Independent variable *A* is ignored. Column marginal totals of ΣX and *n* are found, as shown below, and ss_M is calculated in the familiar fashion. The logic of the *ss* procedure is such that once you understand it, no formula is necessary in order to do the right things.

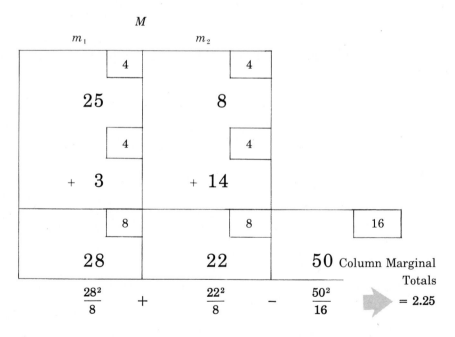

$$ss_M = \sum \frac{(\sum\sum X)_{\text{cols}}}{n} - \frac{(\sum\sum\sum X)^2}{N}$$

Once the resulting ss_M value is placed in the summary table, we can find $ss_{M \times A}$. The principle of additivity requires that ss_c be the total of its component ss sources; we have found two out of the three, so the last ss must be the difference remaining. It is entered into the summary table, and the ss column is complete.

$$ss_A + ss_M + ss_{M \times A} = ss_c$$

SUMMARY TABLE

Source	df	ss	
Cells	3	67.25	67.25
A	1	16.00	−16.00
M	1	2.25	− 2.25
M × A	1	49.00	
			49.00
within cells	12	12.50	
Total	15	79.75	

It should be intuitively clear that the procedures could apply to any two-dimensional experiment. The number of rows or columns could be greater, and the only effect would be in summing over more marginal categories.

Our next step is to calculate the values in the ms (mean square) column, by dividing each ss by its own df. We can ignore the ss_c from now on; it has fulfilled its function. There is an ms value for each of the three between-cells components of overall variance, and there is also an ms value for within cells. We don't care about the ms for the total; it serves no purpose.

The final step in the analysis of variance is to calculate F for each of the between-cells variance sources. Remember,

$$F = \frac{ms_b}{ms_w}$$

The F values are entered into their column, referred to the critical values of the F table, and judged as to their significance. The critical values are found to be 4.75 and 9.33 for the 5% and 1% levels. These F values are located at the conjunction of the 1 df column and the 12 df row in the F table; these dfs are the ones in the summary table.

SUMMARY TABLE

Source		df	ss	ms	F	p
Cells						
	M	1	2.25	2.25	2.16	>0.05
	A	1	16.00	16.00	15.3	<0.01
	A × M	1	49.00	49.00	47.1	<0.01
within cells		12	12.50	1.04		
Total		15	79.75			

Our conclusions are that a significant interaction exists between motivational level and type of athletic skill and that a significant main effect exists for type of athletic skill, but not for motivation overall.

interpretation of interactions

We have a situation in our hypothetical experiment in which type of athletic skill enters into a significant F in two places: as a main effect and as an interaction. How do we interpret this?

The rule is simple: if an interaction is significant, any significant main effect is uninterpretable. This is to safeguard against risk of erroneous or misleading implications from the significant F ratio. Look at Figure 14.4. If, on the basis of its significant main effect, we were to conclude that simple skills get more points than complex skills, which is true overall, we would still be *qualitatively* incorrect when such skills are compared under normal motivation!

The only route to further clarification is to study what happens to one independent variable at only one level of the other independent variable. For example, a t test could be done between simple-task and complex-task scores under normal motivation, and again under very high motivation. Such tests are legitimate only if the overall F is significant. The resulting findings are called **simple effects**, as distinct from **main effects**.

In summary, we see that the first trainer was wrong in his conclusion and naive in his experimental design. Increasing motivational level *does* make a difference—one way or the other.

In our hypothetical experiment we assumed that all cells had the same $n = 4$. Equal cell ns are always a good idea; if N is large they may safely vary somewhat, but they *should* be as nearly equal as possible. Gross inequalities result in some complications that cannot be covered here, except to say that the theoretical bases for computing the interaction suffer if certain cells have a very small n and others have a relatively large n.

And that's all there is to the analysis of variance—at least, all there is about the most basic principles. There are some awesome further developments in more advanced uses, but they do not extend in principle away from what we've just seen.

how to know an interaction
when you meet one

Like most other things in life, interactions can be classified in various ways. An **intrinsic interaction** is one that is really "out there" in the nature of what you're investigating. An **extrinsic interaction** is one that appears merely by accident as a result of random errors in sampling, measurement, or large α value. When an extrinsic interaction is found to be significant, a Type I error has been made. When an intrinsic interaction does not appear as significant in the data, we have a Type II error.

High-order interactions can be mixtures of extrinsic and intrinsic components. Sometimes they are really hard to sort out. The best check is replication; if any interaction persists through replications, it is probably intrinsic.

In complex experiments, interactions can be either **homogeneous** or **heterogeneous**. A homogeneous one is shown in Figure 14.6. Note that the amount of interaction progresses smoothly as you move along the X and Y axes. A heterogeneous interaction is seen in Figure 14.7; in certain places there is no interaction between X and Y, but in other places there is.

For reference purposes, Figure 14.8 shows instances of data in which no interaction is present.

It is very typical of beginning students to think of interactions as undesirable. In fact, a significant intrinsic interaction is a joyous event for most experimenters. After all, even though it is wise to keep experiments simple, Nature usually is not simple at all. Discovery of an interaction provides a ready-made hypothesis (or set of hypotheses) for future experiments.

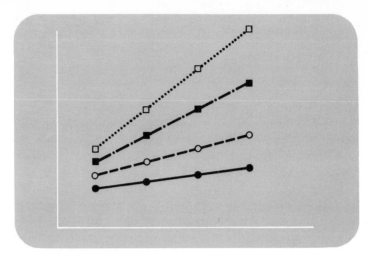

Figure 14.6
Example of a homogeneous interaction.

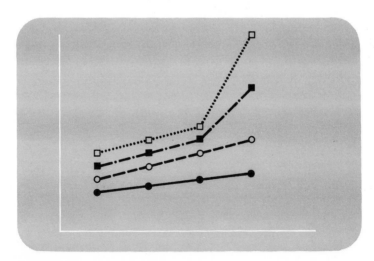

Figure 14.7
Example of a heterogeneous interaction.

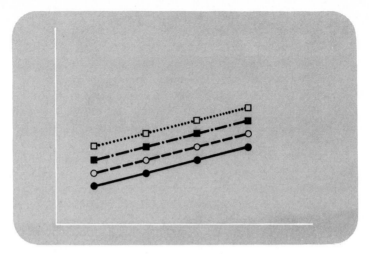

Figure 14.8
Example of 0 interaction and equal X and Y main effects.

word list to memorize out loud on the commuter train

DESIGN/DATA MATRIX

INTERACTION

MAIN EFFECT

SIMPLE EFFECT

INTRINSIC INTERACTION

EXTRINSIC INTERACTION

HOMOGENEOUS INTERACTION

HETEROGENEOUS INTERACTION

self-test on chapter 14

Try to answer all of the following questions without looking ahead to the solutions, which follow. As before, consider 100% the passing grade.

1 An experiment is set up to compare six experimental conditions (variable A) among three age groups (variable B). The total N is 180, and group sizes are equal. The value of $ss_A = 55.0$; $ss_B = 4.0$; $ss_{A \times B} = 50.0$; $ss_w = 162.0$.

  a How many interactions are there?
  b List four ways (not necessarily mutually exclusive) in which interaction may occur.
  c Complete a summary table.
  d For $\alpha = 0.01$, give the critical values of F for the tests that can be made.
  e What conclusion(s) would be drawn?
  f What, if anything, should be done next?

2 If the null hypothesis is true, between what values should the F ratio for variable B of question 1 fall on 99% of all possible tests? What should the mean value be?

3 Why is a negative F ratio impossible?

4 An alternative name for mean square, or ms, is *variance estimate*. Estimate of what?

5 In what way is the denominator of the F test similar to the denominator of the t test?

Answer to Question 1

a There is only one interaction, $A \times B$. Remember, the number of interactions depends upon the number of variables, not the number of levels within the variables.

b Intrinsic, extrinsic, homogeneous, heterogeneous.

c The summary table is as follows.

SUMMARY TABLE

Source	df		ss	ms	F	p
between cells						
A	$A - 1$	5	55.0	11.0	11.0	<0.01
B	$B - 1$	2	4.0	2.0	2.0	>0.05
$A \times B$	$(A - 1)(B - 1)$	10	50.0	5.0	5.0	<0.01
within cells	$N - AB$	162	162.0	1.0		
Total	$N - 1$	179	271.0			

d There are 5 df for A, 2 df for B, and 162 df within cells. The critical F values (from the Table of F) are, therefore,

Variable A: 3.14 (using scaled within-cells df of 150)
Variable B: 4.75
Interaction: 2.44

e Conclusion: a significant interaction exists between A and B.

f The next step is to test simple effects of A. Simple effects of B may also be tested.

Answer to Question 2

With 2 and 150 df, 99% of F ratios will fall *below* the critical value of 4.75 (when the null hypothesis is true). The bottom limit is 0. The mean of any F distribution is 1.0.

Answer to Question 3

It is impossible to have negative F ratios because they are ratios of summed squared values, and squares are always positive.

Answer to Question 4

The term *mean square* describes how *ms* is calculated. The sum of squares is divided by $n - 1$ (minus 1 to correct for the fact that s^2 underestimates σ^2). So the mean square is an unbiased estimate of population variance.

Answer to Question 5

The denominators of the t and F ratios are similar in several ways.

1 They are error estimates, against which the numerator value may be compared.
2 They both are calculated from the variability of the data *within* the groups of the experiment.
3 They are related mathematically. The denominator of t is

$$\sqrt{\frac{n_E s_E^2 + n_C s_C^2}{n_E + n_C - 2}\left(\frac{1}{n_E} + \frac{1}{n_C}\right)} = \sqrt{\frac{ss_E + ss_C}{df_w}\left(\frac{1}{n_E} + \frac{1}{n_C}\right)}$$

and the denominator of F is

$$\frac{ss_1 + ss_2 + \ldots ss_K}{df_w}$$

So the denominator of t is

$$\sqrt{ms_w\left(\frac{1}{n_E} + \frac{1}{n_C}\right)}$$

and for F it is ms_w.

Take full credit for noting any of these similarities, or even getting just the general idea.

ſome ordinary problemſ

1 State the general rule for establishing *df* for between cells, variables (*A, B,* etc.), within cells, and total. How do the first of these relate to the total?
2 Under what circumstances does ΣX^2 equal $(\Sigma X)^2/N$?
3 In this chapter, the topic has been (as it is frequently called) two-way ANOVA. What would be the characteristics of three-way ANOVA?

4 Do a two-way ANOVA of the following data. Construct a summary table.

		Variable A			
		a_1	a_2	a_3	a_4
		12	14	14	12
		16	15	18	19
	b_1	12	16	20	18
		10	14	19	24
Variable B					
		8	15	16	15
		7	18	17	25
	b_2	10	13	20	28
		6	16	18	24

5 Draw two figures in which the results of the preceding question are summarized, the parameter being different in the two figures.

6 Draw appropriate conclusions regarding Question 4 above, using correct F_{crit} values.

7 A manufacturer claims to have discovered a phonograph stylus (Brand A) that will outlast its top-line competitors. In an experimental investigation in which the dependent variable was the number of records played before failure, the following data appeared. Was the manufacturer's claim correct? Could it be expressed any more accurately?

	High Quality HiFi Systems				Economy HiFi Systems			
Brand A	2500	1900	3000	4000	400	500	900	2000
Brand B	800	900	1800	1800	1700	1900	2000	1400
Brand C	1000	2000	2000	2400	2000	1500	1800	1700
Brand D	1000	1100	1200	1300	1300	1200	1200	1400

8 Many students are confused between *interaction* and *confounding*. The two notions are not similar. However, they may well be related. In what way might confounding have occurred in such a way as to justify the claim of our manufacturer in the previous question?

chapter 15
nonparametric statistics

If you are honest and conscientious, you now know *how* to perform *t* tests and analyses of variance. Now we must consider *when* to do them. Unfortunately, sometimes we *cannot* do them legitimately.

As you know, the *t* and *F* tests are examples of **parametric tests**. Recall that a parameter is a characteristic of a population, such as its mean value μ or its standard deviation σ. Another parametric characteristic of a population is the *shape of its frequency distribution*.

The tables of *t* and *F* theoretically apply *only* when the raw scores in an experiment are randomly and independently sampled from population distributions that are *Gaussian* in shape (because only then is μ statistically independent of σ). Moreover, it is assumed that the independent variable does not affect the mean (this is the null hypothesis) or the variance in the population (and hence in the sample) to which it is applied. Fortunately, it is known that population distributions *may* in practice vary somewhat from the Gaussian, provided that they remain roughly bell-shaped and not skewed. Similarly, the variances of the treatment groups (and hence of their hypothetical populations) may in practice be *somewhat* unequal. The *t* and *F* tests are not sensitive to violation of these two assumptions, and yet they remain acutely sensitive to differences between population means. All this is to the good.

Sometimes, however, population distributions differ radically from the Gaussian; data that fall into a J-shaped curve are an example (see Figure 4.2). Sometimes, too, the variances differ very markedly between the two or more cells in an experiment. As a rule of thumb, if the largest variance is more than ten times the smallest, your experiment should not be analyzed by a parametric test. When the assumptions of variance equality or Gaussian shape cannot be justified, we must turn to alternative types of statistical tests that do not require such assumptions. Recall that these are named **nonparametric tests**.

Nonparametric tests have the very valuable feature of not requiring a Gaussian distribution of population scores, because the highest level of scaling is ordinal in such tests. We saw this in the case of the Spearman r_S correlation. The shape of a frequency distribution of *ranks* is indeterminate, unlike the distribution of scores from which the ranks came.

One variety of nonparametric tests rests at the level of nominal or category scaling; in such tests we desire to know the relative frequencies of scores within the categories and the significance of the differences between those relative frequencies. The binomial test, considered in Chapter 7, is an example of such a test.

Frequently, but perhaps not frequently enough, a nonparametric test is adopted when a parametric test *could* perhaps have been used but the parametric assumptions were questionable. Recall that parametric tests are robust—but possibly not robust enough when more than one of the parametric assumptions fail. The α probabilities associated with the t and F tests are based upon the truth of the parametric assumptions; if there is substantial departure from these assumptions, the tabled critical values and α values may be substantially invalid. The worst risk is a Type I error, rejection of the null hypothesis when it is really true, perhaps when the obtained t or F value barely reaches a critical value whose α value is invalid to begin with because of a false parametric assumption. The probability of this particular sort of Type I error cannot be determined accurately, but a nonparametric test avoids the difficulty. However, nonparametric tests must pay a penalty for this advantage.

power efficiency

The penalty is that nonparametric tests are not as powerful as parametric tests. We have met the concept of power before; recall from previous chapters that t and F ratios are the more likely to be significant, the greater the values of n and N. (Recall that increased N serves to diminish the error term denominator, thus making the ratio larger; it also gives more df and a smaller critical value.) The F test in analysis of variance is considered to be the most powerful of all statistical devices; that is, when the null hypothesis is actually not true, the F test manages to reject it using a minimum size of N. The more powerful a test, the smaller the N needed to reject the null hypothesis. The t test, F test, and Pearson r all may be considered maximally efficient for their purposes.

To illustrate the concept of **power efficiency**, we can compare the Pearson r with the Spearman r_S. If a population contains a correlation, and the Pearson r manages to reject the null hypothesis (of no correlation) with a minimum n of 50, the Spearman r_S would need a minimum n of 55 to do so.

Power efficiency is expressed as a proportion or percentage. Suppose that we consider our most powerful test to be (relatively) 100% efficient, and therefore a standard for lesser tests. If we symbolize the most powerful test as Test A and another one as Test B, then, the power efficiency of the latter is

$$\frac{n_A}{n_B}$$

where n_A and n_B are the minimum necessary values of n for rejecting H_0. Thus, for the Spearman r_S the power efficiency is

$$\frac{50}{55} = 0.91$$

The penalty of lower efficiency is not a serious one for the nonparametric test. All we need do is increase the n.

more bad news

Nonparametric tests suffer from an abundance of minor irritating drawbacks other than the major one above. They usually require categorized or rank-ordered data, difficult when the dependent variable was cardinal and the data not originally rank ordered. (Remember the same issue with percentiles?) They are not suitable for large and complex experimental designs. Few computer programs exist for them. (ANOVA is typically done on computers nowadays.) There are prejudices for and against nonparametric tests, prejudices that are not always strictly founded on fact, fair-mindedness, or admission that there are usually several right ways to do anything. Nonparametric tests often have trouble with tied scores (remember the median?). These tests usually cannot be performed on the basis of data reconstructed from values of $\overline{X}$, s, and n, as the t test can be. The formulas for nonparametric tests are sometimes complex and difficult to rationalize (although they do work easily). The tables of critical values are almost as diverse as the total number of nonparametric tests (over 30 of them).

but now the good news

Nonparametric tests
are often appropriate,
sometimes necessary,
usually easy, and
by Gauss,
always there.

In this book, we cannot cover them all. We consider only some particularly useful and necessary ones, including

a tests of categorical frequencies (the binomial test, for example),

b substitutes for the t and F tests,

c measure of correlation (already covered in Chapter 9).

Many excellent textbooks may be found that are solely devoted to an exhaustive range of nonparametric tests.

significance in nonparametric tests

The **test statistics** t, F, and r depend upon probability distributions based upon the Gaussian or Student t distributions. Most of the time, nonparametric tests produce **test statistics** whose probabilities under the null hypothesis are based upon the **binomial** or **chi-square** distributions. The necessary tables are furnished in conjunction with each test to be considered, and the concept of **critical value** applies to nonparametric procedures just as for parametric ones. Likewise, the problems of Type I and Type II errors are similar, and we still need good experimental design.

tests for categorical frequencies

In many types of survey or research, we may be interested in knowing whether scores have fallen in equal numbers into several categories (after allowing for the effects of mere chance). For example, we might be interested in the relative frequencies with which women and men are granted honorary university degrees. The null hypothesis says that there is no systematic difference in treatment of the sexes. But suppose that, in one year in North America, it happens that 1326 men and 474 women are honored in this way. Is this a result of equal treatment of the sexes, the greater number of men receiving degrees being a product only of chance, or is it evidence that one sex is favored over the other?

In Chapter 7 we considered the binomial test, and it could give us the probability of such data, assuming the null hypothesis. Unfortunately, the binomial test is limited to testing between two samples. Therefore, we will consider a technique that can compare any number of categories in terms of their frequency counts.

the χ^2 test

The symbol χ is the Greek letter *chi* (pronounced "kye"). The term χ^2 (chi-square) is applied to one of the basic probability distributions in statistics, comparable in importance to the Gaussian, Student t, and F distributions. The χ^2 test is based upon this distribution, just as the t and F tests are based on their distributions.

To visualize the χ^2 distribution, imagine that you have a barrel filled with beads. Half the beads are white, half are black, and they are well mixed up. Thus if you blindly reached in and pulled out a given number of the beads, you would expect half of them to be white and half of them black. Or you could have several colors of beads, each color in any given quantity. As long as they were well mixed and you sampled blindly, you would *expect* to get numbers of each color proportional to the number of each color in the barrel. But of course, this expectation would almost never be exactly met. The *obtained* frequency of each color would usually not match the *expected* frequencies exactly.

The χ^2 test furnishes us with the probability that any obtained frequency arose by chance, assuming (this is the null hypothesis) that the barrel actually contains the expected proportaions of each color.

In real life, the barrel is Nature and the beads are events. A white bead might be a certain event, the black bead the absence of that event.

In our example on honorary degrees, there are 1800 degrees given, and the null hypothesis says that half of them should have been given to men, the other half to women (the proportions of the two sexes in the population being approximately equal). This is like saying that there *should* be equal numbers of black and white beads in the barrel. So we want to know what the probability is of finding 1326 white beads and 474 black ones in a sample of 1800, when the barrel contains equal numbers of both color.

If many, many samples of equal size are randomly drawn from the barrel, a χ^2 distribution will be formed from the following statistic. statistic.

$$\chi^2 = \frac{\sum (f_o - f_e)^2}{f_e} \quad \text{or} \quad \sum \frac{(f_o - f_e)^2}{f_e}$$

Table 15.A
SOME CRITICAL VALUES OF χ^2

	Level of significance for a directional test				
	0.10	0.05	0.025	0.01	0.005
	Level of significance for a nondirectional test				
df	0.20	0.10	0.05	0.02	0.01
1	1.64	2.71	3.84	5.41	6.64
2	3.22	4.60	5.99	7.82	9.21
3	4.64	6.25	7.82	9.84	11.34
4	5.99	7.78	9.49	11.67	13.28
5	7.29	9.24	11.07	13.39	15.09

The symbols f_o and f_e stand for *observed* and *expected* frequencies, and the summations are over all categories whose frequencies we are measuring (in this case, degrees given to men and degrees given to women). Like the t and F distributions, the χ^2 distribution can have many shapes, depending upon the df, which we'll consider in a moment. Some typical χ^2 distributions are shown in Figure 15.1.

The question about male and female honorary degrees can now be settled. All that is necessary is to substitute numerical values into the formula for χ^2.

$$\chi^2 = \frac{(1326 - 900)^2}{900} + \frac{(474 - 900)^2}{900} = 403.28$$

Our obtained value of 403.28 must now be evaluated in the same way as a t or F ratio, by comparison with a table of critical values of χ^2, given in Table 15.A. (A complete Table of Critical Values of χ^2 appears in the Appendix.) It is used just as a t table is: the probability value at the top of each column refers to the likelihood of the values of χ^2 listed below being reached or exceeded merely by chance. The left-hand column is our old friend df.

The df value for χ^2 is different from those involved in parametric tests, where df is closely tied to n, the number of scores. With the χ^2 test, df is tied to the number of *categories* in which frequencies of scores are merely the dependent variable. Above, there are two parenthesized

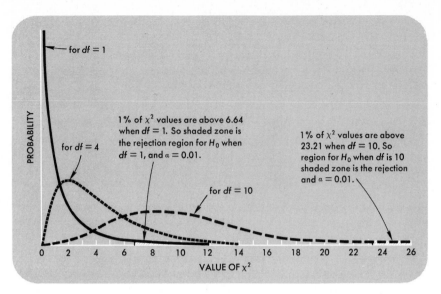

PROBABILITY

for df = 1

for df = 4

1% of χ^2 values are above 6.64 when df = 1. So shaded zone is the rejection region for H_0 when df = 1, and $\alpha = 0.01$.

1% of χ^2 values are above 23.21 when df = 10. So region for H_0 when df is 10 shaded zone is the rejection and $\alpha = 0.01$.

for df = 10

VALUE OF χ^2

Figure 15.1
Several χ^2 distributions. There is a different distribution for every value of df; the mean of each is equal to its df. As df rises, the distribution moves to the right along the abscissa; there is thus no upper limit to the abscissa. A χ^2 value that falls into the rejection region may justify rejecting the null hypothesis. Relate this figure to the information in Table 15.A.

categories in the formula, one for males and the other for females. Therefore, we start with 2 df. But as before 1 df is lost whenever a constraint is placed upon freedom to vary. If there are 1326 males and a total of 1800, the female frequency is not free. It *must* be 474; otherwise the value of f_e could not be ascertained as $(1326 + 474)/2 = 900$. So 1 df is lost, and we have one remaining. Referring to Table 15.A, we find that the null hypothesis may be rejected at the 0.01 level if χ^2 equals or exceeds 6.64. (Look at the table and see for yourself.) Our obtained value of 403.28 exceeds 6.64 quite handsomely (this often happens with χ^2, much more than with t and F tests), so we reject the null hypothesis. We may rest comfortably (or uncomfortably) with the conclusion that men are more eminent (or officially perceived that way) than women.

The foregoing is an example of the one-sample χ^2 test. We can also perform a multi-sample χ^2 test.

Suppose that at a certain college there are three sections of the statistics class in the Psychology Department. Each section contains students who were abitrarily assigned to it at registration. The sections have three different instructors, and some students feel that Pass and Fail grades are not equivalent among the three sections. The picture is as follows.

	Section 1	Section 2	Section 3
Pass	9	7	14
Fail	8	11	1

In this instance we have six categories from three independent groups. The null hypothesis is that the three sections differ only by chance in their pass–fail frequencies.

In the multi-sample test, the most difficult step is to find the expected frequencies for each cell, and even that is not too difficult.

The procedure is to sum the observed frequencies toward the margins of the design/data matrix and then find the total N from the marginal totals.

	Section 1	Section 2	Section 3	Marginal total
Pass	9	7	14	30
Fail	8	11	1	20
Marginal total	17	18	15	50 Grand total

Then, to find the expected frequency in any cell, merely multiply its marginal totals together and divide by the grand total. For example, the marginal totals for *Pass* and *Section 1* are 30 and 17; since the grand total is 50, we have $(30 \times 17)/50 = 10.2$. This value, 10.2, is the expected frequency (f_e) for the upper left-hand cell. Expected frequencies are found for the remaining cells by the same principle.

Within each cell, we next find the difference between f_o and f_e and square it. The result is then divided by f_e, giving

$$\frac{(f_o - f_e)^2}{f_e}$$

For the upper left-hand cell, therefore, we get

$$\frac{(9 - 10.2)^2}{10.2} = 0.141$$

The resulting values, six of them in this instance, are then totaled. And that's all there is to the calculation. The value of χ^2 for our example is 10.67. (As practice, check this with your own calculations.)

Finally, we decide on the value of α, determine df, ascertain the critical χ^2 value, and compare the latter with our obtained χ^2. If the obtained χ^2 equals or exceeds the tabled value, we reject the null hypothesis.

In determining df, we see that there are six categories or cells. Since the value of N must be used in determining the f_e values, we lose 1 df. Then there are several marginal *column* totals (three in this example), and we lose another df for each of them except one (which has already lost 1 df in the calculation of N and cannot lose it again). Thus, two more df are lost. Finally, there are several marginal *row* totals (two in this case), and each loses 1 df except for one of them, which already lost 1 df in calculating N. (N is equally well calculated by summing row totals or column totals.) Four degrees of freedom have been lost altogether, and the principle extends to any number of rows and columns. There are 2 df remaining from the original six. The quick way to remember the df for any such χ^2 data matrix is as

$$df = (R - 1)(C - 1)$$

where R stands for the number of rows and C for the number of columns. So for our present example,

$$df = (2 - 1)(3 - 1) = 2$$

Looking at either the Table of Critical Values of χ^2 in the Appendix or Table 15.A, we see that for $\alpha = 0.01$ and $df = 2$, the critical value is 9.21. Our obtained χ^2 is 10.67, so we reject the null hypothesis and decide that the three classes were *not* being treated similarly.

It may seem obvious that grades in Section 3 are generally too high. However, the test has not *located* the difference precisely in the data matrix; it has merely indicated that a significant deviation is there *somewhere*. We can, however, sometimes get a good inkling from just inspecting the matrix.

The formula for *df* is not, obviously, blindly followed when only one row is present, because *df* cannot be 0. (If $R - 1 = 1 - 1 = 0$, then *df* would be 0 too, and that is not possible.) The *df* value is actually easier to calculate under such conditions. Suppose, for example, that you wonder whether a course (Physics 150) is more popular with males than with females. You find 75 males and 35 females on the class list. So χ^2 is calculated as

$$\chi^2 = \frac{(75 - 55)^2}{55} + \frac{(35 - 55)^2}{55} = 14.54$$

and this is significant with 1 *df*. Remember the basic rule: **df is given by the number of categories in which frequency may freely vary**, once the values of f_e are known. So in this simpler instance, $df = C - 1$ (columns minus 1).

There are three more precautions to be observed in using the χ^2 test. (1) It must be used on raw *frequencies* only, not on percentages or cardinal data. Remember, it is a test at the nominal scaling level, although if rank-order predictions are verified for the cells of a data matrix whose χ^2 is significant, ordinal connotations are possible. (2) For the test to be applicable, the *expected* frequency, f_e, must be 5 or greater in every cell. In our last sample experiment, the smallest f_e was 6.00. (3) The crux of proper use of χ^2 is that the frequencies must have fallen *independently* into the data matrix; no score in any category may relate in any way to any other score, in the same category or not.

Basically, the χ^2 test decides whether it is likely that the observed frequencies in two or more categories occurred by chance if all frequencies were the result of independent random sampling in the same population. Therefore, being sensitive to *difference* but not *direction* of difference, it is inherently two-tailed. However, if the user of the χ^2 test has a directional hypothesis, he or she can retain H_0 when χ^2 is "significant" in the wrong direction. This would allow some area to be added to the rejection region of the χ^2 distribution, applicable when the direction of difference between frequencies turns out as predicted. This one-tailed test should be used only with one-sample comparisons, like the one comparing men and women in the awarding of honorary degrees. More complex comparisons do not allow a clear either/or prediction of the direction of a difference. The Table of Critical Values of χ^2 gives the one-tailed critical values of χ^2.

the sign test

The sign test is an alternative to the repeated-measures or within-subjects t test, useful when the parametric assumptions are in doubt and also when you want a quick and easy test of significance between sets of measures that are *related*.

Suppose that a marriage counselor, after years of treating troubled couples, begins to wonder whether one of the partners is typically helped more than the other by the counseling. Possibly the success of counseling depends upon the wife (or the husband) being the one for whom the treatment works best. It would be nice to find out.

The *sign* test is so called because it uses plus and minus signs as its data. It's very useful when we can rank-order within pairs of related observations without actually being able to attach cardinal values. Thus, within each couple there will be a plus sign (+) assigned to the *more* improved member and a minus sign (−) assigned to the *less* improved member. If the null hypothesis is true, meaning that the husbands are generally as adequately assisted as their wives (and vice versa), we would expect about half the plus signs to be attached to each sex. However, if the null hypothesis is false (if the therapy helps one spouse more than the other), then more plus signs should be found in the favored group.

The real issue is, of course, how to interpret the relative frequencies of plus and minus signs as actually found. Do they differ by chance alone, or do they indicate a true difference between the groups?

This is one of the easiest of the nonparametric tests, because we need only (1) count the number of pairs in which some difference is observed, and (2) count the total number of plus signs in the group that has fewer of them. Let's give the symbol g to the latter number. Suppose that our counselor looks at the 20 couples seen during one week. In 14 of these pairs, the wife has benefited more than the husband. In two instances, the counselor cannot decide who benefited more, so the total number (N) drops to 18. Thus 14 out of 18 wives score a plus, and $g = 4$. The probability of this happening is the same as that of tossing 18 coins and getting 14 heads, and the possible outcomes are binomially distributed around a mean of 9. We could work out this probability directly with the binomial expansion, but a table could be used if available; see Table H in the Appendix. The total N is shown down the left, and g (the number of plus signs) is shown along the top. We see that the probability of finding only four plus signs ($18 - 14$) for husbands is 0.015 if the null hypothesis is true. So our counselor decides instead that the wives tend to be helped much more than the husbands.

The sign test may be one- or two-tailed. If the former, the direction of difference between frequencies must be predicted in advance. Table H is set up for the one-tailed situation; usually this test is an *ad hoc* one, in which a difference is observed first and wondered about second. For a two-tailed test, the probability values in the body of Table H would be doubled. So the clinician's p is really 0.03, since no prediction of which spouse tended to benefit more was made.

The power efficiency of the sign test is at least 0.63, and it can be greater when N (the number of pairs) is small. Thus, the sign test requires half as many scores again as the t test if it's to reject the null hypothesis as readily.

the binomial test again

In Chapter 7, the binomial test was touched upon briefly. Now we expand a bit on the earlier, intuitive, explanation.

Let's return to County Kayo for a worked illustration of the binomial test. Around a given village, there are 20 leprechauns altogether, 15 male and 5 female. An elf chances by, in dire straits because his shoes need mending. Now, leprechauns are unfriendly to elves, but they are good at fixing shoes; so our elf blows his horn loudly on the off chance that a leprechaun might be willing to assist him. Three female leprechauns and one male leprechaun appear in answer. Our elf wonders whether this indicates a greater friendliness on the part of the women than of the men. The binomial test can tell him.

In statistical terms, the question is: How many combinations of three females and one male are there, out of the total number of possible 4-case samples, (assuming the null hypothesis, which says that the two sexes are equally friendly)? The probability of a male offering to help (p_m) is theoretically 15/20, and p_f is 5/20. If the number of 3-female and 1-male samples is too small a percentage of all possible combinations, we might then conclude that it is too improbable to be consistent with the null hypothesis.

We let P stand for the probability, under H_0, of one event (a male offering to help), and Q will stand for the probability of the other event (a female offering to help). Then g is the value of the less frequent event, and $(N - g)$ is the value of the more frequent one. The formula for the probability of getting g events (males) out of N total (females plus males) is

$$p_g = \left(\frac{N!}{g!(N-g)!}\right)(P^g)(Q^{(N-g)})$$

With values substituted into this, we have

$$p_g = \left(\frac{4 \times 3 \times 2 \times 1}{1 \times 3 \times 2 \times 1}\right)\left(\frac{15}{20}\right)^1\left(\frac{5}{20}\right)^3$$

$$= 4 \times 0.75 \times 0.0156$$

$$= 0.0468$$

There is just less than a 5% chance of getting exactly three females and one male under the null hypothesis.

Our elf considers this and decides that the question is not yet fully answered. The *real* question is whether a result this extreme *or more so* could have happened by chance. So he, and we, should add the probability of the more extreme outcome, 4 females and no males. We would use the above formula, getting

$$p_g = \left(\frac{4!}{(0!)(4!)}\right)\left(\frac{15}{20}\right)^0\left(\frac{5}{20}\right)^4$$

$$= \frac{4 \times 3 \times 2 \times 1}{(1)(4 \times 3 \times 2 \times 1)} \times 1 \times 0.0039$$

(remembering that 0! is equal to 1 and that anything raised to the power 0 is also 1)

$$= 1 \times 1 \times 0.0039$$

$$= 0.0039$$

When the two probabilities (of getting g events or less out of N possibilities) are added, we have a total $p_g = 0.0507$. Our elf decides that H_0 cannot be rejected at $\alpha = 0.05$, and certainly not at 0.01.

When $P = Q$, the calculations are easier. Table H in the Appendix does the work for you, for N values to 20 and all possible g values, given N.

using table H

Suppose that in another village in County Kayo, our elf knows that there are equal numbers of male and female leprechauns. When he goes there and blows his horn, 6 females and 2 males stick their heads out of their thatched cottages. The probability of 2 males or less in an N of 8, when P and Q are equal, is 0.145 (Find this value in Table H in the Appendix.) The null hypothesis cannot be rejected in this second village.

The binomial test is useful when f_e is too small for the χ^2 test. In this latest example, an N of 8 could provide an f_e value of only 4 in each of the two groups, male and female.

The major problem with the binomial test is that with large values of N, huge values result from taking the factorial. The χ^2 test is therefore preferable when total frequencies are large.

the Mann–Whitney *U* test

The **Mann–Whitney *U* Test** is an alternative to the t test for independent groups; it is extremely power-efficient (0.95) and popular when there is any doubt about the parametric assumptions necessary for the t test. Its only drawback is the difficulty of ranking large quantities of scores. It may be used for testing the probability that two sets of ordinally scaled scores came from populations having the same mean value μ. When this probability is too small, the null hypothesis is rejected.

The procedure is very easy in principle and in practice. Let's imagine that we have two samples, A and B, randomly drawn from two parent populations, A and B. The first step is to rank-order all the scores, lining them up in a vertical column and keeping track of which group each score is from. If n_A is 5 and the A scores are 2, 6, 7, 9, and 14, and n_B is 4 and the B scores are 8, 10, 10, and 15, you would arrange the picture like this.

Scores	
A	B
2	
6	
7	
	8
9	
	10
	10
14	
	15

Notice that the scores are ranked from the smallest at the top to the largest at the bottom.

Next, it is merely necessary to decide which group of scores generally ranks lower; in this example, group A does. Finally, we count the number of times a score in the other group, B, ranks lower than a score in group A. The resulting total is the statistic U.

Scores

A	B	
2		
6		
7		
	8	Lower than 2 A scores (9 and 14)
9		
	10	Lower than 1 A score (14)
	10	Lower than 1 A score (14)
14	15	$4 = U$

Now if the null hypothesis is true, we would expect the two sets of scores to lie approximately side by side, so that an A score falls lower than a B score about as often as a B score falls lower than an A score. If there are no ties between scores in opposite groups, $n_A n_B$ gives the total number of times that a score in one group falls below a score in the opposite group. We would expect the value of U to be $n_A n_B/2$, calculating it either way (A relative to B or B relative to A).

In this example the value of U is 4. It ought to be 10, or $(5 \times 4)/2$. Is the discrepancy accidental? The last step is to refer to a Table of U Values (Table I in the Appendix.). If your obtained U value is *equal to* or **less** *than* the tabled value for your values of n_A and n_B, then the null hypothesis may be rejected.

Sometimes we find that our obtained U value is very large when it should (by inspection of the ranked data) have been small. The problem is that we did the comparisons the wrong way around. The statistic so obtained is called U' ("U prime"), not U. U may easily be calculated from U', since

$$U = n_A n_B - U'$$

The preceding example involves two problems that you may have noticed already. First, the task of counting the number of *preceded scores* (the number of times a B score precedes an A score) is onerous and gets worse as N increases. Second, the Table of U stops with n_A and $n_B = 20$, and quite often an experiment has larger ns than that.

calculation of U with large n_A and n_B

The value of U can be calculated directly if we know the values of n_A and n_B and if the overall ranks are known for one of the groups (preferably the overall higher-ranking one). In our previous example, group B seems to be the higher group. So we enter the overall ranks for both groups and total them for group B, getting the value ΣR_B

		Scores		
Rank	A	B	Rank	
1	2			
2	6			
3	7			
		8	4	
5	9			
		10	6.5	
		10	6.5	
8	14			
		15	9	
			26 $= \Sigma R_B$	

The value of U can be calculated as

$$U = n_A n_B + \frac{n_B(n_B + 1)}{2} - R_B$$

which we see to be

$$U = 5 \times 4 + \frac{4 \times (4 + 1)}{2} - 26$$

$$= 20 + 10 - 26 = 4$$

(If U' arises by mistake from this formula, simply subtract it from $n_A n_B$ as above to get U.)

interpreting the significance of U with large n_A or n_B

When either n_A or n_B is larger than 20, we run off the table. All is not lost, however. Despite the absence of our parametric assumptions about populations A and B, the values of U fall into a Gaussian distribution when the null hypothesis is true and n_A and n_B are large enough. As hinted above, the mean value of the distribution is $(n_A n_B)/2$. The standard deviation of the Gaussian distribution is difficult to rationalize briefly, but it works out as

$$s_U = \sqrt{\frac{(n_A)(n_B)(n_A + n_B + 1)}{12}}$$

With this information we can easily perform a z test, with the numerator as the difference between U and $(n_A n_B)/2$ and with the denominator as the standard deviation. We can then use Table 8.A, as for any z test.

what to do about tied scores

When any A score is tied with a B score, we are in an ambiguous situation, because U requires us to count the number of times a B score is *smaller than* an A score, or an A score *smaller than* a B score. There is argument about how to handle this problem, but here is a good option.

step 1

Consider tied scores to be slightly different in favor of Group A. If several ties exist, favor Group A consistently. Calculate U as shown above, and determine its significance.

ORIGINAL PICTURE

Score

Rank	A	B	Rank	
1	2			
2.5	3	3	2.5	Tied
4	4			
5	6			
6.5	7	7	6.5	Tied
		8	8	
		10	9	
10	15			
		16	11	

Favoring A

Rank	A	B	Rank
1	2		
2	3	3	3
4	4		
5	6		
6	7	7	7
		8	8
		10	9
10	15		
		16	11

$$\sum R_\mathrm{B} = 38$$

$$U = 6 \times 5 + \frac{5 \times (5 + 1)}{2} - 38 = 7$$
[not significant]

step 2

Repeat Step 1, but this time favor Group B consistently. Calculate U and test its significance.

Favoring B

Rank	A	B	Rank
1	2		
3	3	3	2
4	4		
5	6		
7	7	7	6
		8	8
		10	9
10	15		
		16	11

$$\sum R_{\mathrm{B}} = 36$$

$$U = 6 \times 5 + \frac{5 \times (5 + 1)}{2} - 36 = 9 \text{ [not significant]}$$

step 3

If neither of the obtained U values is significant, the null hypothesis may be retained unambiguously. If both of the obtained U values are significant, the null hypothesis may be rejected unambiguously. If one U is significant and the other not, you're out of luck, but you may wish to venture to Step 4.

step 4

Average the two U values. If the mean U value is significant, reject the null hypothesis.

how to use Table H

Determine the value of the Mann–Whitney U, making sure that it is U and not U'. Then determine the values of n_A and n_B and locate the proper intersection in the Table. If your obtained U is *equal to or smaller than* the critical value, you may reject H_0 at the associated level of significance.

Kruskal–Wallis one-way analysis of variance

Just as the t test must give way to the F test and ANOVA when C (the number of separate experimental cells) exceeds two, the Mann–Whitney U test must give way to the **Kruskal–Wallis test** when the number of groups is larger than two, when the groups are randomly and independently sampled, and when doubt as to the parametric assumptions renders the F test undesirable. The power efficiency of the Kruskal–Wallis test is about 0.95. It is based on the χ^2 distribution and requires at least ordinally scaled data. Such ordinal scaling may be applied to cardinal raw data, shown vertically rank-ordered here:

GROUP	A		B		C		D		E	
	X	Rank	X	Rank	X	Rank	X	Rank	X	Rank
	16									
	19									
	20									
					23					
					25					
							26			
			28				27			
			34							
	35									
									36	
									38	
	40									
	45									
			47							
					48					
					49					
							51			
									55	
			57							
			58							
			59							
							64			
					67					
					69					
							73			
							75		77	
									79	
									80	

The Kruskal–Wallis test is best visualized by an example. Suppose that our experiment has involved five groups of subjects, with $n = 6$ in each group, so that $N = 30$. (The Kruskal–Wallis test is unsatisfactory if n is less than 5.) The scores from the five groups are rank ordered in a way similar to the Mann–Whitney procedure. It is then easy to rank-order overall, as shown to the right of each column.

GROUP	A		B		C		D		E	
	X	Rank	X	Rank	X	Rank	X	Rank	X	Rank
	16	1								
	19	2								
	20	3								
					23	4				
					25	5				
							26	6		
			28	8			27	7		
			34	9						
	35	10								
									36	11
									38	12
	40	13								
	45	14								
			47	15						
					48	16				
					49	17				
							51	18		
									55	19
			57	20						
			58	21						
			59	22						
							64	23		
					67	24				
					69	25				
							73	26		
							75	27	77	28
									79	29
									80	30
$\sum R =$	43		95		91		107		129	
$(\sum R)^2 =$	1849		9025		8281		11449		16641	
$\dfrac{(\sum R)^2}{n} =$	308.16		1504.16		1380.16		1908.16		2773.50	

The ranks are then totaled to give ΣR, and each such total is squared, as shown, and divided by its n. The resulting quotients are added to give $\Sigma (\Sigma R)^2/n$.

$$\Sigma \frac{(\Sigma R)^2}{n} = 308.16 + 1504.16 + 1380.16 + 1908.16 + 2773.50$$

$$= 7874.15$$

The Kruskal–Wallis statistic is called H and is calculated as

$$H = \frac{12}{N(N + 1)} \Sigma \frac{(\Sigma R)^2}{n} - 3(N + 1)$$

Now, this formula is almost impossible to intuit, but it works out so that H equals 0 when the values of ΣR are equal, as they are when the null hypothesis is true. With greater departure of H from 0, the probability increases that the groups are not drawn from the same population. The significance of H is determined by comparing it with the Table of Critical Values of χ^2 for $C - 1$ df. In the present instance, there are five groups,, hence 4 df.

The value N in the formula refers to the total N, 30 in the example. The values of 12 and 3 (which never vary) are placed where shown in the equation for H.

Working the formula for H in our present example, we find that

$$H = \frac{12}{30(31)} \times 7874.15 - 3(30 + 1)$$

$$= 8.60$$

The critical value of χ^2 for 4 df and $\alpha = 0.05$ is 9.49, which has not been reached by our obtained H. Hence, we cannot reject the null hypothesis for our worked example.

what to do about tied values

When ties occur, the rank assigned to each is the mean rank for all the tied scores. For example, if the raw scores 24, 24, and 24 would normally rank 9th, 10th, and 11th, they all get the mean rank of $(9 + 10 + 11)/3 = 10$.

A tie does not greatly affect H. Usually you can ignore the problem, unless more than 25% of all scores are tied in some way or other. Even then, the effect is to reduce H slightly from what it should have been; if it is significant, it would also have been significant if the ties had been taken care of in some way.

summary

So far, we have managed to find a nonparametric test to parallel every parametric one that we've covered in this book. The repeated-measures *t* test (or within-subjects *t* test) is paralleled by the sign test. The independent-measures *t* test is paralleled by the Mann–Whitney *U* Test. The one-way analysis of variance and *F* test are paralleled by the Kruskal–Wallis test. The Pearson *r* is paralleled by the Spearman r_S. It's nice to note too that the nonparametric tests discussed here are accompanied by others (which space considerations bar from this book) giving great choice and flexibility to the experimenter.

However, we are left with a problem outstanding. What nonparametric test will serve as alternative to the two-way (or multi-way) ANOVA considered in Chapter 14?

Alas: There are no widely accepted, well-known, or hazard-free methods for looking nonparametrically at two experiments simultaneously while also looking at the interaction. This is disappointing, but remember that parametric analysis of variance *is* robust in the face of suspect parametric assumptions and that if necessary the main effects and simple effects can still be back-up tested by the Mann–Whitney or Kruskal–Wallis techniques.

terminology for chapter 15

PARAMETRIC ASSUMPTIONS

PARAMETRIC TESTS

NONPARAMETRIC TESTS

POWER EFFICIENCY

TEST STATISTIC (such as *r, t,* and *F* for parametric tests and r_S, *U*, and *H* for nonparametric tests)

CHI-SQUARE TEST

THE SIGN TEST

THE MANN–WHITNEY *U* TEST

THE KRUSKAL–WALLIS ONE-WAY ANALYSIS OF VARIANCE

THE BINOMIAL TEST (again!)

self-test on chapter 15

This is a short-answer Self-test. Fill in the blanks with the appropriate answers from your well-filled memory.

1 ＿＿＿＿＿ A ＿＿ test is one that requires that the population of scores form a Gaussian distribution.

2 ＿＿＿＿＿ A ＿＿ test is one that allows the population distribution to fall into any shape.

3 ＿＿＿＿＿ You ascertain that a given parametric test needs a minimum of 50 scores (one each from 50 subjects) to reject the null hypothesis. An equivalent nonparametric test requires a minimum of 60 scores. Therefore, the power efficiency of the latter test has the value ＿＿.

4 ＿＿＿＿＿ You wish to know whether fillies or colts more often win the Irish Sweepstakes. A proper test would be the ＿＿ test with

5 ＿＿＿＿＿ degrees of freedom.

6 ＿＿＿＿＿ A nutrition researcher decides to compare two types of diet, using experimental rats. The researcher uses litter mate pairs, one member getting one diet and the other member getting the other. The proper nonparametric test would be the ＿＿.

7 ＿＿＿＿＿ The correct answer to Question 6 depends upon the fact that the two samples are ＿＿.

8 ＿＿＿＿＿ If the nutrition researcher had decided to use three random samples of rats to compare three diets, the correct nonparametric test would be the ＿＿.

9 ＿＿＿＿＿ If the χ^2 test or a test based upon the χ^2 distribution is used and the null hypothesis is true, we would expect the test statistic to equal ＿＿.

10 ＿＿＿＿＿ The value of *df* in the χ^2 and Kruskal–Wallis tests is determined from the number of ＿＿.

11 ＿＿＿＿＿ A primate researcher takes two groups of chimpanzees at random and administers an experimental treatment to one group and a control treatment to the other. The best choice of nonparametric test would be the ＿＿.

SCORING AND EVALUATION of Self-Test questions

1 Parametric

2 Nonparametric

3 0.83 or 83%. Power efficiency is determined as n_A/n_B or 50/60 = 0.83.

4 Chi-square test. This is the one for comparing frequencies within categories.

5 1 *df*. Two categories, one degree of reedom.

6 Sign test. This is the one to use when you have related data or within-subjects data or, at any rate, some kind of pairing between the two groups.

7 Related. If the two samples were independent, the Mann–Whitney test would be used.

8 Kruskal–Wallis. With more than two independent groups, that's the choice.

9 0 or merely random departure from 0.

10 Groups (or categories). Not the number of subjects.

11 Mann–Whitney *U* test for two independent groups.

12 Four large department stores report the following incidences of shoplifting over a two-week period: 50, 81, 27, and 42. Perform a x^2 test and decide whether a significant difference exists among the stores, using $\alpha = 0.05$.

_____ = x^2

_____ = critical value of x^2

( ) Retain null hypothesis
( ) Reject null hypothesis

13 A buying agent, wishing to compare two brands of pocket calculator, assembles a panel of 9 statistics students and has each student work a problem on each machine. The problems are known to be exactly equivalent in difficulty, and a counterbalancing procedure is used. The data are the times (in minutes) required to work each problem correctly:

Student	1	2	3	4	5	6	7	8	9
Brand A	9	15	16	10	8	17	21	14	15
Brand B	13	13	18	19	11	18	21	15	17

Perform a sign test, and come to a conclusion with $\alpha = 0.05$.

31.08
7.82
Reject

Answer to Question 12

$$x^2 = \sum \frac{(f_o - f_e)^2}{f_e} \qquad f_e = [50 + 81 + 27 + 42]/4 = 50$$

$$= \frac{(50 - 50)^2 + (81 - 50)^2 + (27 - 50)^2 + (42 - 50)^2}{50}$$

$$= \frac{0 + 961 + 529 + 64}{50} = \frac{1554}{50} = 31.08$$

$df = C - 1 = 4 - 1 = 3$

Critical value of x^2 for $\alpha = 0.05$ is 7.82

$31.08 > 7.82$, so reject null hypothesis

Reject H_0 **Answer to Question 13**

Sign Test: Pair	1	2	3	4	5	6	7	8	9
A		+					Tied		
B	+		+	+	+	+		+	+

$N = 8, n_{A+} = 1 = g$

$p = 0.035 < \alpha$

Reject H_0 at $\alpha = 0.05$

14 An educator compares two random groups of third-year students to determine whether a programmed textbook is *worse* than an ordinary textbook. Group P uses the programmed one, and Group C gets the ordinary one. The data are the final test scores on a common examination.

Group P: 57   78   45   33   52   57   28   76
Group C: 77   86   34   85   60   87   94   79   56   90

Perform a Mann–Whitney U Test, and come to a conclusion using the 0.025 level.

_____ $= U$.
_____ $=$ Critical value for U.
( ) Retain null hypothesis.
( ) Reject null hypothesis.

Answer to Question 14

Mann–Whitney U test:

Score

Rank	P	C	Rank
1	28		
2	33		
		34	3
4	45		
5	52		
		56	6
7	57		
8	57		
		60	9
10	76		
		77	11
12	78		
		79	13
	— —	85	14
$\sum R_P = 49$		86	15
		87	16
		90	17
		94	18
$n_P = 8$		$n_C = 10$	

13
17
Reject

$$U' = n_P n_C + \frac{n_P(n_P + 1)}{2} - \sum R_P$$

$$= 8 \times 10 + \frac{8(8 + 1)}{2} - 49$$

$$= 80 + 36 - 49$$

$$= 67$$

$$U = n_P n_C - U'$$

$$= 80 - 67 = 13$$

Critical value $= 17$ for $\alpha = 0.025$

$$U < U_{crit}, \text{ so reject } H_0$$

15 Four different types of buried-metal detectors are being compared among four groups of leprechauns, each group using one of the four types. The dependent variable is the amount of gold recovered, and we want to know whether any significant differences exist among the four types of detector. Use the Kruskal–Wallis test to find out. The data are

Machine A:	68	36	78	69	50	91
Machine B:	79	65	93	98	64	62
Machine C:	48	51	70	44	68	68
Machine D:	43	64	52	39	*	27

(*Note:* * refers to Sean Bogthumper's score, which was lost because he dropped his machine into Cuthbert's Chasm. Carry on regardless.)

_____ = H.
_____ = df.
_____ = critical value of H for $\alpha = 0.05$.
_____ = Reject or retain null hypothesis.

Answer to Question 15

Kruskal–Walis test:

	A		B		C		D	
	X	R	X	R	X	R	X	R
							27	1
	36	2						
							39	3
							43	4
					44	5		
					48	6		
	50	7						
					51	8		
							52	9
			62	10				
			64	11.5			64	11.5
			65	13				
	68	15			68	15		
					68	15		
	69	17						
					70	18		
	78	19						
			79	20				
	91	21						
			93	22				
			98	23				

$N = 23$

$\sum R =$	81	99.5	67	28.5
$(\sum R)^2 =$	6561	9900.25	4489	812.25
$(\sum R)^2/n =$	1093.5	1650.04	748.16	162.45

$$\sum (\sum R)^2/N = 3654.15$$

$$H = \frac{12}{N(N+1)} \sum \frac{(\sum R)^2}{n} - 3(N+1)$$

7.438

$df = 3 \ (C - 1 = 4 - 1)$

7.82 (See Table of χ^2)

Retain null hypothesis

$$= \frac{12}{23 \times 24} \times 3654.15 - 3 \times 24$$

$$= 7.438$$

SCORING YOUR PERFORMANCE

Let us remain with our need for perfection, except for the odd rounding error.

ſome ordinary problemſ

1 We know that parametric tests are always more efficient than non-parametric ones when both are appropriate for a given problem. Why is this?

2 We have seen that *power efficiency* refers to a test's capability of rejecting the null hypothesis when it is false. How does this fact relate to β, the probability of a Type II error?

3 Hollingshead and Redlich found the following data on the relationship between social class and incidence of schizophrenia in the city of New Haven.[1]

Class Level	Proportion of population	Number of patients per 100,000 population
I	0.03	}111
II	0.09	
III	0.21	168
IV	0.49	300
V	0.18	895

Test whether prevalence of schizophrenia is related to class level.

4 An ophthalmologist is studying the effect of a new treatment for retinal disease. The disease in question is one that affects both eyes, and the researcher treats only one eye (left or right selected at random) and compares the results with the untreated "control" eye. Using a nonparametric procedure, test whether the resulting data show any significant effect. The data are

Patient	Treated	Control
H. J.	26	32
B. W.	44	41
D. O.	30	39
P. D.	58	62
S. F.	35	38
J. L.	25	31
S. W.	28	29
A. B.	47	49
C. E.	29	35

[1] A. B. Hollingshead and F. Redlich, *Social Class and Mental Illness,* New York: Wiley, 1958.

5 In Question 4, how many more patients would be required to allow H_0 to be rejected at $\alpha = 0.01$, assuming that all such fresh patients showed the same trend in results?

6 Perform a nonparametric test for *independent* samples on the data of Question 4. Is there a significant difference? Explain any discrepancy with the proper test that was done above.

7 A nursing supervisor asks the hospital psychologist to check on an observation made by many nurses, namely, that men make better patients than women (from a nursing point of view). The available data are ratings (made on a 5-point rating scale in which 5 is "excellent" and 1 is "terrible"). The psychologist finds that the data look like this.

Men	Women
16 score 5	12 score 5
23 score 4	20 score 4
11 score 3	19 score 3
9 score 2	21 score 2
9 score 1	18 score 1

What should the psychologist decide?

8 A department store delivery-service superintendent gets a year-end breakdown on the amount of down time among the store's fleet of delivery trucks. Is there a significant difference among the different makes of vehicle?

Hunkajunk Vans:	48	24	36	89	36	77	20	89	80
Bloke's Wagons:	19	35	46	48	25	25	18		
Murkk Panels:	79	51	98	69	90	64	95	103	81
Kreisler's Carriages:	85	86	96	97	63	100	92		
Stanley Steamers:	5	11	9	17	0	16	11		

9 An experimental psychologist is preparing stimulus materials for an experiment in color vision, intending to use color slides for projection onto a screen. In comparing two brands of color film (A and B), the psychologist finds that they rank (from best to worst) as follows in their fidelity of color reproduction (tied when bracketed).

A A B A [B B] A B A [A A] B B B
[B A] B B B

Is either film preferable to the other?

10 An experimenter finds that a nonparametric test and a parametric test are equally appropriate for analyzing the results of an experiment. Size of N is unimportant to the experiment. Which choice of test is preferable?

11 A statistician finds that with the t test the null hypothesis can be rejected using 40 subjects. How many subjects would be needed in order to reject H_0 with the same alpha and the Mann–Whitney U test?

12 In performing the U test, the statistician finds a value of U (or is it U'?) of 615 when $n_1 = 30$ and $n_2 = 50$. What can be concluded?

epilogue

One problem with ending a book is that the topic does not end when the course does. The subject matter goes on, yet the book and its reader must recognize the end of the term and conclude as gracefully as possible. The problem is how. Too many issues have been stirred around to permit a really graceful exit of basic statistics from your life. These remarks are an effort to bring some sort of closure to the subject of statistics, at least for the time being.

where does statistics go from here?

There are several directions in which statistical thinking can be pursued by those with the need or inclination. An early step would be to back up the intuitive quality of the concepts developed here with their formal mathematical backbone. It is proper to begin with intuition, just as Archimedes, Gauss, and Gosset did, but the solid mathematical foundation is necessary for progress into advanced statistical methods.

There are many such methods, and we can describe some of them. You have already learned how to run an experiment with two independent variables and one dependent variable, but imagine an experiment to ascertain the effects of *any* number of independent variables on one dependent variable. Some powerful experiments can be conducted using such *high-order univariate analysis of variance.* (The word *univariate* refers to the fact that just one dependent variable is involved in the experimental design.) Even more powerful experiments can be run using experimental designs that have any number of independent variables *and* any number of dependent variables. The latter, due to the effects of the former, may be correlated in various complex ways, while at the same time interactions may exist among the independent variables. The statistical method needed for such experiments is called *multivariate analysis of variance* (MANOVA) and requires matrix algebra and the use of a computer. Some experimental problems require a MANOVA approach, and so will you if you become interested in such problems.

Our treatment of correlation has been limited to the relationship between two sets of scores. However, a complex *multiple correlation* may exist among more than two sets, and advanced statistics is needed in such situations as well.

As a final example of where statistics can go from here, there is the field of research called *factor analysis*. We can imagine this by considering the fact that individuals possess a very large (if not infinite) number of physical and behavioral characteristics that can be measured. Such characteristics can be called *factors* of personality. But they are probably not all independent of each other: some are correlated with others, the strength of correlation and coefficient of determination being measures of how much any number of factors have in common. Sorting the personality in terms of factor groupings requires familiarity with multiple correlation, although the computations are done nowadays by computer. The list of areas served by advanced statistics could be extended greatly.

advanced statistics and advanced experimentation

It is important to distinguish between the substance and the role of statistics. Since it is a branch of mathematics, there are mathematical statisticians who specialize in it, and they have provided innovations, improvements, and refinements of technique that stand experimenters in good stead. But such contributions do have one drawback: The statistician would often rather do "good mathematics" than provide the experimenter with the simplest possible method of drawing a conclusion from an experiment. The statistician would never be nonchalant about whether an experiment is being done with $\alpha = 0.03$ when it is believed that the 0.01 level applies. (Such discrepancies can occur from failure of parametric assumptions.) If the exact value of α cannot be ascertained, the statistician might even decide that the experiment should not be done at all. However, the experimenter merely wants an *adequately* low value of α in order to get on with the job of answering empirical questions without overkill. As a rule, the empirical researcher should concentrate on the question at hand and use the simplest statistics that will suffice to answer it.

Amusing things sometimes happen in experimentation. An experimenter may test a directional hypothesis and find that the data come out the wrong way around. When this happens you don't need to do any statistics; you just have to look at the data. But experimenters have been known to grind through the statistics anyway, even though no decision-making process is involved. Such activity indicates a confusion between means and ends.

Advanced experimentation is defined not by statistical complexity but by the creativeness and quality of the thought that went into the hypotheses being examined. Science's most far-reaching discoveries have not involved statistics at all. Michael Faraday discovered electromagnetic induction without statistics, complex design, or intricate apparatus; in psychology, William James, Ivan Pavlov, and B. F. Skinner accomplished much without highly intricate experiments. At the same time we must recognize that the conception of any scientific hypothesis demands (at least potentially) a methodology complex enough to test it. With powerful statistics, complex hypotheses can now be formulated.

statistics and science in a cultural context

The impersonality of mathematics, statistics, and research often obscures the fact that these are all human activities carried on within a human cultural context. The social motives underlying them are diverse—sometimes worthy and sometimes perhaps not—but the goals of scientists are generally related to the goals and preoccupations of their society. There have been instances of sick societies aided and abetted by sick science and sick statistics; the racial theories of the Nazis are the most chilling example of modern times. In some respects, science in our own society has the status of a game, similar to the games of politics, big business, even war, in which the essential ingredient is people interacting with each other in certain ways and for certain purposes. The word *game* is not intended to be pejorative; it simply implies that certain rules are in effect to guide the interaction. In contemporary behavioral science, as we've seen, the rules are heavily statistical; by and large this seems to be desirable. Another rule is that scientific findings must be carefully checked before being announced; another is that the announcement must take place clearly, fairly, and publicly. These are excellent rules, but they place added responsibility on the scientist to acquire the verbal and writing skills that are needed.

Science, which includes statistical thinking, differs from most other games in that it involves logic and a basic desire to get at the truth. You may not plan to be a scientist or a statistician, but you need not debar yourself from retaining and developing the logical clarity and critical judgment offered by statistical thinking. These are abilities that every thoughtful person needs.

statistical tables

Table A
AREAS UNDER THE GAUSSIAN CURVE

z			z			z		
0.00	.0000	.5000	0.55	.2088	.2912	1.10	.3643	.1357
0.01	.0040	.4960	0.56	.2123	.2877	1.11	.3665	.1335
0.02	.0080	.4920	0.57	.2157	.2843	1.12	.3686	.1314
0.03	.0120	.4880	0.58	.2190	.2810	1.13	.3708	.1292
0.04	.0160	.4840	0.59	.2224	.2776	1.14	.3729	.1271
0.05	.0199	.4801	0.60	.2257	.2743	1.15	.3749	.1251
0.06	.0239	.4761	0.61	.2291	.2709	1.16	.3770	.1230
0.07	.0279	.4721	0.62	.2324	.2676	1.17	.3790	.1210
0.08	.0319	.4681	0.63	.2357	.2643	1.18	.3810	.1190
0.09	.0359	.4641	0.64	.2389	.2611	1.19	.3830	.1170
0.10	.0398	.4602	0.65	.2422	.2578	1.20	.3849	.1151
0.11	.0438	.4562	0.66	.2454	.2546	1.21	.3869	.1131
0.12	.0478	.4522	0.67	.2486	.2514	1.22	.3888	.1112
0.13	.0517	.4483	0.68	.2517	.2483	1.23	.3907	.1093
0.14	.0557	.4443	0.69	.2549	.2451	1.24	.3925	.1075
0.15	.0596	.4404	0.70	.2580	.2420	1.25	.3944	.1056
0.16	.0636	.4364	0.71	.2611	.2389	1.26	.3962	.1038
0.17	.0675	.4325	0.72	.2642	.2358	1.27	.3980	.1020
0.18	.0714	.4286	0.73	.2673	.2327	1.28	.3997	.1003
0.19	.0753	.4247	0.74	.2704	.2296	1.29	.4015	.0985
0.20	.0793	.4207	0.75	.2734	.2266	1.30	.4032	.0968
0.21	.0832	.4168	0.76	.2764	.2236	1.31	.4049	.0951
0.22	.0871	.4129	0.77	.2794	.2206	1.32	.4066	.0934
0.23	.0910	.4090	0.78	.2823	.2177	1.33	.4082	.0918
0.24	.0948	.4052	0.79	.2852	.2148	1.34	.4099	.0901
0.25	.0987	.4013	0.80	.2881	.2119	1.35	.4115	.0885
0.26	.1026	.3974	0.81	.2910	.2090	1.36	.4131	.0869
0.27	.1064	.3936	0.82	.2939	.2061	1.37	.4147	.0853
0.28	.1103	.3897	0.83	.2967	.2033	1.38	.4162	.0838
0.29	.1141	.3859	0.84	.2995	.2005	1.39	.4177	.0823
0.30	.1179	.3821	0.85	.3023	.1977	1.40	.4192	.0808
0.31	.1217	.3783	0.86	.3051	.1949	1.41	.4207	.0793
0.32	.1255	.3745	0.87	.3078	.1922	1.42	.4222	.0778
0.33	.1293	.3707	0.88	.3106	.1894	1.43	.4236	.0764
0.34	.1331	.3669	0.89	.3133	.1867	1.44	.4251	.0749
0.35	.1368	.3632	0.90	.3159	.1841	1.45	.4265	.0735
0.36	.1406	.3594	0.91	.3186	.1814	1.46	.4279	.0721
0.37	.1443	.3557	0.92	.3212	.1788	1.47	.4292	.0708
0.38	.1480	.3520	0.93	.3238	.1762	1.48	.4306	.0694
0.39	.1517	.3483	0.94	.3264	.1736	1.49	.4319	.0681
0.40	.1554	.3446	0.95	.3289	.1711	1.50	.4332	.0668
0.41	.1591	.3409	0.96	.3315	.1685	1.51	.4345	.0655
0.42	.1628	.3372	0.97	.3340	.1660	1.52	.4357	.0643
0.43	.1664	.3336	0.98	.3365	.1635	1.53	.4370	.0630
0.44	.1700	.3300	0.99	.3389	.1611	1.54	.4382	.0618
0.45	.1736	.3264	1.00	.3413	.1587	1.55	.4394	.0606
0.46	.1772	.3228	1.01	.3438	.1562	1.56	.4406	.0594
0.47	.1808	.3192	1.02	.3461	.1539	1.57	.4418	.0582
0.48	.1844	.3156	1.03	.3485	.1515	1.58	.4429	.0571
0.49	.1879	.3121	1.04	.3508	.1492	1.59	.4441	.0559
0.50	.1915	.3085	1.05	.3531	.1469	1.60	.4452	.0548
0.51	.1950	.3050	1.06	.3554	.1446	1.61	.4463	.0537
0.52	.1985	.3015	1.07	.3577	.1423	1.62	.4474	.0526
0.53	.2019	.2981	1.08	.3599	.1401	1.63	.4484	.0516
0.54	.2054	.2946	1.09	.3621	.1379	1.64	.4495	.0505

The table shows the area of the shaded portions relative to the area of the whole distribution. The shaded areas pertain only to the upper half of the distribution. To find the area between $-z$ and $+z$, or beyond $-z$ and $+z$, double the figures given in the table.

z	0 z	0 z	z	0 z	0 z	z	0 z	0 z
1.65	.4505	.0495	2.22	.4868	.0132	2.79	.4974	.0026
1.66	.4515	.0485	2.23	.4871	.0129	2.80	.4974	.0026
1.67	.4525	.0475	2.24	.4875	.0125	2.81	.4975	.0025
1.68	.4535	.0465	2.25	.4878	.0122	2.82	.4976	.0024
1.69	.4545	.0455	2.26	.4881	.0119	2.83	.4977	.0023
1.70	.4554	.0446	2.27	.4884	.0116	2.84	.4977	.0023
1.71	.4564	.0436	2.28	.4887	.0113	2.85	.4978	.0022
1.72	.4573	.0427	2.29	.4890	.0110	2.86	.4979	.0021
1.73	.4582	.0418	2.30	.4893	.0107	2.87	.4979	.0021
1.74	.4591	.0409	2.31	.4896	.0104	2.88	.4980	.0020
1.75	.4599	.0401	2.32	.4898	.0102	2.89	.4981	.0019
1.76	.4608	.0392	2.33	.4901	.0099	2.90	.4981	.0019
1.77	.4616	.0384	2.34	.4904	.0096	2.91	.4982	.0018
1.78	.4625	.0375	2.35	.4906	.0094	2.92	.4982	.0018
1.79	.4633	.0367	2.36	.4909	.0091	2.93	.4983	.0017
1.80	.4641	.0359	2.37	.4911	.0089	2.94	.4984	.0016
1.81	.4649	.0351	2.38	.4913	.0087	2.95	.4984	.0016
1.82	.4656	.0344	2.39	.4916	.0084	2.96	.4985	.0015
1.83	.4664	.0336	2.40	.4918	.0082	2.97	.4985	.0015
1.84	.4671	.0329	2.41	.4920	.0080	2.98	.4986	.0014
1.85	.4678	.0322	2.42	.4922	.0078	2.99	.4986	.0014
1.86	.4686	.0314	2.43	.4925	.0075	3.00	.4987	.0013
1.87	.4693	.0307	2.44	.4927	.0073	3.01	.4987	.0013
1.88	.4699	.0301	2.45	.4929	.0071	3.02	.4987	.0013
1.89	.4706	.0294	2.46	.4931	.0069	3.03	.4988	.0012
1.90	.4713	.0287	2.47	.4932	.0068	3.04	.4988	.0012
1.91	.4719	.0281	2.48	.4934	.0066	3.05	.4989	.0011
1.92	.4726	.0274	2.49	.4936	.0064	3.06	.4989	.0011
1.93	.4732	.0268	2.50	.4938	.0062	3.07	.4989	.0011
1.94	.4738	.0262	2.51	.4940	.0060	3.08	.4990	.0010
1.95	.4744	.0256	2.52	.4941	.0059	3.09	.4990	.0010
1.96	.4750	.0250	2.53	.4943	.0057	3.10	.4990	.0010
1.97	.4756	.0244	2.54	.4945	.0055	3.11	.4991	.0009
1.98	.4761	.0239	2.55	.4946	.0054	3.12	.4991	.0009
1.99	.4767	.0233	2.56	.4948	.0052	3.13	.4991	.0009
2.00	.4772	.0228	2.57	.4949	.0051	3.14	.4992	.0008
2.01	.4778	.0222	2.58	.4951	.0049	3.15	.4992	.0008
2.02	.4783	.0217	2.59	.4952	.0048	3.16	.4992	.0008
2.03	.4788	.0212	2.60	.4953	.0047	3.17	.4992	.0008
2.04	.4793	.0207	2.61	.4955	.0045	3.18	.4993	.0007
2.05	.4798	.0202	2.62	.4956	.0044	3.19	.4993	.0007
2.06	.4803	.0197	2.63	.4957	.0043	3.20	.4993	.0007
2.07	.4808	.0192	2.64	.4959	.0041	3.21	.4993	.0007
2.08	.4812	.0188	2.65	.4960	.0040	3.22	.4994	.0006
2.09	.4817	.0183	2.66	.4961	.0039	3.23	.4994	.0006
2.10	.4821	.0179	2.67	.4962	.0038	3.24	.4994	.0006
2.11	.4826	.0174	2.68	.4963	.0037	3.25	.4994	.0006
2.12	.4830	.0170	2.69	.4964	.0036	3.30	.4995	.0005
2.13	.4834	.0166	2.70	.4965	.0035	3.35	.4996	.0004
2.14	.4838	.0162	2.71	.4966	.0034	3.40	.4997	.0003
2.15	.4842	.0158	2.72	.4967	.0033	3.45	.4997	.0003
2.16	.4846	.0154	2.73	.4968	.0032	3.50	.4998	.0002
2.17	.4850	.0150	2.74	.4969	.0031	3.60	.4998	.0002
2.18	.4854	.0146	2.75	.4970	.0030	3.70	.4999	.0001
2.19	.4857	.0143	2.76	.4971	.0029	3.80	.4999	.0001
2.20	.4861	.0139	2.77	.4972	.0028	3.90	.49995	.00005
2.21	.4864	.0136	2.78	.4973	.0027	4.00	.49997	.00003

SOURCE: Richard P. Runyon and Audrey Haber, *Fundamentals of Behavioral Statistics,* Second Edition (1971), Reading, Mass.: Addison-Wesley Publishing Co. Artwork from Robert B. McCall, *Fundamental Statistics for Psychology,* Second Edition (1975), New York: Harcourt Brace Jovanovich, Inc.

Table B
CRITICAL VALUES OF r FOR 5% AND 1% LEVELS OF SIGNIFICANCE

n	5%	1%	n	5%	1%
3	0.997	0.999	38	0.320	0.413
4	0.950	0.990	39	0.316	0.408
5	0.878	0.959	40	0.312	0.403
6	0.811	0.917	41	0.308	0.398
7	0.754	0.874	42	0.304	0.393
8	0.707	0.834	43	0.301	0.389
9	0.666	0.798	44	0.297	0.384
10	0.632	0.765	45	0.294	0.380
11	0.602	0.735	46	0.291	0.376
12	0.576	0.708	47	0.288	0.372
13	0.553	0.684	48	0.284	0.368
14	0.532	0.661	49	0.281	0.364
15	0.514	0.641	50	0.279	0.361
16	0.497	0.623	55	0.166	0.345
17	0.482	0.606	60	0.254	0.330
18	0.468	0.590	65	0.244	0.317
19	0.456	0.575	70	0.235	0.306
20	0.444	0.561	75	0.227	0.296
21	0.433	0.549	80	0.220	0.286
22	0.423	0.537	85	0.213	0.278
23	0.413	0.526	90	0.207	0.270
24	0.404	0.515	95	0.202	0.263
25	0.396	0.505	100	0.195	0.256
26	0.388	0.496	125	0.170	0.230
27	0.331	0.487	150	0.159	0.210
28	0.374	0.478	175	0.148	0.194
29	0.367	0.470	200	0.138	0.181
30	0.361	0.463	300	0.113	0.148
31	0.355	0.456	400	0.098	0.128
32	0.349	0.449	500	0.088	0.115
33	0.344	0.442	600	0.080	0.105
34	0.339	0.436	700	0.074	0.097
35	0.334	0.430	800	0.070	0.091
36	0.329	0.424	900	0.065	0.086
37	0.325	0.418	1000	0.062	0.081

Table C
CRITICAL VALUES OF r_S FOR 5% AND 1% LEVELS OF SIGNIFICANCE

n	5%	1%
5	1.000	—
6	0.886	1.000
7	0.786	0.929
8	0.738	0.881
9	0.683	0.833
10	0.648	0.794
12	0.591	0.777
14	0.544	0.715
16	0.506	0.665
18	0.475	0.625
20	0.450	0.591
22	0.428	0.562
24	0.409	0.537
26	0.392	0.515
28	0.377	0.496
30	0.364	0.478

This table is taken from E. G. Olds, The 5 percent significance levels of sums of squares of rank differences and a correction. *Annals of Mathematical Statistics* (1949), **20,** 117–118; and Olds, Distribution of sums of squares of rank differences for small numbers of individuals. *Annals of Mathematical Statistics* (1938), **9,** 133–148. This table is much abbreviated from the original.

The obtained value of r_S is significant if it is greater than or equal to the value listed in the table.

This table is taken from Table IX of James E. Wert, Charles O. Neidt, and J. Stanley Ahmann, *Statistical Methods in Educational and Psychological Research* (1954), New York: Appleton-Century-Crofts.

The obtained value of r is significant if it is greater than or equal to the value listed in the table.

Table D

TRANSFORMATION OF r TO Z

r	z_r	r	z_r	r	z_r	r	z_r	r	z_r
.000	.000	.200	.203	.400	.424	.600	.693	.800	1.099
.005	.005	.205	.208	.405	.430	.605	.701	.805	1.113
.010	.010	.210	.213	.410	.436	.610	.709	.810	1.127
.015	.015	.215	.218	.415	.442	.615	.717	.815	1.142
.020	.020	.220	.224	.420	.448	.620	.725	.820	1.157
.025	.025	.225	.229	.425	.454	.625	.733	.825	1.172
.030	.030	.230	.234	.430	.460	.630	.741	.830	1.188
.035	.035	.235	.239	.435	.466	.635	.750	.835	1.204
.040	.040	.240	.245	.440	.472	.640	.758	.840	1.221
.045	.045	.245	.250	.445	.478	.645	.767	.845	1.238
.050	.050	.250	.255	.450	.485	.650	.775	.850	1.256
.055	.055	.255	.261	.455	.491	.655	.784	.855	1.274
.060	.060	.260	.266	.460	.497	.660	.793	.860	1.293
.065	.065	.265	.271	.465	.504	.665	.802	.865	1.313
.070	.070	.270	.277	.470	.510	.670	.811	.870	1.333
.075	.075	.275	.282	.475	.517	.675	.820	.875	1.354
.080	.080	.280	.288	.480	.523	.680	.829	.880	1.376
.085	.085	.285	.293	.485	.530	.685	.838	.885	1.398
.090	.090	.290	.299	.490	.536	.690	.848	.890	1.422
.095	.095	.295	.304	.495	.543	.695	.858	.895	1.447
.100	.100	.300	.310	.500	.549	.700	.867	.900	1.472
.105	.105	.305	.315	.505	.556	.705	.877	.905	1.499
.110	.110	.310	.321	.510	.563	.710	.887	.910	1.528
.115	.116	.315	.326	.515	.570	.715	.897	.915	1.557
.120	.121	.320	.332	.520	.576	.720	.908	.920	1.589
.125	.126	.325	.337	.525	.583	.725	.918	.925	1.623
.130	.131	.330	.343	.530	.590	.730	.929	.930	1.658
.135	.136	.335	.348	.535	.597	.735	.940	.935	1.697
.140	.141	.340	.354	.540	.604	.740	.950	.940	1.738
.145	.146	.345	.360	.545	.611	.745	.962	.945	1.783
.150	.151	.350	.365	.550	.618	.750	.973	.950	1.832
.155	.156	.355	.371	.555	.626	.755	.984	.955	1.886
.160	.161	.360	.377	.560	.633	.760	.996	.960	1.946
.165	.167	.365	.383	.565	.640	.765	1.008	.965	2.014
.170	.172	.370	.388	.570	.648	.770	1.020	.970	2.092
.175	.177	.375	.394	.575	.655	.775	1.033	.975	2.185
.180	.182	.380	.400	.580	.662	.780	1.045	.980	2.298
.185	.187	.385	.406	.585	.670	.785	1.058	.985	2.443
.190	.192	.390	.412	.590	.678	.790	1.071	.990	2.647
.195	.198	.395	.418	.595	.685	.795	1.085	.995	2.994

SOURCE: From *Statistical Methods,* Second Edition, by Allen L. Edwards. Copyright © 1967 by Allen L. Edwards. First edition copyright 1954 by Allen L. Edwards under the title *Statistical Methods for the Behavioral Sciences.* Reprinted by permission of Holt, Rinehart and Winston.

Table E
CRITICAL VALUES OF t

	Level of significance for one-tailed test					
	.10	.05	.025	.01	.005	.0005
	Level of significance for two-tailed test					
df	.20	.10	.05	.02	.01	.001
1	3.078	6.314	12.706	31.821	63.657	636.619
2	1.886	2.920	4.303	6.965	9.925	31.598
3	1.638	2.353	3.182	4.541	5.841	12.941
4	1.533	2.132	2.776	3.747	4.604	8.610
5	1.476	2.015	2.571	3.365	4.032	6.859
6	1.440	1.943	2.447	3.143	3.707	5.959
7	1.415	1.895	2.365	2.998	3.499	5.405
8	1.397	1.860	2.306	2.896	3.355	5.041
9	1.383	1.833	2.262	2.821	3.250	4.781
10	1.372	1.812	2.228	2.764	3.169	4.587
11	1.363	1.796	2.201	2.718	3.106	4.437
12	1.356	1.782	2.179	2.681	3.055	4.318
13	1.350	1.771	2.160	2.650	3.012	4.221
14	1.345	1.761	2.145	2.624	2.977	4.140
15	1.341	1.753	2.131	2.602	2.947	4.073
16	1.337	1.746	2.120	2.583	2.921	4.015
17	1.333	1.740	2.110	2.567	2.898	3.965
18	1.330	1.734	2.101	2.552	2.878	3.922
19	1.328	1.729	2.093	2.539	2.861	3.883
20	1.325	1.725	2.086	2.528	2.845	3.850
21	1.323	1.721	2.080	2.518	2.831	3.819
22	1.321	1.717	2.074	2.508	2.819	3.792
23	1.319	1.714	2.069	2.500	2.807	3.767
24	1.318	1.711	2.064	2.492	2.797	3.745
25	1.316	1.708	2.060	2.485	2.787	3.725
26	1.315	1.706	2.056	2.479	2.779	3.707
27	1.314	1.703	2.052	2.473	2.771	3.690
28	1.313	1.701	2.048	2.467	2.763	3.674
29	1.311	1.699	2.045	2.462	2.756	3.659
30	1.310	1.697	2.042	2.457	2.750	3.646
40	1.303	1.684	2.021	2.423	2.704	3.551
60	1.296	1.671	2.000	2.390	2.660	3.460
120	1.289	1.658	1.980	2.358	2.617	3.373
∞	1.282	1.645	1.960	2.326	2.576	3.291

t curves for $df = 1$ and $df = \infty$:

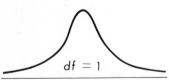

df = 1

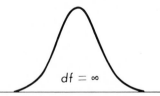

df = ∞

SOURCE: Richard P. Runyon and Audrey Haber, *Fundamentals of Behavioral Statistics,* Second Edition (1971), Reading, Mass.: Addison-Wesley Publishing Co. Taken from Table III of Fisher and Yates, *Statistical Tables for Biological, Agricultural and Medical Research,* published by Longman Group, Ltd., London (previously published by Oliver and Boyd, Ltd., Edinburgh), and by permission of the authors and publishers.

The obtained value of t is significant if it is greater than or equal to the value listed in the table.

Table F
CRITICAL VALUES OF F

The obtained F is significant at a given level if it is equal to or greater than the value shown in the table.
0.05 (light row) and 0.01 (dark row) points for the distribution of F

Degrees of freedom for greater mean square (top value = 0.05, bottom value = 0.01)

Lesser df	1	2	3	4	5	6	7	8	9	10	11	12	14	16	20	24	30	40	50	75	100	200	500	∞
1	161 / 4052	200 / 4999	216 / 5403	225 / 5625	230 / 5764	234 / 5859	237 / 5928	239 / 5981	241 / 6022	242 / 6056	243 / 6082	244 / 6106	245 / 6142	246 / 6169	248 / 6208	249 / 6234	250 / 6258	251 / 6286	252 / 6302	253 / 6323	253 / 6334	254 / 6352	254 / 6361	254 / 6366
2	18.51 / 98.49	19.00 / 99.01	19.16 / 99.17	19.25 / 99.25	19.30 / 99.30	19.33 / 99.33	19.36 / 99.34	19.37 / 99.36	19.38 / 99.38	19.39 / 99.40	19.40 / 99.41	19.41 / 99.42	19.42 / 99.43	19.43 / 99.44	19.44 / 99.45	19.45 / 99.46	19.46 / 99.47	19.47 / 99.48	19.47 / 99.48	19.48 / 99.49	19.49 / 99.49	19.49 / 99.49	19.50 / 99.50	19.50 / 99.50
3	10.13 / 34.12	9.55 / 30.81	9.28 / 29.46	9.12 / 28.71	9.01 / 28.24	8.94 / 27.91	8.88 / 27.67	8.84 / 27.49	8.81 / 27.34	8.78 / 27.23	8.76 / 27.13	8.74 / 27.05	8.71 / 26.92	8.69 / 26.83	8.66 / 26.69	8.64 / 26.60	8.62 / 26.50	8.60 / 26.41	8.58 / 26.30	8.57 / 26.27	8.56 / 26.23	8.54 / 26.18	8.54 / 26.14	8.53 / 26.12
4	7.71 / 21.20	6.94 / 18.00	6.59 / 16.69	6.39 / 15.98	6.26 / 15.52	6.16 / 15.21	6.09 / 14.98	6.04 / 14.80	6.00 / 14.66	5.96 / 14.54	5.93 / 14.45	5.91 / 14.37	5.87 / 14.24	5.84 / 14.15	5.80 / 14.02	5.77 / 13.93	5.74 / 13.83	5.71 / 13.74	5.70 / 13.69	5.68 / 13.61	5.66 / 13.57	5.65 / 13.52	5.64 / 13.48	5.63 / 13.46
5	6.61 / 16.26	5.79 / 13.27	5.41 / 12.06	5.19 / 11.39	5.05 / 10.97	4.95 / 10.67	4.88 / 10.45	4.82 / 10.27	4.78 / 10.15	4.74 / 10.05	4.70 / 9.96	4.68 / 9.89	4.64 / 9.77	4.60 / 9.68	4.56 / 9.55	4.53 / 9.47	4.50 / 9.38	4.46 / 9.29	4.44 / 9.24	4.42 / 9.17	4.40 / 9.13	4.38 / 9.07	4.37 / 9.04	4.36 / 9.02
6	5.99 / 13.74	5.14 / 10.92	4.76 / 9.78	4.53 / 9.15	4.39 / 8.75	4.28 / 8.47	4.21 / 8.26	4.15 / 8.10	4.10 / 7.98	4.06 / 7.87	4.03 / 7.79	4.00 / 7.72	3.96 / 7.60	3.92 / 7.52	3.87 / 7.39	3.84 / 7.31	3.81 / 7.23	3.77 / 7.14	3.75 / 7.09	3.72 / 7.02	3.71 / 6.99	3.69 / 6.94	3.68 / 6.90	3.67 / 6.88
7	5.59 / 12.25	4.74 / 9.55	4.35 / 8.45	4.12 / 7.85	3.97 / 7.46	3.87 / 7.19	3.79 / 7.00	3.73 / 6.84	3.68 / 6.71	3.63 / 6.62	3.60 / 6.54	3.57 / 6.47	3.52 / 6.35	3.49 / 6.27	3.44 / 6.15	3.41 / 6.07	3.38 / 5.98	3.34 / 5.90	3.32 / 5.85	3.29 / 5.78	3.28 / 5.75	3.25 / 5.70	3.24 / 5.67	3.23 / 5.65
8	5.32 / 11.26	4.46 / 8.65	4.07 / 7.59	3.84 / 7.01	3.69 / 6.63	3.58 / 6.37	3.50 / 6.19	3.44 / 6.03	3.39 / 5.91	3.34 / 5.82	3.31 / 5.74	3.28 / 5.67	3.23 / 5.56	3.20 / 5.48	3.15 / 5.36	3.12 / 5.28	3.08 / 5.20	3.05 / 5.11	3.03 / 5.06	3.00 / 5.00	2.98 / 4.96	2.96 / 4.91	2.94 / 4.88	2.93 / 4.86
9	5.12 / 10.56	4.26 / 8.02	3.86 / 6.99	3.63 / 6.42	3.48 / 6.06	3.37 / 5.80	3.29 / 5.62	3.23 / 5.47	3.18 / 5.35	3.13 / 5.26	3.10 / 5.18	3.07 / 5.11	3.02 / 5.00	2.98 / 4.92	2.93 / 4.80	2.90 / 4.73	2.86 / 4.64	2.82 / 4.56	2.80 / 4.51	2.77 / 4.45	2.76 / 4.41	2.73 / 4.36	2.72 / 4.33	2.71 / 4.31
10	4.96 / 10.04	4.10 / 7.56	3.71 / 6.55	3.48 / 5.99	3.33 / 5.64	3.22 / 5.39	3.14 / 5.21	3.07 / 5.06	3.02 / 4.95	2.97 / 4.85	2.94 / 4.78	2.91 / 4.71	2.86 / 4.60	2.82 / 4.52	2.77 / 4.41	2.74 / 4.33	2.70 / 4.25	2.67 / 4.17	2.64 / 4.12	2.61 / 4.05	2.59 / 4.01	2.56 / 3.96	2.55 / 3.93	2.54 / 3.91
11	4.84 / 9.65	3.98 / 7.20	3.59 / 6.22	3.36 / 5.67	3.20 / 5.32	3.09 / 5.07	3.01 / 4.88	2.95 / 4.74	2.90 / 4.63	2.86 / 4.54	2.82 / 4.46	2.79 / 4.40	2.74 / 4.29	2.70 / 4.21	2.65 / 4.10	2.61 / 4.02	2.57 / 3.94	2.53 / 3.86	2.50 / 3.80	2.47 / 3.74	2.45 / 3.70	2.42 / 3.66	2.41 / 3.62	2.40 / 3.60
12	4.75 / 9.33	3.88 / 6.93	3.49 / 5.95	3.26 / 5.41	3.11 / 5.06	3.00 / 4.82	2.92 / 4.65	2.85 / 4.50	2.80 / 4.39	2.76 / 4.30	2.72 / 4.22	2.69 / 4.16	2.64 / 4.05	2.60 / 3.98	2.54 / 3.86	2.50 / 3.78	2.46 / 3.70	2.42 / 3.61	2.40 / 3.56	2.36 / 3.49	2.35 / 3.46	2.32 / 3.41	2.31 / 3.38	2.30 / 3.36
13	4.67 / 9.07	3.80 / 6.70	3.41 / 5.74	3.18 / 5.20	3.02 / 4.86	2.92 / 4.62	2.84 / 4.44	2.77 / 4.30	2.72 / 4.19	2.67 / 4.10	2.63 / 4.02	2.60 / 3.96	2.55 / 3.85	2.51 / 3.78	2.46 / 3.67	2.42 / 3.59	2.38 / 3.51	2.34 / 3.42	2.32 / 3.37	2.28 / 3.30	2.26 / 3.27	2.24 / 3.21	2.22 / 3.18	2.21 / 3.16
14	4.60 / 8.86	3.74 / 6.51	3.34 / 5.56	3.11 / 5.03	2.96 / 4.69	2.85 / 4.46	2.77 / 4.28	2.70 / 4.14	2.65 / 4.03	2.60 / 3.94	2.56 / 3.86	2.53 / 3.80	2.48 / 3.70	2.44 / 3.62	2.39 / 3.51	2.35 / 3.43	2.31 / 3.34	2.27 / 3.26	2.24 / 3.21	2.21 / 3.14	2.19 / 3.11	2.16 / 3.06	2.14 / 3.02	2.13 / 3.00
15	4.54 / 8.68	3.68 / 6.36	3.29 / 5.42	3.06 / 4.89	2.90 / 4.56	2.79 / 4.32	2.70 / 4.14	2.64 / 4.00	2.59 / 3.89	2.55 / 3.80	2.51 / 3.73	2.48 / 3.67	2.43 / 3.56	2.39 / 3.48	2.33 / 3.36	2.29 / 3.29	2.25 / 3.20	2.21 / 3.12	2.18 / 3.07	2.15 / 3.00	2.12 / 2.97	2.10 / 2.92	2.08 / 2.89	2.07 / 2.87

Degrees of freedom for lesser mean square

SOURCE: Richard P. Runyon and Audrey Haber, *Fundamentals of Behavioral Statistics,* Second Edition (1971), Reading, Mass.: Addison-Wesley Publishing Co. Taken from G. W. Snedecor and William G. Cochran, *Statistical Methods,* 6th edition, Ames, Iowa: Iowa State University Press © 1967.

The obtained F is significant at a given level if it is equal to or greater than the value shown in the table.
0.05 (light row) and 0.01 (dark row) points for the distribution of F

Degrees of freedom for greater mean square

df	1	2	3	4	5	6	7	8	9	10	11	12	14	16	20	24	30	40	50	75	100	200	500	∞
16	4.49 / 8.53	3.63 / 6.23	3.24 / 5.29	3.01 / 4.77	2.85 / 4.44	2.74 / 4.20	2.66 / 4.03	2.59 / 3.89	2.54 / 3.78	2.49 / 3.69	2.45 / 3.61	2.42 / 3.55	2.37 / 3.45	2.33 / 3.37	2.28 / 3.25	2.24 / 3.18	2.20 / 3.10	2.16 / 3.01	2.13 / 2.96	2.09 / 2.89	2.07 / 2.86	2.04 / 2.80	2.02 / 2.77	2.01 / 2.75
17	4.45 / 8.40	3.59 / 6.11	3.20 / 5.18	2.96 / 4.67	2.81 / 4.34	2.70 / 4.10	2.62 / 3.93	2.55 / 3.79	2.50 / 3.68	2.45 / 3.59	2.41 / 3.52	2.38 / 3.45	2.33 / 3.35	2.29 / 3.27	2.23 / 3.16	2.19 / 3.08	2.15 / 3.00	2.11 / 2.92	2.08 / 2.86	2.04 / 2.79	2.02 / 2.76	1.99 / 2.70	1.97 / 2.67	1.96 / 2.65
18	4.41 / 8.28	3.55 / 6.01	3.16 / 5.09	2.93 / 4.58	2.77 / 4.25	2.66 / 4.01	2.58 / 3.85	2.51 / 3.71	2.46 / 3.60	2.41 / 3.51	2.37 / 3.44	2.34 / 3.37	2.29 / 3.27	2.25 / 3.19	2.19 / 3.07	2.15 / 3.00	2.11 / 2.91	2.07 / 2.83	2.04 / 2.78	2.00 / 2.71	1.98 / 2.68	1.95 / 2.62	1.93 / 2.59	1.92 / 2.57
19	4.38 / 8.18	3.52 / 5.93	3.13 / 5.01	2.90 / 4.50	2.74 / 4.17	2.63 / 3.94	2.55 / 3.77	2.48 / 3.63	2.43 / 3.52	2.38 / 3.43	2.34 / 3.36	2.31 / 3.30	2.26 / 3.19	2.21 / 3.12	2.15 / 3.00	2.11 / 2.92	2.07 / 2.84	2.02 / 2.76	2.00 / 2.70	1.96 / 2.63	1.94 / 2.60	1.91 / 2.54	1.90 / 2.51	1.88 / 2.49
20	4.35 / 8.10	3.49 / 5.85	3.10 / 4.94	2.87 / 4.43	2.71 / 4.10	2.60 / 3.87	2.52 / 3.71	2.45 / 3.56	2.40 / 3.45	2.35 / 3.37	2.31 / 3.30	2.28 / 3.23	2.23 / 3.13	2.18 / 3.05	2.12 / 2.94	2.08 / 2.86	2.04 / 2.77	1.99 / 2.69	1.96 / 2.63	1.92 / 2.56	1.90 / 2.53	1.87 / 2.47	1.85 / 2.44	1.84 / 2.42
21	4.32 / 8.02	3.47 / 5.78	3.07 / 4.87	2.84 / 4.37	2.68 / 4.04	2.57 / 3.81	2.49 / 3.65	2.42 / 3.51	2.37 / 3.40	2.32 / 3.31	2.28 / 3.24	2.25 / 3.17	2.20 / 3.07	2.15 / 2.99	2.09 / 2.88	2.05 / 2.80	2.00 / 2.72	1.96 / 2.63	1.93 / 2.58	1.89 / 2.51	1.87 / 2.47	1.84 / 2.42	1.82 / 2.38	1.81 / 2.36
22	4.30 / 7.94	3.44 / 5.72	3.05 / 4.82	2.82 / 4.31	2.66 / 3.99	2.55 / 3.76	2.47 / 3.59	2.40 / 3.45	2.35 / 3.35	2.30 / 3.26	2.26 / 3.18	2.23 / 3.12	2.18 / 3.02	2.13 / 2.94	2.07 / 2.83	2.03 / 2.75	1.98 / 2.67	1.93 / 2.58	1.91 / 2.53	1.87 / 2.46	1.84 / 2.42	1.81 / 2.37	1.80 / 2.33	1.78 / 2.31
23	4.28 / 7.88	3.42 / 5.66	3.03 / 4.76	2.80 / 4.26	2.64 / 3.94	2.53 / 3.71	2.45 / 3.54	2.38 / 3.41	2.32 / 3.30	2.28 / 3.21	2.24 / 3.14	2.20 / 3.07	2.14 / 2.97	2.10 / 2.89	2.04 / 2.78	2.00 / 2.70	1.96 / 2.62	1.91 / 2.53	1.88 / 2.48	1.84 / 2.41	1.82 / 2.37	1.79 / 2.32	1.77 / 2.28	1.76 / 2.26
24	4.26 / 7.82	3.40 / 5.61	3.01 / 4.72	2.78 / 4.22	2.62 / 3.90	2.51 / 3.67	2.43 / 3.50	2.36 / 3.36	2.30 / 3.25	2.26 / 3.17	2.22 / 3.09	2.18 / 3.03	2.13 / 2.93	2.09 / 2.85	2.02 / 2.74	1.98 / 2.66	1.94 / 2.58	1.89 / 2.49	1.86 / 2.44	1.82 / 2.36	1.80 / 2.33	1.76 / 2.27	1.74 / 2.23	1.73 / 2.21
25	4.24 / 7.77	3.38 / 5.57	2.99 / 4.68	2.76 / 4.18	2.60 / 3.86	2.49 / 3.63	2.41 / 3.46	2.34 / 3.32	2.28 / 3.21	2.24 / 3.13	2.20 / 3.05	2.16 / 2.99	2.11 / 2.89	2.06 / 2.81	2.00 / 2.70	1.96 / 2.62	1.92 / 2.54	1.87 / 2.45	1.84 / 2.40	1.80 / 2.32	1.77 / 2.29	1.74 / 2.23	1.72 / 2.19	1.71 / 2.17
26	4.22 / 7.72	3.37 / 5.53	2.98 / 4.64	2.74 / 4.14	2.59 / 3.82	2.47 / 3.59	2.39 / 3.42	2.32 / 3.29	2.27 / 3.17	2.22 / 3.09	2.18 / 3.02	2.15 / 2.96	2.10 / 2.86	2.05 / 2.77	1.99 / 2.66	1.95 / 2.58	1.90 / 2.50	1.85 / 2.41	1.82 / 2.36	1.78 / 2.28	1.76 / 2.25	1.72 / 2.19	1.70 / 2.15	1.69 / 2.13
27	4.21 / 7.68	3.35 / 5.49	2.96 / 4.60	2.73 / 4.11	2.57 / 3.79	2.46 / 3.56	2.37 / 3.39	2.30 / 3.26	2.25 / 3.14	2.20 / 3.06	2.16 / 2.98	2.13 / 2.93	2.08 / 2.83	2.03 / 2.74	1.97 / 2.63	1.93 / 2.55	1.88 / 2.47	1.84 / 2.38	1.80 / 2.33	1.76 / 2.25	1.74 / 2.21	1.71 / 2.16	1.68 / 2.12	1.67 / 2.10
28	4.20 / 7.64	3.34 / 5.45	2.95 / 4.57	2.71 / 4.07	2.56 / 3.76	2.44 / 3.53	2.36 / 3.36	2.29 / 3.23	3.24 / 3.11	2.19 / 3.03	2.15 / 2.95	2.12 / 2.90	2.06 / 2.80	2.02 / 2.71	1.96 / 2.60	1.91 / 2.52	1.87 / 2.44	1.81 / 2.35	1.78 / 2.30	1.75 / 2.22	1.72 / 2.18	1.69 / 2.13	1.67 / 2.09	1.65 / 2.06
29	4.18 / 7.60	3.33 / 5.42	2.93 / 4.54	2.70 / 4.04	2.54 / 3.73	2.43 / 3.50	2.35 / 3.33	2.28 / 3.20	2.22 / 3.08	2.18 / 3.00	2.14 / 2.92	2.10 / 2.87	2.05 / 2.77	2.00 / 2.68	1.94 / 2.57	1.90 / 2.49	1.85 / 2.41	1.80 / 2.32	1.77 / 2.27	1.73 / 2.19	1.71 / 2.15	1.68 / 2.10	1.65 / 2.06	1.64 / 2.03
30	4.17 / 7.56	3.32 / 5.39	2.92 / 4.51	2.69 / 4.02	2.53 / 3.70	2.42 / 3.47	2.34 / 3.30	2.27 / 3.17	2.21 / 3.06	2.16 / 2.98	2.12 / 2.90	2.09 / 2.84	2.04 / 2.74	1.99 / 2.66	1.93 / 2.55	1.89 / 2.47	1.84 / 2.38	1.79 / 2.29	1.76 / 2.24	1.72 / 2.16	1.69 / 2.13	1.66 / 2.07	1.64 / 2.03	1.62 / 2.01

Degrees of freedom for lesser mean square

(continued)

0.05 (light row) and 0.01 (dark row) points for the distribution of F

Degrees of freedom for greater mean square

Each cell lists the 0.05 point (light/upper value) and the 0.01 point (dark/lower value).

df (lesser)	1	2	3	4	5	6	7	8	9	10	11	12	14	16	20	24	30	40	50	75	100	200	500	∞
32	4.15 / 7.50	3.30 / 5.34	2.90 / 4.46	2.67 / 3.97	2.51 / 3.66	2.40 / 3.42	2.32 / 3.25	2.25 / 3.12	2.19 / 3.01	2.14 / 2.94	2.10 / 2.86	2.07 / 2.80	2.02 / 2.70	1.97 / 2.62	1.91 / 2.51	1.86 / 2.42	1.82 / 2.34	1.76 / 2.25	1.74 / 2.20	1.69 / 2.12	1.67 / 2.08	1.64 / 2.02	1.61 / 1.98	1.59 / 1.96
34	4.13 / 7.44	3.28 / 5.29	2.88 / 4.42	2.65 / 3.93	2.49 / 3.61	2.38 / 3.38	2.30 / 3.21	2.23 / 3.08	2.17 / 2.97	2.12 / 2.89	2.08 / 2.82	2.05 / 2.76	2.00 / 2.66	1.95 / 2.58	1.89 / 2.47	1.84 / 2.38	1.80 / 2.30	1.74 / 2.21	1.71 / 2.15	1.67 / 2.08	1.64 / 2.04	1.61 / 1.98	1.59 / 1.94	1.57 / 1.91
36	4.11 / 7.39	3.26 / 5.25	2.86 / 4.38	2.63 / 3.89	2.48 / 3.58	2.36 / 3.35	2.28 / 3.18	2.21 / 3.04	2.15 / 2.94	2.10 / 2.86	2.06 / 2.78	2.03 / 2.72	1.98 / 2.62	1.93 / 2.54	1.87 / 2.43	1.82 / 2.35	1.78 / 2.26	1.72 / 2.17	1.69 / 2.12	1.65 / 2.04	1.62 / 2.00	1.59 / 1.94	1.56 / 1.90	1.55 / 1.87
38	4.10 / 7.35	3.25 / 5.21	2.85 / 4.34	2.62 / 3.86	2.46 / 3.54	2.35 / 3.32	2.26 / 3.15	2.19 / 3.02	2.14 / 2.91	2.09 / 2.82	2.05 / 2.75	2.02 / 2.69	1.96 / 2.59	1.92 / 2.51	1.85 / 2.40	1.80 / 2.32	1.76 / 2.22	1.71 / 2.14	1.67 / 2.08	1.63 / 2.00	1.60 / 1.97	1.57 / 1.90	1.54 / 1.86	1.53 / 1.84
40	4.08 / 7.31	3.23 / 5.18	2.84 / 4.31	2.61 / 3.83	2.45 / 3.51	2.34 / 3.29	2.25 / 3.12	2.18 / 2.99	2.12 / 2.88	2.07 / 2.80	2.04 / 2.73	2.00 / 2.66	1.95 / 2.56	1.90 / 2.49	1.84 / 2.37	1.79 / 2.29	1.74 / 2.20	1.69 / 2.11	1.66 / 2.05	1.61 / 1.97	1.59 / 1.94	1.55 / 1.88	1.53 / 1.84	1.51 / 1.81
42	4.07 / 7.27	3.22 / 5.15	2.83 / 4.29	2.59 / 3.80	2.44 / 3.49	2.32 / 3.26	2.24 / 3.10	2.17 / 2.96	2.11 / 2.86	2.06 / 2.77	2.02 / 2.70	1.99 / 2.64	1.94 / 2.54	1.89 / 2.46	1.82 / 2.35	1.78 / 2.26	1.73 / 2.17	1.68 / 2.08	1.64 / 2.02	1.60 / 1.94	1.57 / 1.91	1.54 / 1.85	1.51 / 1.80	1.49 / 1.78
44	4.06 / 7.24	3.21 / 5.12	2.82 / 4.26	2.58 / 3.78	2.43 / 3.46	2.31 / 3.24	2.23 / 3.07	2.16 / 2.94	2.10 / 2.84	2.05 / 2.75	2.01 / 2.68	1.98 / 2.62	1.92 / 2.52	1.88 / 2.44	1.81 / 2.32	1.76 / 2.24	1.72 / 2.15	1.66 / 2.06	1.63 / 2.00	1.58 / 1.92	1.56 / 1.88	1.52 / 1.82	1.50 / 1.78	1.48 / 1.75
46	4.05 / 7.21	3.20 / 5.10	2.81 / 4.24	2.57 / 3.76	2.42 / 3.44	2.30 / 3.22	2.22 / 3.05	2.14 / 2.92	2.09 / 2.82	2.04 / 2.73	2.00 / 2.66	1.97 / 2.60	1.91 / 2.50	1.87 / 2.42	1.80 / 2.30	1.75 / 2.22	1.71 / 2.13	1.65 / 2.04	1.62 / 1.98	1.57 / 1.90	1.54 / 1.86	1.51 / 1.80	1.48 / 1.76	1.46 / 1.72
48	4.04 / 7.19	3.19 / 5.08	2.80 / 4.22	2.56 / 3.74	2.41 / 3.42	2.30 / 3.20	2.21 / 3.04	2.14 / 2.90	2.08 / 2.80	2.03 / 2.71	1.99 / 2.64	1.96 / 2.58	1.90 / 2.48	1.86 / 2.40	1.79 / 2.28	1.74 / 2.20	1.70 / 2.11	1.64 / 2.02	1.61 / 1.96	1.56 / 1.88	1.53 / 1.84	1.50 / 1.78	1.47 / 1.73	1.45 / 1.70
50	4.03 / 7.17	3.18 / 5.06	2.79 / 4.20	2.56 / 3.72	2.40 / 3.41	2.29 / 3.18	2.20 / 3.02	2.13 / 2.88	2.07 / 2.78	2.02 / 2.70	1.98 / 2.62	1.95 / 2.56	1.90 / 2.46	1.85 / 2.39	1.78 / 2.26	1.74 / 2.18	1.69 / 2.10	1.63 / 2.00	1.60 / 1.94	1.55 / 1.86	1.52 / 1.82	1.48 / 1.76	1.46 / 1.71	1.44 / 1.68
55	4.02 / 7.12	3.17 / 5.01	2.78 / 4.16	2.54 / 3.68	2.38 / 3.37	2.27 / 3.15	2.18 / 2.98	2.11 / 2.85	2.05 / 2.75	2.00 / 2.66	1.97 / 2.59	1.93 / 2.53	1.88 / 2.43	1.83 / 2.35	1.76 / 2.23	1.72 / 2.15	1.67 / 2.06	1.61 / 1.96	1.58 / 1.90	1.52 / 1.82	1.50 / 1.78	1.46 / 1.71	1.43 / 1.66	1.41 / 1.64
60	4.00 / 7.08	3.15 / 4.98	2.76 / 4.13	2.52 / 3.65	2.37 / 3.34	2.25 / 3.12	2.17 / 2.95	2.10 / 2.82	2.04 / 2.72	1.99 / 2.63	1.95 / 2.56	1.92 / 2.50	1.86 / 2.40	1.81 / 2.32	1.75 / 2.20	1.70 / 2.12	1.65 / 2.03	1.59 / 1.93	1.56 / 1.87	1.50 / 1.79	1.48 / 1.74	1.44 / 1.68	1.41 / 1.63	1.39 / 1.60
65	3.99 / 7.04	3.14 / 4.95	2.75 / 4.10	2.51 / 3.62	2.36 / 3.31	2.24 / 3.09	2.15 / 2.93	2.08 / 2.79	2.02 / 2.70	1.98 / 2.61	1.94 / 2.54	1.90 / 2.47	1.85 / 2.37	1.80 / 2.30	1.73 / 2.18	1.68 / 2.09	1.63 / 2.00	1.57 / 1.90	1.54 / 1.84	1.49 / 1.76	1.46 / 1.71	1.42 / 1.64	1.39 / 1.60	1.37 / 1.56
70	3.98 / 7.01	3.13 / 4.92	2.74 / 4.08	2.50 / 3.60	2.35 / 3.29	2.23 / 3.07	2.14 / 2.91	2.07 / 2.77	2.01 / 2.67	1.97 / 2.59	1.93 / 2.51	1.89 / 2.45	1.84 / 2.35	1.79 / 2.28	1.72 / 2.15	1.67 / 2.07	1.62 / 1.98	1.56 / 1.88	1.53 / 1.82	1.47 / 1.74	1.45 / 1.69	1.40 / 1.62	1.37 / 1.56	1.35 / 1.53
80	3.96 / 6.96	3.11 / 4.88	2.72 / 4.04	2.48 / 3.56	2.33 / 3.25	2.21 / 3.04	2.12 / 2.87	2.05 / 2.74	1.99 / 2.64	1.95 / 2.55	1.91 / 2.48	1.88 / 2.41	1.82 / 2.32	1.77 / 2.24	1.70 / 2.11	1.65 / 2.03	1.60 / 1.94	1.54 / 1.84	1.51 / 1.78	1.45 / 1.70	1.42 / 1.65	1.38 / 1.57	1.35 / 1.52	1.32 / 1.49

Degrees of freedom for lesser mean square

0.05 (light row) and 0.01 (dark row) points for the distribution of F

Degrees of freedom for greater mean square

	1	2	3	4	5	6	7	8	9	10	11	12	14	16	20	24	30	40	50	75	100	200	500	∞
100	3.94 / 6.90	3.09 / 4.82	2.70 / 3.98	2.46 / 3.51	2.30 / 3.20	2.19 / 2.99	2.10 / 2.82	2.03 / 2.69	1.97 / 2.59	1.92 / 2.51	1.88 / 2.43	1.85 / 2.36	1.79 / 2.26	1.75 / 2.19	1.68 / 2.06	1.63 / 1.98	1.57 / 1.89	1.51 / 1.79	1.48 / 1.73	1.42 / 1.64	1.39 / 1.59	1.34 / 1.51	1.30 / 1.46	1.28 / 1.43
125	3.92 / 6.84	3.07 / 4.78	2.68 / 3.94	2.44 / 3.47	2.29 / 3.17	2.17 / 2.95	2.08 / 2.79	2.01 / 2.65	1.95 / 2.56	1.90 / 2.47	1.86 / 2.40	1.83 / 2.33	1.77 / 2.23	1.72 / 2.15	1.65 / 2.03	1.60 / 1.94	1.55 / 1.85	1.49 / 1.75	1.45 / 1.68	1.39 / 1.59	1.36 / 1.54	1.31 / 1.46	1.27 / 1.40	1.25 / 1.37
150	3.91 / 6.81	3.06 / 4.75	2.67 / 3.91	2.43 / 3.44	2.27 / 3.13	2.16 / 2.92	2.07 / 2.76	2.00 / 2.62	1.94 / 2.53	1.89 / 2.44	1.85 / 2.37	1.82 / 2.30	1.76 / 2.20	1.71 / 2.12	1.64 / 2.00	1.59 / 1.91	1.54 / 1.83	1.47 / 1.72	1.44 / 1.66	1.37 / 1.56	1.34 / 1.51	1.29 / 1.43	1.25 / 1.37	1.22 / 1.33
200	3.89 / 6.76	3.04 / 4.71	2.65 / 3.88	2.41 / 3.41	2.26 / 3.11	2.14 / 2.90	2.05 / 2.73	1.98 / 2.60	1.92 / 2.50	1.87 / 2.41	1.83 / 2.34	1.80 / 2.28	1.74 / 2.17	1.69 / 2.09	1.62 / 1.97	1.57 / 1.88	1.52 / 1.79	1.45 / 1.69	1.42 / 1.62	1.35 / 1.53	1.32 / 1.48	1.26 / 1.39	1.22 / 1.33	1.19 / 1.28
400	3.86 / 6.70	3.02 / 4.66	2.62 / 3.83	2.39 / 3.36	2.23 / 3.06	2.12 / 2.85	2.03 / 2.69	1.96 / 2.55	1.90 / 2.46	1.85 / 2.37	1.81 / 2.29	1.78 / 2.23	1.72 / 2.12	1.67 / 2.04	1.60 / 1.92	1.54 / 1.84	1.49 / 1.74	1.42 / 1.64	1.38 / 1.57	1.32 / 1.47	1.28 / 1.42	1.22 / 1.32	1.16 / 1.24	1.13 / 1.19
1000	3.85 / 6.66	3.00 / 4.62	2.61 / 3.80	2.38 / 3.34	2.22 / 3.04	2.10 / 2.82	2.02 / 2.66	1.95 / 2.53	1.89 / 2.43	1.84 / 2.34	1.80 / 2.26	1.76 / 2.20	1.70 / 2.09	1.65 / 2.01	1.58 / 1.89	1.53 / 1.81	1.47 / 1.71	1.41 / 1.61	1.36 / 1.54	1.30 / 1.44	1.26 / 1.38	1.19 / 1.28	1.13 / 1.19	1.08 / 1.11
∞	3.84 / 6.64	2.99 / 4.60	2.60 / 3.78	2.37 / 3.32	2.21 / 3.02	2.09 / 2.80	2.01 / 2.64	1.94 / 2.51	1.88 / 2.41	1.83 / 2.32	1.79 / 2.24	1.75 / 2.18	1.69 / 2.07	1.64 / 1.99	1.57 / 1.87	1.52 / 1.79	1.46 / 1.69	1.40 / 1.59	1.35 / 1.52	1.28 / 1.41	1.24 / 1.36	1.17 / 1.25	1.11 / 1.15	1.00 / 1.00

Degrees of freedom for lesser mean square

Table G

CRITICAL VALUES OF χ^2

	Level of significance for a directional test					
	.10	.05	.025	.01	.005	.0005
	Level of significance for a non-directional test					
df	.20	.10	.05	.02	.01	.001
1	1.64	2.71	3.84	5.41	6.64	10.83
2	3.22	4.60	5.99	7.82	9.21	13.82
3	4.64	6.25	7.82	9.84	11.34	16.27
4	5.99	7.78	9.49	11.67	13.28	18.46
5	7.29	9.24	11.07	13.39	15.09	20.52
6	8.56	10.64	12.59	15.03	16.81	22.46
7	9.80	12.02	14.07	16.62	18.48	24.32
8	11.03	13.36	15.51	18.17	20.09	26.12
9	12.24	14.68	16.92	19.68	21.67	27.88
10	13.44	15.99	18.31	21.16	23.21	29.59
11	14.63	17.28	19.68	22.62	24.72	31.26
12	15.81	18.55	21.03	24.05	26.22	32.91
13	16.98	19.81	22.36	25.47	27.69	34.53
14	18.15	21.06	23.68	26.87	29.14	36.12
15	19.31	22.31	25.00	28.26	30.58	37.70
16	20.46	23.54	26.30	29.63	32.00	39.29
17	21.62	24.77	27.59	31.00	33.41	40.75
18	22.76	25.99	28.87	32.35	34.80	42.31
19	23.90	27.20	30.14	33.69	36.19	43.82
20	25.04	28.41	31.41	35.02	37.57	45.32
21	26.17	29.62	32.67	36.34	38.93	46.80
22	27.30	30.81	33.92	37.66	40.29	48.27
23	28.43	32.01	35.17	38.97	41.64	49.73
24	29.55	33.20	36.42	40.27	42.98	51.18
25	30.68	34.38	37.65	41.57	44.31	52.62
26	31.80	35.56	38.88	42.86	45.64	54.05
27	32.91	36.74	40.11	44.14	46.96	55.48
28	34.03	37.92	41.34	45.42	48.28	56.89
29	35.14	39.09	42.69	46.69	49.59	58.30
30	36.25	40.26	43.77	47.96	50.89	59.70
32	38.47	42.59	46.19	50.49	53.49	62.49
34	40.68	44.90	48.60	53.00	56.06	65.25
36	42.88	47.21	51.00	55.49	58.62	67.99
38	45.08	49.51	53.38	57.97	61.16	70.70
40	47.27	51.81	55.76	60.44	63.69	73.40
44	51.64	56.37	60.48	65.34	68.71	78.75
48	55.99	60.91	65.17	70.20	73.68	84.04
52	60.33	65.42	69.83	75.02	78.62	89.27
56	64.66	69.92	74.47	79.82	83.51	94.46
60	68.97	74.40	79.08	84.58	88.38	99.61

SOURCE: Table G is taken from Table IV of Fisher and Yates, *Statistical Tables for Biological, Agricultural and Medical Research,* published by Longman Group, Ltd., London (previously published by Oliver and Boyd, Ltd., Edinburgh), and by permission of the authors and publishers.

The obtained value of χ^2 is significant if it is greater than or equal to the value listed in the table.

Table H

PROBABILITIES ASSOCIATED WITH VALUES AS SMALL AS OBSERVED VALUES OF *g* IN THE BINOMIAL TEST AND SIGN TEST

This table is valid only when $P = Q = 0.5$. The decimal points have been omitted; each p value is to three decimal places.

g → N ↓	0	1	2	3	4	5	6	7	8	9	10
5	031	188	500								
6	016	109	344	656							
7	008	062	227	500							
8	004	035	145	363	637						
9	002	020	090	254	500						
10	001	011	055	172	377	623					
11		006	033	113	274	500					
12		003	019	073	194	387	613				
13		002	011	046	133	291	500				
14		001	006	029	090	212	395	605			
15			004	018	059	151	304	500			
16			002	011	038	105	227	402	598		
17			001	006	025	072	166	315	500		
18			001	004	015	048	119	240	407	593	
19				002	010	032	084	180	324	500	
20				001	006	021	058	132	252	412	588

See note below.

All values in this zone (N = 14–20, left columns) are 000 or very nearly so.

0.01 level—reject H_0 0.05 level—reject H_0 retain H_0?

If *g* value for any N is too large, you have chosen the wrong value as *g*; interchange *g* with $(N - g)$. This is not permissible, however, if the hypothesis predicted that *g* would be smaller than $(N - g)$ and it turned out not to be.

From *Statistical Inference* by Helen M. Walker and Joseph Lev. Copyright, 1953, by Holt, Rinehart and Winston, Inc. Reprinted by permission of Holt, Rinehart and Winston.

Table I

CRITICAL VALUES OF U

n_A	3	4	5	6	7	8	9	10	11	12	13	14	15	16	17	18	19	20
n_B	—	—	0	1	1	2	2	3	3	4	4	5	5	6	6	7	7	8
3	—	—	—	—	—	—	0	0	0	1	1	1	2	2	2	2	3	3
4	—	0	1	2	3	4	4	5	6	7	8	9	10	11	11	12	13	14
	—	—	—	0	0	1	1	2	2	3	3	4	5	5	6	6	7	8
5			2	3	5	6	7	8	9	11	12	13	14	15	17	18	19	20
			0	1	1	2	3	4	5	6	7	7	8	9	10	11	12	13
6				5	6	8	10	11	13	14	16	17	19	21	22	24	25	27
				2	3	4	5	6	7	9	10	11	12	13	15	16	17	18
7					8	10	12	14	16	18	20	22	24	26	28	30	32	34
					4	6	7	9	10	12	13	15	16	18	19	21	22	24
8						13	15	17	19	22	24	26	29	31	34	36	38	41
						7	9	11	13	15	17	18	20	22	24	26	28	30
9							17	20	23	26	28	31	34	37	39	42	45	48
							11	13	16	18	20	22	24	27	29	31	33	36
10								23	26	29	33	36	39	42	45	48	52	55
								16	18	21	24	26	29	31	34	37	39	42
11									30	33	37	40	44	47	51	55	58	62
									21	24	27	30	33	36	39	42	45	48
12										37	41	45	49	53	57	61	65	69
										27	31	34	37	41	44	47	51	54
13											45	50	54	59	63	67	72	76
											34	38	42	45	49	53	56	60
14												55	59	64	67	74	78	83
												42	46	50	54	58	63	67
15													64	70	75	80	85	90
													51	55	60	64	69	73
16														75	81	86	92	98
														60	65	70	74	79
17															87	93	99	105
															70	75	81	86
18																99	106	112
																81	87	92
19																	113	119
																	93	99
20																		127
																		105

The dashes indicate that no decision is possible at the stated value of alpha.

This table gives the critical values of U for the 0.05 level (upper figures) and 0.01 level (lower figures) with a two-tailed test. For a one-tailed test, these probabilities are halved.

Table I is adapted and condensed from H. B. Mann and D. R. Whitney, On a test of whether one or two random variables is stochastically larger than the other. *Annals of Mathematical Statistics* (1947), **18**, 50–60.

The obtained value of U is *not* significant if it falls between the two values listed in the table; otherwise, it is significant.

Table J

TABLE OF SQUARES AND SQUARE ROOTS

N	N^2	$\sqrt{N}$	$1/N$	N	N^2	$\sqrt{N}$	$1/N$
1	1	1.0000	1.000000	41	1681	6.4031	.024390
2	4	1.4142	.500000	42	1764	6.4807	.023810
3	9	1.7321	.333333	43	1849	6.5574	.023256
4	16	2.0000	.250000	44	1936	6.6332	.022727
5	25	2.2361	.200000	45	2025	6.7082	.022222
6	36	2.4495	.166667	46	2116	6.7823	.021739
7	49	2.6458	.142857	47	2209	6.8557	.021277
8	64	2.8284	.125000	48	2304	6.9282	.020833
9	81	3.0000	.111111	49	2401	7.0000	.020408
10	100	3.1623	.100000	50	2500	7.0711	.020000
11	121	3.3166	.090909	51	2601	7.1414	.019608
12	144	3.4641	.083333	52	2704	7.2111	.019231
13	169	3.6056	.076923	53	2809	7.2801	.018868
14	196	3.7417	.071429	54	2916	7.3485	.018519
15	225	3.8730	.066667	55	3025	7.4162	.018182
16	256	4.0000	.062500	56	3136	7.4833	.017857
17	289	4.1231	.058824	57	3249	7.5498	.017544
18	324	4.2426	.055556	58	3364	7.6158	.017241
19	361	4.3589	.052632	59	3481	7.6811	.016949
20	400	4.4721	.050000	60	3600	7.7460	.016667
21	441	4.5826	.047619	61	3721	7.8102	.016393
22	484	4.6904	.045455	62	3844	7.8740	.016129
23	529	4.7958	.043478	63	3969	7.9373	.015873
24	576	4.8990	.041667	64	4096	8.0000	.015625
25	625	5.0000	.040000	65	4225	8.0623	.015385
26	676	5.0990	.038462	66	4356	8.1240	.015152
27	729	5.1962	.037037	67	4489	8.1854	.014925
28	784	5.2915	.035714	68	4624	8.2462	.014706
29	841	5.3852	.034483	69	4761	8.3066	.014493
30	900	5.4772	.033333	70	4900	8.3666	.014286
31	961	5.5678	.032258	71	5041	8.4261	.014085
32	1024	5.6569	.031250	72	5184	8.4853	.013889
33	1089	5.7446	.030303	73	5329	8.5440	.013699
34	1156	5.8310	.029412	74	5476	8.6023	.013514
35	1225	5.9161	.028571	75	5625	8.6603	.013333
36	1296	6.0000	.027778	76	5776	8.7178	.013158
37	1369	6.0828	.027027	77	5929	8.7750	.012987
38	1444	6.1644	.026316	78	6084	8.8318	.012821
39	1521	6.2450	.025641	79	6241	8.8882	.012658
40	1600	6.3246	.025000	80	6400	8.9443	.012500

SOURCE: J. W. Dunlap and A. K. Kurtz, *Handbook of Statistical Monographs, Tables, and Formulas* (1932), New York: World Book Company, as used in A. L. Edwards, *Statistical Methods for the Behavioral Sciences* (1954), New York: Holt, Rinehart and Winston.

N	N^2	$\sqrt{N}$	$1/N$	N	N^2	$\sqrt{N}$	$1/N$
81	6561	9.0000	.012346	121	14641	11.0000	.00826446
82	6724	9.0554	.012195	122	14884	11.0454	.00819672
83	6889	9.1104	.012048	123	15129	11.0905	.00813008
84	7056	9.1652	.011905	124	15376	11.1355	.00806452
85	7225	9.2195	.011765	125	15625	11.1803	.00800000
86	7396	9.2736	.011628	126	15876	11.2250	.00793651
87	7569	9.3274	.011494	127	16129	11.2694	.00787402
88	7744	9.3808	.011364	128	16384	11.3137	.00781250
89	7921	9.4340	.011236	129	16641	11.3578	.00775194
90	8100	9.4868	.011111	130	16900	11.4018	.00769231
91	8281	9.5394	.010989	131	17161	11.4455	.00763359
92	8464	9.5917	.010870	132	17424	11.4891	.00757576
93	8649	9.6437	.010753	133	17689	11.5326	.00751880
94	8836	9.6954	.010638	134	17956	11.5758	.00746269
95	9025	9.7468	.010526	135	18225	11.6190	.00740741
96	9216	9.7980	.010417	136	18496	11.6619	.00735294
97	9409	9.8489	.010309	137	18769	11.7047	.00729927
98	9604	9.8995	.010204	138	19044	11.7473	.00724638
99	9801	9.9499	.010101	139	19321	11.7898	.00719424
100	10000	10.0000	.010000	140	19600	11.8322	.00714286
101	10201	10.0499	.00990099	141	19881	11.8743	.00709220
102	10404	10.0995	.00980392	142	20164	11.9164	.00704225
103	10609	10.1489	.00970874	143	20449	11.9583	.00699301
104	10816	10.1980	.00961538	144	20736	12.0000	.00694444
105	11025	10.2470	.00952381	145	21025	12.0416	.00689655
106	11236	10.2956	.00943396	146	21316	12.0830	.00684932
107	11449	10.3441	.00934579	147	21609	12.1244	.00680272
108	11664	10.3923	.00925926	148	21904	12.1655	.00675676
109	11881	10.4403	.00917431	149	22201	12.2066	.00671141
110	12100	10.4881	.00909091	150	22500	12.2474	.00666667
111	12321	10.5357	.00900901	151	22801	12.2882	.00662252
112	12544	10.5830	.00892857	152	23104	12.3288	.00657895
113	12769	10.6301	.00884956	153	23409	12.3693	.00653595
114	12996	10.6771	.00877193	154	23716	12.4097	.00649351
115	13225	10.7238	.00869565	155	24025	12.4499	.00645161
116	13456	10.7703	.00862069	156	24336	12.4900	.00641026
117	13689	10.8167	.00854701	157	24649	12.5300	.00636943
118	13924	10.8628	.00847458	158	24964	12.5698	.00632911
119	14161	10.9087	.00840336	159	25281	12.6095	.00628931
120	14400	10.9545	.00833333	160	25600	12.6491	.00625000

N	N^2	$\sqrt{N}$	$1/N$	N	N^2	$\sqrt{N}$	$1/N$
161	25921	12.6886	.00621118	201	40401	14.1774	.00497512
162	26244	12.7279	.00617284	202	40804	14.2127	.00495050
163	26569	12.7671	.00613497	203	41209	14.2478	.00492611
164	26896	12.8062	.00609756	204	41616	14.2829	.00490196
165	27225	12.8452	.00606061	205	42025	14.3178	.00487805
166	27556	12.8841	.00602410	206	42436	14.3527	.00485437
167	27889	12.9228	.00598802	207	42849	14.3875	.00483092
168	28224	12.9615	.00595238	208	43264	14.4222	.00480769
169	28561	13.0000	.00591716	209	43681	14.4568	.00478469
170	28900	13.0384	.00588235	210	44100	14.4914	.00476190
171	29241	13.0767	.00584795	211	44521	14.5258	.00473934
172	29584	13.1149	.00581395	212	44944	14.5602	.00471698
173	29929	13.1529	.00578035	213	45369	14.5945	.00469484
174	30276	13.1909	.00574713	214	45796	14.6287	.00467290
175	30625	13.2288	.00571429	215	46225	14.6629	.00465116
176	30976	13.2665	.00568182	216	46656	14.6969	.00462963
177	31329	13.3041	.00564972	217	47089	14.7309	.00460829
178	31684	13.3417	.00561798	218	47524	14.7648	.00458716
179	32041	13.3791	.00558659	219	47961	14.7986	.00456621
180	32400	13.4164	.00555556	220	48400	14.8324	.00454545
181	32761	13.4536	.00552486	221	48841	14.8661	.00452489
182	33124	13.4907	.00549451	222	49284	14.8997	.00450450
183	33489	13.5277	.00546448	223	49729	14.9332	.00448430
184	33856	13.5647	.00543478	224	50176	14.9666	.00446429
185	34225	13.6015	.00540541	225	50625	15.0000	.00444444
186	34596	13.6382	.00537634	226	51076	15.0333	.00442478
187	34969	13.6748	.00534759	227	51529	15.0665	.00440529
188	35344	13.7113	.00531915	228	51984	15.0997	.00438596
189	35721	13.7477	.00529101	229	52441	15.1327	.00436681
190	36100	13.7840	.00526316	230	52900	15.1658	.00434783
191	36481	13.8203	.00523560	231	53361	15.1987	.00432900
192	36864	13.8564	.00520833	232	53824	15.2315	.00431034
193	37249	13.8924	.00518135	233	54289	15.2643	.00429185
194	37636	13.9284	.00515464	234	54756	15.2971	.00427350
195	38025	13.9642	.00512821	235	55225	15.3297	.00425532
196	38416	14.0000	.00510204	236	55696	15.3623	.00423729
197	38809	14.0357	.00507614	237	56169	15.3948	.00421941
198	39204	14.0712	.00505051	238	56644	15.4272	.00420168
199	39601	14.1067	.00502513	239	57121	15.4596	.00418410
200	40000	14.1421	.00500000	240	57600	15.4919	.00416667

N	N^2	$\sqrt{N}$	$1/N$	N	N^2	$\sqrt{N}$	$1/N$
241	58081	15.5242	.00414938	281	78961	16.7631	.00355872
242	58564	15.5563	.00413223	282	79524	16.7929	.00354610
243	59049	15.5885	.00411523	283	80089	16.8226	.00353357
244	59536	15.6205	.00409836	284	80656	16.8523	.00352113
245	60025	15.6525	.00408163	285	81225	16.8819	.00350877
246	60516	15.6844	.00406504	286	81796	16.9115	.00349650
247	61009	15.7162	.00404858	287	82369	16.9411	.00348432
248	61504	15.7480	.00403226	288	82944	16.9706	.00347222
249	62001	15.7797	.00401606	289	83521	17.0000	.00346021
250	62500	15.8114	.00400000	290	84100	17 0294	.00344828
251	63001	15.8430	.00398406	291	84681	17.0587	.00343643
252	63504	15.8745	.00396825	292	85264	17.0880	.00342466
253	64009	15.9060	.00395257	293	85849	17.1172	.00341297
254	64516	15.9374	.00393701	294	86436	17.1464	.00340136
255	65025	15.9687	.00392157	295	87025	17.1756	.00338983
256	65536	16.0000	.00390625	296	87616	17.2047	.00337838
257	66049	16.0312	.00389105	297	88209	17.2337	.00336700
258	66564	16.0624	.00387597	298	88804	17.2627	.00335570
259	67081	16.0935	.00386100	299	89401	17.2916	.00334448
260	67600	16.1245	.00384615	300	90000	17.3205	.00333333
261	68121	16.1555	.00383142	301	90601	17.3494	.00332226
262	68644	16.1864	.00381679	302	91204	17.3781	.00331126
263	69169	16.2173	.00380228	303	91809	17.4069	.00330033
264	69696	16.2481	.00378788	304	92416	17.4356	.00328947
265	70225	16.2788	.00377358	305	93025	17.4642	.00327869
266	70756	16.3095	.00375940	306	93636	17.4929	.00326797
267	71289	16.3401	.00374532	307	94249	17.5214	.00325733
268	71824	16.3707	.00373134	308	94864	17.5499	.00324675
269	72361	16.4012	.00371747	309	95481	17.5784	.00323625
270	72900	16.4317	.00370370	310	96100	17.6068	.00322581
271	73441	16.4621	.00369004	311	96721	17.6352	.00321543
272	73984	16.4924	.00367647	312	97344	17.6635	.00320513
273	74529	16.5227	.00366300	313	97969	17.6918	.00319489
274	75076	16.5529	.00364964	314	98596	17.7200	.00318471
275	75625	16.5831	.00363636	315	99225	17.7482	.00317460
276	76176	16.6132	.00362319	316	99856	17.7764	.00316456
277	76729	16.6433	.00361011	317	100489	17.8045	.00315457
278	77284	16.6733	.00359712	318	101124	17.8326	.00314465
279	77841	16.7033	.00358423	319	101761	17.8606	.00313480
280	78400	16.7332	.00357143	320	102400	17.8885	.00312500

N	N²	√N	1/N	N	N²	√N	1/N
321	103041	17.9165	.00311526	361	130321	19.0000	.00277008
322	103684	17.9444	.00310559	362	131044	19.0263	.00276243
323	104329	17.9722	.00309598	363	131769	19.0526	.00275482
324	104976	18.0000	.00308642	364	132496	19.0788	.00274725
325	105625	18.0278	.00307692	365	133225	19.1050	.00273973
326	106276	18.0555	.00306748	366	133956	19.1311	.00273224
327	106929	18.0831	.00305810	367	134689	19.1572	.00272480
328	107584	18.1108	.00304878	368	135424	19.1833	.00271739
329	108241	18.1384	.00303951	369	136161	19.2094	.00271003
330	108900	18.1659	.00303030	370	136900	19.2354	.00270270
331	109561	18.1934	.00302115	371	137641	19.2614	.00269542
332	110224	18.2209	.00301205	372	138384	19.2873	.00268817
333	110889	18.2483	.00300300	373	139129	19.3132	.00268097
334	111556	18.2757	.00299401	374	139876	19.3391	.00267380
335	112225	18.3030	.00298507	375	140625	19.3649	.00266667
336	112896	18.3303	.00297619	376	141376	19.3907	.00265957
337	113569	18.3576	.00296736	377	142129	19.4165	.00265252
338	114244	18.3848	.00295858	378	142884	19.4422	.00264550
339	114921	18.4120	.00294985	379	143641	19.4679	.00263852
340	115600	18.4391	.00294118	380	144400	19.4936	.00263158
341	116281	18.4662	.00293255	381	145161	19.5192	.00262467
342	116964	18.4932	.00292398	382	145924	19.5448	.00261780
343	117649	18.5203	.00291545	383	146689	19.5704	.00261097
344	118336	18.5472	.00290698	384	147456	19.5959	.00260417
345	119025	18.5742	.00289855	385	148225	19.6214	.00259740
346	119716	18.6011	.00289017	386	148996	19.6469	.00259067
347	120409	18.6279	.00288184	387	149769	19.6723	.00258398
348	121104	18.6548	.00287356	388	150544	19.6977	.00257732
349	121801	18.6815	.00286533	389	151321	19.7231	.00257069
350	122500	18.7083	.00285714	390	152100	19.7484	.00256410
351	123201	18.7350	.00284900	391	152881	19.7737	.00255754
352	123904	18.7617	.00284091	392	153664	19.7990	.00255102
353	124609	18.7883	.00283286	393	154449	19.8242	.00254453
354	125316	18.8149	.00282486	394	155236	19.8494	.00253807
355	126025	18.8414	.00281690	395	156025	19.8746	.00253165
356	126736	18.8680	.00280899	396	156816	19.8997	.00252525
357	127449	18.8944	.00280112	397	157609	19.9249	.00251889
358	128164	18.9209	.00279330	398	158404	19.9499	.00251256
359	128881	18.9473	.00278552	399	159201	19.9750	.00250627
360	129000	18.9737	.00277778	400	160000	20.0000	.00250000

N	N²	$\sqrt{N}$	1/N	N	N²	$\sqrt{N}$	1/N
401	160801	20.0250	.00249377	441	194481	21.0000	.00226757
402	161604	20.0499	.00248756	442	195364	21.0238	.00226244
403	162409	20.0749	.00248139	443	196249	21.0476	.00225734
404	163216	20.0998	.00247525	444	197136	21.0713	.00225225
405	164025	20.1246	.00246914	445	198025	21.0950	.00224719
406	164836	20.1494	.00246305	446	198916	21.1187	.00224215
407	165649	20.1742	.00245700	447	199809	21.1424	.00223714
408	166464	20.1990	.00245098	448	200704	21.1660	.00223214
409	167281	20.2237	.00244499	449	201601	21.1896	.00222717
410	168100	20.2485	.00243902	450	202500	21.2132	.00222222
411	168921	20.2731	.00243309	451	203401	21.2368	.00221729
412	169744	20.2978	.00242718	452	204304	21.2603	.00221239
413	170569	20.3224	.00242131	453	205209	21.2838	.00220751
414	171396	20.3470	.00241546	454	206116	21.3073	.00220264
415	172225	20.3715	.00240964	455	207025	21.3307	.00219780
416	173056	20.3961	.00240385	456	207936	21.3542	.00219298
417	173889	20.4206	.00239808	457	208849	21.3776	.00218818
418	174724	20.4450	.00239234	458	209764	21.4009	.00218341
419	175561	20.4695	.00238663	459	210681	21.4243	.00217865
420	176400	20.4939	.00238095	460	211600	21.4476	.00217391
421	177241	20.5183	.00237530	461	212521	21.4709	.00216920
422	178084	20.5426	.00236967	462	213444	21.4942	.00216450
423	178929	20.5670	.00236407	463	214369	21.5174	.00215983
424	179776	20.5913	.00235849	464	215296	21.5407	.00215517
425	180625	20.6155	.00235294	465	216225	21.5639	.00215054
426	181476	20.6398	.00234742	466	217156	21.5870	.00214592
427	182329	20.6640	.00234192	467	218089	21.6102	.00214133
428	183184	20.6882	.00233645	468	219024	21.6333	.00213675
429	184041	20.7123	.00233100	469	219961	21.6564	.00213220
430	184900	20.7364	.00232558	470	220900	21.6795	.00212766
431	185761	20.7605	.00232019	471	221841	21.7025	.00212314
432	186624	20.7846	.00231481	472	222784	21.7256	.00211864
433	187489	20.8087	.00230947	473	223729	21.7486	.00211416
434	188356	20.8327	.00230415	474	224676	21.7715	.00210970
435	189225	20.8567	.00229885	475	225625	21.7945	.00210526
436	190096	20.8806	.00229358	476	226576	21.8174	.00210084
437	190969	20.9045	.00228833	477	227529	21.8403	.00209644
438	191844	20.9284	.00228311	478	228484	21.8632	.00209205
439	192721	20.9523	.00227790	479	229441	21.8861	.00208768
440	193600	20.9762	.00227273	480	230400	21.9089	.00208333

N	N^2	$\sqrt{N}$	$1/N$	N	N^2	$\sqrt{N}$	$1/N$
481	231361	21.9317	.00207900	521	271441	22.8254	.00191939
482	232324	21.9545	.00207469	522	272484	22.8473	.00191571
483	233289	21.9773	.00207039	523	273529	22.8692	.00191205
484	234256	22.0000	.00206612	524	274576	22.8910	.00190840
485	235225	22.0227	.00206186	525	275625	22.9129	.00190476
486	236196	22.0454	.00205761	526	276676	22.9347	.00190114
487	237169	22.0681	.00205339	527	277729	22.9565	.00189753
488	238144	22.0907	.00204918	528	278784	22.9783	.00189394
489	239121	22.1133	.00204499	529	279841	23.0000	.00189036
490	240100	22.1359	.00204082	530	280900	23.0217	.00188679
491	241081	22.1585	.00203666	531	281961	23.0434	.00188324
492	242064	22.1811	.00203252	532	283024	23.0651	.00187970
493	243049	22.2036	.00202840	533	284089	23.0868	.00187617
494	244036	22.2261	.00202429	534	285156	23.1084	.00187266
495	245025	22.2486	.00202020	535	286225	23.1301	.00186916
496	246016	22.2711	.00201613	536	287296	23.1517	.00186567
497	247009	22.2935	.00201207	537	288369	23.1733	.00186220
498	248004	22.3159	.00200803	538	289444	23.1948	.00185874
499	249001	22.3383	.00200401	539	290521	23.2164	.00185529
500	250000	22.3607	.00200000	540	291600	23.2379	.00185185
501	251001	22.3830	.00199601	541	292681	23.2594	.00184843
502	252004	22.4054	.00199203	542	293764	23.2809	.00184502
503	253009	22.4277	.00198807	543	294849	23.3024	.00184162
504	254016	22.4499	.00198413	544	295936	23.3238	.00183824
505	255025	22.4722	.00198020	545	297025	23.3452	.00183486
506	256036	22.4944	.00197628	546	298116	23.3666	.00183150
507	257049	22.5167	.00197239	547	299209	23.3880	.00182815
508	258064	22.5389	.00196850	548	300304	23.4094	.00182482
509	259081	22.5610	.00196464	549	301401	23.4307	.00182149
510	260100	22.5832	.00196078	550	302500	23.4521	.00181818
511	261121	22.6053	.00195695	551	303601	23.4734	.00181488
512	262144	22.6274	.00195312	552	304704	23.4947	.00181159
513	263169	22.6495	.00194932	553	305809	23.5160	.00180832
514	264196	22.6716	.00194553	554	306916	23.5372	.00180505
515	265225	22.6936	.00194175	555	308025	23.5584	.00180180
516	266256	22.7156	.00193798	556	309136	23.5797	.00179856
517	267289	22.7376	.00193424	557	310249	23.6008	.00179533
518	268324	22.7596	.00193050	558	311364	23.6220	.00179211
519	269361	22.7816	.00192678	559	312481	23.6432	.00178891
520	270400	22.8035	.00192308	560	313600	23.6643	.00178571

N	N²	$\sqrt{N}$	1/N	N	N²	$\sqrt{N}$	1/N
561	314721	23.6854	.00178253	601	361201	24.5153	.00166389
562	315844	23.7065	.00177936	602	362404	24.5357	.00166113
563	316969	23.7276	.00177620	603	363609	24.5561	.00165837
564	318096	23.7487	.00177305	604	364816	24.5764	.00165563
565	319225	23.7697	.00176991	605	366025	24.5967	.00165289
566	320356	23.7908	.00176678	606	367236	24.6171	.00165017
567	321489	23.8118	.00176367	607	368449	24.6374	.00164745
568	322624	23.8328	.00176056	608	369664	24.6577	.00164474
569	323761	23.8537	.00175747	609	370881	24.6779	.00164204
570	324900	23.8747	.00175439	610	372100	24.6982	.00163934
571	326041	23.8956	.00175131	611	373321	24.7184	.00163666
572	327184	23.9165	.00174825	612	374544	24.7386	.00163399
573	328329	23.9374	.00174520	613	375769	24.7588	.00163132
574	329476	23.9583	.00174216	614	376996	24.7790	.00162866
575	330625	23.9792	.00173913	615	378225	24.7992	.00162602
576	331776	24.0000	.00173611	616	379456	24.8193	.00162338
577	332929	24.0208	.00173310	617	380689	24.8395	.00162075
578	334084	24.0416	.00173010	618	381924	24.8596	.00161812
579	335241	24.0624	.00172712	619	383161	24.8797	.00161551
580	336400	24.0832	.00172414	620	384400	24.8998	.00161290
581	337561	24.1039	.00172117	621	385641	24.9199	.00161031
582	338724	24.1247	.00171821	622	386884	24.9399	.00160772
583	339889	24.1454	.00171527	623	388129	24.9600	.00160514
584	341056	24.1661	.00171233	624	389376	24.9800	.00160256
585	342225	24.1868	.00170940	625	390625	25.0000	.00160000
586	343396	24.2074	.00170648	626	391876	25.0200	.00159744
587	344569	24.2281	.00170358	627	393129	25.0400	.00159490
588	345744	24.2487	.00170068	628	394384	25.0599	.00159236
589	346921	24.2693	.00169779	629	395641	25.0799	.00158983
590	348100	24.2899	.00169492	630	396900	25.0998	.00158730
591	349281	24.3105	.00169205	631	398161	25.1197	.00158479
592	350464	24.3311	.00168919	632	399424	25.1396	.00158228
593	351649	24.3516	.00168634	633	400689	25.1595	.00157978
594	352836	24.3721	.00168350	634	401956	25.1794	.00157729
595	354025	24.3926	.00168067	635	403225	25.1992	.00157480
596	355216	24.4131	.00167785	636	404496	25.2190	.00157233
597	356409	24.4336	.00167504	637	405769	25.2389	.00156986
598	357604	24.4540	.00167224	638	407044	25.2587	.00156740
599	358801	24.4745	.00166945	639	408321	25.2784	.00156495
600	360000	24.4949	.00166667	640	409600	25.2982	.00156250

N	N^2	$\sqrt{N}$	$1/N$	N	N^2	$\sqrt{N}$	$1/N$
641	410881	25.3180	.00156006	681	463761	26.0960	.00146843
642	412164	25.3377	.00155763	682	465124	26.1151	.00146628
643	413449	25.3574	.00155521	683	466489	26.1343	.00146413
644	414736	25.3772	.00155280	684	467856	26.1534	.00146199
645	416025	25.3969	.00155039	685	469225	26.1725	.00145985
646	417316	25.4165	.00154799	686	470596	26.1916	.00145773
647	418609	25.4362	.00154560	687	471969	26.2107	.00145560
648	419904	25.4558	.00154321	688	473344	26.2298	.00145349
649	421201	25.4755	.00154083	689	474721	26.2488	.00145138
650	422500	25.4951	.00153846	690	476100	26.2679	.00144928
651	423801	25.5147	.00153610	691	477481	26.2869	.00144718
652	425104	25.5343	.00153374	692	478864	26.3059	.00144509
653	426409	25.5539	.00153139	693	480249	26.3249	.00144300
654	427716	25.5734	.00152905	694	481636	26.3439	.00144092
655	429025	25.5930	.00152672	695	483025	26.3629	.00143885
656	430336	25.6125	.00152439	696	484416	26.3818	.00143678
657	431649	25.6320	.00152207	697	485809	26.4008	.00143472
658	432964	25.6515	.00151976	698	487204	26.4197	.00143266
659	434281	25.6710	.00151745	699	488601	26.4386	.00143062
660	435600	25.6905	.00151515	700	490000	26.4575	.00142857
661	436921	25.7099	.00151286	701	491401	26.4764	.00142653
662	438244	25.7294	.00151057	702	492804	26.4953	.00142450
663	439569	25.7488	.00150830	703	494209	26.5141	.00142248
664	440896	25.7682	.00150602	704	495616	26.5330	.00142045
665	442225	25.7876	.00150376	705	497025	26.5518	.00141844
666	443556	25.8070	.00150150	706	498436	26.5707	.00141643
667	444889	25.8263	.00149925	707	499849	26.5895	.00141443
668	446224	25.8457	.00149701	708	501264	26.6083	.00141243
669	447561	25.8650	.00149477	709	502681	26.6271	.00141044
670	448900	25.8844	.00149254	710	504100	26.6458	.00140845
671	450241	25.9037	.00149031	711	505521	26.6646	.00140647
672	451584	25.9230	.00148810	712	506944	26.6833	.00140449
673	452929	25.9422	.00148588	713	508369	26.7021	.00140252
674	454276	25.9615	.00148368	714	509796	26.7208	.00140056
675	455625	25.9808	.00148148	715	511225	26.7395	.00139860
676	456976	26.0000	.00147929	716	512656	26.7582	.00139665
677	458329	26.0192	.00147710	717	514089	26.7769	.00139470
678	459684	26.0384	.00147493	718	515524	26.7955	.00139276
679	461041	26.0576	.00147275	719	516961	26.8142	.00139082
680	462400	26.0768	.00147059	720	518400	26.8328	.00138889

N	N^2	$\sqrt{N}$	$1/N$	N	N^2	$\sqrt{N}$	$1/N$
721	519841	26.8514	.00138696	761	579121	27.5862	.00131406
722	521284	26.8701	.00138504	762	580644	27.6043	.00131234
723	522729	26.8887	.00138313	763	582169	27.6225	.00131062
724	524176	26.9072	.00138122	764	583696	27.6405	.00130890
725	525625	26.9258	.00137931	765	585225	27.6586	.00130719
726	527076	26.9444	.00137741	766	586756	27.6767	.00130548
727	528529	26.9629	.00137552	767	588289	27.6948	.00130378
728	529984	26.9815	.00137363	768	589824	27.7128	.00130208
729	531441	27.0000	.00137174	769	591361	27.7308	.00130039
730	532900	27.0185	.00136986	770	592900	27.7489	.00129870
731	534361	27.0370	.00136799	771	594441	27.7669	.00129702
732	535824	27.0555	.00136612	772	595984	27.7849	.00129534
733	537289	27.0740	.00136426	773	597529	27.8029	.00129366
734	538756	27.0924	.00136240	774	599076	27.8209	.00129199
735	540225	27.1109	.00136054	775	600625	27.8388	.00129032
736	541696	27.1293	.00135870	776	602176	27.8568	.00128866
737	543169	27.1477	.00135685	777	603729	27.8747	.00128700
738	544644	27.1662	.00135501	778	605284	27.8927	.00128535
739	546121	27.1846	.00135318	779	606841	27.9106	.00128370
740	547600	27.2029	.00135135	780	608400	27.9285	.00128205
741	549081	27.2213	.00134953	781	609961	27.9464	.00128041
742	550564	27.2397	.00134771	782	611524	27.9643	.00127877
743	552049	27.2580	.00134590	783	613089	27.9821	.00127714
744	553536	27.2764	.00134409	784	614656	28.0000	.00127551
745	555025	27.2947	.00134228	785	616225	28.0179	.00127389
746	556516	27.3130	.00134048	786	617796	28.0357	.00127226
747	558009	27.3313	.00133869	787	619369	28.0535	.00127065
748	559504	27.3496	.00133690	788	620944	28.0713	.00126904
749	561001	27.3679	.00133511	789	622521	28.0891	.00126743
750	562500	27.3861	.00133333	790	624100	28.1069	.00126582
751	564001	27.4044	.00133156	791	625681	28.1247	.00126422
752	565504	27.4226	.00132979	792	627264	28.1425	.00126263
753	567009	27.4408	.00132802	793	628849	28.1603	.00126103
754	568516	27.4591	.00132626	794	630436	28.1780	.00125945
755	570025	27.4773	.00132450	795	632025	28.1957	.00125786
756	571536	27.4955	.00132275	796	633616	28.2135	.00125628
757	573049	27.5136	.00132100	797	635209	28.2312	.00125471
758	574564	27.5318	.00131926	798	636804	28.2489	.00125313
759	576081	27.5500	.00131752	799	638401	28.2666	.00125156
760	577600	27.5681	.00131579	800	640000	28.2843	.00125000

N	N^2	$\sqrt{N}$	$1/N$	N	N^2	$\sqrt{N}$	$1/N$
801	641601	28.3019	.00124844	841	707281	29.0000	.00118906
802	643204	28.3196	.00124688	842	708964	29.0172	.00118765
803	644809	28.3373	.00124533	843	710649	29.0345	.00118624
804	646416	28.3549	.00124378	844	712336	29.0517	.00118483
805	648025	28.3725	.00124224	845	714025	29.0689	.00118343
806	649636	28.3901	.00124069	846	715716	29.0861	.00118203
807	651249	28.4077	.00123916	847	717409	29.1033	.00118064
808	652864	28.4253	.00123762	848	719104	29.1204	.00117925
809	654481	28.4429	.00123609	849	720801	29.1376	.00117786
810	656100	28.4605	.00123457	850	722500	29.1548	.00117647
811	657721	28.4781	.00123305	851	724201	29.1719	.00117509
812	659344	28.4956	.00123153	852	725904	29.1890	.00117371
813	660969	28.5132	.00123001	853	727609	29.2062	.00117233
814	662596	28.5307	.00122850	854	729316	29.2233	.00117096
815	664225	28.5482	.00122699	855	731025	29.2404	.00116959
816	665856	28.5657	.00122549	856	732736	29.2575	.00116822
817	667489	28.5832	.00122399	857	734449	29.2746	.00116686
818	669124	28.6007	.00122249	858	736164	29.2916	.00116550
819	670761	28.6182	.00122100	859	737881	29.3087	.00116414
820	672400	28.6356	.00121951	860	739600	29.3258	.00116279
821	674041	28.6531	.00121803	861	741321	29.3428	.00116144
822	675684	28.6705	.00121655	862	743044	29.3598	.00116009
823	677329	28.6880	.00121507	863	744769	29.3769	.00115875
824	678976	28.7054	.00121359	864	746496	29.3939	.00115741
825	680625	28.7228	.00121212	865	748225	29.4109	.00115607
826	682276	28.7402	.00121065	866	749956	29.4279	.00115473
827	683929	28.7576	.00120919	867	751689	29.4449	.00115340
828	685584	28.7750	.00120773	868	753424	29.4618	.00115207
829	687241	28.7924	.00120627	869	755161	29.4788	.00115075
830	688900	28.8097	.00120482	870	756900	29.4958	.00114943
831	690561	28.8271	.00120337	871	758641	29.5127	.00114811
832	692224	28.8444	.00120192	872	760384	29.5296	.00114679
833	693889	28.8617	.00120048	873	762129	29.5466	.00114548
834	695556	28.8791	.00119904	874	763876	29.5635	.00114416
835	697225	28.8964	.00119760	875	765625	29.5804	.00114286
836	698896	28.9137	.00119617	876	767376	29.5973	.00114155
837	700569	28.9310	.00119474	877	769129	29.6142	.00114025
838	702244	28.9482	.00119332	878	770884	29.6311	.00113895
839	703921	28.9655	.00119190	879	772641	29.6479	.00113766
840	705600	28.9828	.00119048	880	774400	29.6648	.00113636

N	N^2	$\sqrt{N}$	$1/N$	N	N^2	$\sqrt{N}$	$1/N$
881	776161	29.6816	.00113507	921	848241	30.3480	.00108578
882	777924	29.6985	.00113379	922	850084	30.3645	.00108460
883	779689	29.7153	.00113250	923	851929	30.3809	.00108342
884	781456	29.7321	.00113122	924	853776	30.3974	.00108225
885	783225	29.7489	.00112994	925	855625	30.4138	.00108108
886	784996	29.7658	.00112867	926	857476	30.4302	.00107991
887	786769	29.7825	.00112740	927	859329	30.4467	.00107875
888	788544	29.7993	.00112613	928	861184	30.4631	.00107759
889	790321	29.8161	.00112486	929	863041	30.4795	.00107643
890	792100	29.8329	.00112360	930	864900	30.4959	.00107527
891	793881	29.8496	.00112233	931	866761	30.5123	.00107411
892	795664	29.8664	.00112108	932	868624	30.5287	.00107296
893	797449	29.8831	.00111982	933	870489	30.5450	.00107181
894	799236	29.8998	.00111857	934	872356	30.5614	.00107066
895	801025	29.9166	.00111732	935	874225	30.5778	.00106952
896	802816	29.9333	.00111607	936	876096	30.5941	.00106838
897	804609	29.9500	.00111483	937	877969	30.6105	.00106724
898	806404	29.9666	.00111359	938	879844	30.6268	.00106610
899	808201	29.9833	.00111235	939	881721	30.6431	.00106496
900	810000	30.0000	.00111111	940	883600	30.6594	.00106383
901	811801	30.0167	.00110988	941	885481	30.6757	.00106270
902	813604	30.0333	.00110865	942	887364	30.6920	.00106157
903	815409	30.0500	.00110742	943	889249	30.7083	.00106045
904	817216	30.0666	.00110619	944	891136	30.7246	.00105932
905	819025	30.0832	.00110497	945	893025	30.7409	.00105820
906	820836	30.0998	.00110375	946	894916	30.7571	.00105708
907	822649	30.1164	.00110254	947	896809	30.7734	.00105597
908	824464	30.1330	.00110132	948	898704	30.7896	.00105485
909	826281	30.1496	.00110011	949	900601	30.8058	.00105374
910	828100	30.1662	.00109890	950	902500	30.8221	.00105263
911	829921	30.1828	.00109769	951	904401	30.8383	.00105152
912	831744	30.1993	.00109649	952	906304	30.8545	.00105042
913	833569	30.2159	.00109529	953	908209	30.8707	.00104932
914	835396	30.2324	.00109409	954	910116	30.8869	.00104822
915	837225	30.2490	.00109290	955	912025	30.9031	.00104712
916	839056	30.2655	.00109170	956	913936	30.9192	.00104603
917	840889	30.2820	.00109051	957	915849	30.9354	.00104493
918	842724	30.2985	.00108932	958	917764	30.9516	.00104384
919	844561	30.3150	.00108814	959	919681	30.9677	.00104275
920	846400	30.3315	.00108696	960	921600	30.9839	.00104167

N	N^2	$\sqrt{N}$	$1/N$	N	N^2	$\sqrt{N}$	$1/N$
961	923521	31.0000	.00104058	981	962361	31.3209	.00101937
962	925444	31.0161	.00103950	982	964324	31.3369	.00101833
963	927369	31.0322	.00103842	983	966289	31.3528	.00101729
964	929296	31.0483	.00103734	984	968256	31.3688	.00101626
965	931225	31.0644	.00103627	985	970225	31.3847	.00101523
966	933156	31.0805	.00103520	986	972196	31.4006	.00101420
967	935089	31.0966	.00103413	987	974169	31.4166	.00101317
968	937024	31.1127	.00103306	988	976144	31.4325	.00101215
969	938961	31.1288	.00103199	989	978121	31.4484	.00101112
970	940900	31.1448	.00103093	990	980100	31.4643	.00101010
971	942841	31.1609	.00102987	991	982081	31.4802	.00100908
972	944784	31.1769	.00102881	992	984064	31.4960	.00100806
973	946729	31.1929	.00102775	993	986049	31.5119	.00100705
974	948676	31.2090	.00102669	994	988036	31.5278	.00100604
975	950625	31.2250	.00102564	995	990025	31.5436	.00100503
976	952576	31.2410	.00102459	996	992016	31.5595	.00100402
977	954529	31.2570	.00102354	997	994009	31.5753	.00100301
978	956484	31.2730	.00102249	998	996004	31.5911	.00100200
979	958441	31.2890	.00102145	999	998001	31.6070	.00100100
980	960400	31.3050	.00102041	1000	1000000	31.6228	.00100000

answers to selected problems

Chapter 4

1 $\Sigma X = 952$

14 b 3

   c $\overline{X} = 4.02857$;

median = 3.38889

Chapter 5

6 $s = 3.0388$

15 a $s = 9.662$

     $QD = 9.25$

Chapter 6

1 186.76

3 z value of first score is -1.8316

Chapter 7

4 25

6 a 0.0009765

  b 0.2460937

  c 0.0976562

8 a 0.25

  b 6.4×10^{-5}

  c 9.7×10^{-4}

9 a <0.00003

  b 0.3413

  c 0.1359

  d 0.0456

12 0.171875

Chapter 9

7 $r = 0.9849$

11 c $r = -0.3117$

   d $r = 0.8564$, $r_S = 0.9941$

Chapter 10

2 $r = 0.667$

4 b $r = 0.7307$

  c $r^2 = 0.5339$; $k^2 = 0.4660$

  d $s_X = 3.364867$; $s_Y = 2.3460912$

  e $Y' = 9.47$

  g 1.6015

  h 6.331, 12.608

Chapter 11

3 c $z = +1.667$

5 $z = +4.07$

8 $z = 0.3069$

Chapter 12

8 $t = 2.121$

Chapter 13

1 $F = 22.197$

2 0.788

8 a 3.83

  b 3.54

  c 5.01

Chapter 14

4 $F = 15.936$ for A; $F = 0.02$ for B;

   $F = 3.0396$ for $A \times B$ interaction

Chapter 15

3 $\chi^2 = 1830.5368$ for $df = 4$

4 $p = 0.0107$

6 $U = 26$ favoring T, 28 favoring C

7 $\chi^2 = 10.714$ for 5 df

8 $H = 25.72$ for 4 df

9 $U = 23$ or 24

489

glossary of symbols

Symbol	Name	Formula
X, Y	Scores	
n	Number of scores in sample	
N	Number of scores in all samples combined	
$\sum$	Summation	
$\bar{X}, \bar{Y}$	Arithmetic mean for a sample	$\left.\begin{array}{l}\bar{X} = \dfrac{\sum X}{n} \\[2mm] \bar{Y} = \dfrac{\sum Y}{n}\end{array}\right\} \cong \dfrac{\sum (\text{frequency} \times \text{midpoint})}{n}$
μ (mu)	Mean for a population	$\mu \cong \hat{\mu} = \bar{X}$
$\hat{\mu}$ (mu circumflex)	Circumflex indicates a *best estimate* of a parameter	
s	Standard deviation for a sample	$s = \sqrt{\dfrac{\sum (X - \bar{X})^2}{n}} = \sqrt{\dfrac{\sum X^2}{n} - \bar{X}^2}$
s^2	Variance for a sample	$s^2 \cong \dfrac{\sum (\text{frequency} \times \text{midpoint}^2)}{n}$
$\sigma, \hat{\sigma}$ (sigma)	Standard deviation for a population	$\sigma \cong \hat{\sigma} = s\sqrt{\dfrac{n}{n-1}}$
		$\hat{\sigma} = \sqrt{\dfrac{\sum (X - \bar{X})^2}{n-1}} = \sqrt{\dfrac{n \sum X^2 - (\sum X)^2}{n(n-1)}}$
σ^2	Variance for a population	
Range	Largest score minus smallest score plus 1	$L - S + 1$
QD	Quartile deviation	$QD = \dfrac{Q_3 - Q_1}{2}$
z	z score	$z = \dfrac{X - \bar{X}}{s}$, so $X = \bar{X} + sz$
T	T score	$T = 10(z + 5)$

Symbol	Name	Formula
P	Percentile point	$P = X_{LL} + IW \dfrac{(P - f_{LL})}{f_I}$
PR	Percentile rank	$PR = \dfrac{(\text{Total of scores less than } X) + 0.5}{n} \times 100$
$\sigma_{\bar{X}}, \hat{\sigma}_{\bar{X}}$	Standard error of the mean	$\sigma_{\bar{X}} \cong \hat{\sigma}_{\bar{X}} = \dfrac{\hat{\sigma}}{\sqrt{n}} = \dfrac{s}{\sqrt{n-1}}$
$z_{\bar{X}}$	z test statistic	$z_{\bar{X}} = \dfrac{\bar{X} - \mu}{\sigma_{\bar{X}}} \cong \dfrac{(\bar{X} - \mu)\sqrt{n}}{\hat{\sigma}}$
r	Pearson product–moment coefficient of correlation	$r = \dfrac{\sum z_X z_Y}{n} = \dfrac{(\sum XY/n) - \bar{X}\bar{Y}}{s_X s_Y}$ $r = \dfrac{[(\sum XY)/n] - \bar{X}\bar{Y}}{\sqrt{[(\sum X^2)/n - \bar{X}^2][(\sum Y^2)/n - \bar{Y}^2]}}$
ρ (rho)	Population value of Pearson coefficient	
r_S	Spearman rank-order coefficient of correlation	$r_S = 1 - \dfrac{6\sum D^2}{n(n^2 - 1)}$
z'_Y	Predicted z score on Y given X	$z'_Y = r z_X$
Y'	Predicted raw score on Y given X	$Y' = \bar{Y} + \dfrac{r s_Y (X - \bar{X})}{s_X}$
$s_{Y \cdot X}$	Standard error of the estimate of Y given X	$s_{Y \cdot X} = s_Y \sqrt{1 - r^2}$
D	Difference score	$D = X_1 - X_2$
$\bar{D}$	Difference between two means, or mean difference for a sample	$\bar{D} = \bar{X}_1 - \bar{X}_2 \qquad \bar{D} = \sum D/n$
$\sigma_{\bar{D}}, \hat{\sigma}_{\bar{D}}$	Standard error of the difference between two means	$\sigma_{\bar{D}} \cong \hat{\sigma}_{\bar{D}} = \sqrt{\dfrac{s_E^2}{n_E - 1} + \dfrac{s_C^2}{n_C - 1}}$
$z_{\bar{D}}$	Statistic for the z test (H_0 that $\mu = 0$)	$z_{\bar{D}} = \dfrac{\bar{D}}{\hat{\sigma}_{\bar{D}}}$
H_0	Null hypothesis, usually that $\mu_E = \mu_C$, and thus that $\mu_D = 0$	
μ_D	Difference between two means, or mean difference for a population	$\mu_D = \mu_E - \mu_C$ $\mu_D \cong \hat{\mu}_D = \bar{D}$

Symbol	Name	Formula
Z	Z transformation of r	$Z = 0.5 \log_e \dfrac{1 + r}{1 - r}$
$\sigma_Z, \hat{\sigma}_Z$	Standard error of Z	$\sigma_Z \cong \hat{\sigma}_Z = s_Z = \dfrac{1}{\sqrt{n - 3}}$
$\sigma_{Z_1 - Z_2}$	Standard error of the difference between two Z values	$\sigma_{Z_1 - Z_2} \cong \hat{\sigma}_{Z_1 - Z_2} = \sqrt{\dfrac{1}{n_1 - 3} + \dfrac{1}{n_2 - 3}}$
z	z test statistic for the difference between two correlations	$z = \dfrac{Z_1 - Z_2}{\hat{\sigma}_{Z_1 - Z_2}}$
t	Student t statistic to test H_0 that $\mu_D = 0$	
	For matched-subjects design or repeated-measures design:	$t = \dfrac{\bar{D}}{\sqrt{\dfrac{n \sum D^2 - (\sum D)^2}{n^2(n - 1)}}}$
	For independent groups of equal size:	$t = \dfrac{\bar{D}}{\sqrt{\dfrac{s_E{}^2}{n_E - 1} + \dfrac{s_C{}^2}{n_C - 1}}}$
	For independent groups of unequal size:	$t = \dfrac{\bar{D}}{\sqrt{\dfrac{n_E s_E{}^2 + n_C s_C{}^2}{n_E + n_C - 2} \left(\dfrac{1}{n_E} + \dfrac{1}{n_C} \right)}}$
df	Degrees of freedom	
F	F statistic to test H_0 that $\mu_1 = \mu_2 = \mu_n$	$F = \dfrac{ms_b}{ms_w} = \dfrac{ss_b / df_b}{ss_w / df_w}$
ss	Sum of squares	
ms	Mean square	
η (eta)	Correlation coefficient	$\eta = \sqrt{\dfrac{ss_b}{ss_t}}$
χ^2	Chi-square statistic	$\chi^2 = \dfrac{\sum (f_o - f_e)^2}{f_e}$

index

Page references are to the first page when a topic spans several pages.

A 6
B 7
C 8
D 9
E 0
F 1
G 2
H 3
I 4
J 5